T0203133

# Lecture Notes in Computer Science 13252

More information about this series at  https://link.springer.com/bookseries/558

Kyungmin Bae (Ed.)

# Rewriting Logic
# and Its Applications

14th International Workshop, WRLA 2022
Munich, Germany, April 2–3, 2022
Revised Selected Papers

 Springer

*Editor*
Kyungmin Bae 🄳
Department of Computer Science and Engineering
Pohang University of Science and Technology
Pohang, Korea (Republic of)

ISSN 0302-9743 ISSN 1611-3349 (electronic)
Lecture Notes in Computer Science
ISBN 978-3-031-12440-2 ISBN 978-3-031-12441-9 (eBook)
https://doi.org/10.1007/978-3-031-12441-9

This Springer imprint is published by the registered company Springer Nature Switzerland AG
The registered company address is: Gewerbestrasse 11, 6330 Cham, Switzerland

# Preface

This volume contains the formal proceedings of the 14th International Workshop on Rewriting Logic and its Applications (WRLA 2022), held as a satellite event of the European Joint Conferences on Theory and Practice of Software (ETAPS 2022) in Munich, Germany, during April 2–3, 2022.

Rewriting logic is a natural model of computation and an expressive semantic framework for concurrency, parallelism, communication, and interaction. It can be used for specifying a wide range of systems and languages in various application fields. It also has good properties as a metalogical framework for representing logics. Over the years, several languages based on rewriting logic have been designed and implemented. The aim of the workshop is to bring together researchers with a common interest in rewriting logic and its applications and to give them the opportunity to present their recent works, discuss future research directions, and exchange ideas.

The previous meetings were held in Asilomar, USA (1996), Pont-a-Mousson, France (1998), Kanazawa, Japan (2000), Pisa, Italy (2002), Barcelona, Spain (2004), Vienna, Austria (2006), Budapest, Hungary (2008), Paphos, Cyprus (2010), Tallinn, Estonia (2012), Grenoble, France (2014), Eindhoven, the Netherlands (2016), Thessaloniki, Greece (2018), and online as a virtual event (2020).

This year, we received 13 submissions. Each was reviewed by at least three Program Committee members. After extensive discussions, the Program Committee decided to accept 11 papers for presentation at the workshop and nine papers for inclusion in these proceedings. This volume also includes two invited papers by Gwen Salaün and Sebastian Mödersheim, two invited tutorials by Santiago Escobar and Rubén Rubio, and an invited experience report by Peter Csaba Ölveczky.

We sincerely thank all the authors of papers submitted to the workshop, and the invited speakers for kindly accepting to contribute to WRLA 2022. We are grateful to the members of the Program Committee and the subreviewers for their careful work in the review process. We also thank the members of the WRLA steering committee for their valuable suggestions. Finally, we express our gratitude to all members of the local organization team of ETAPS 2022, whose work made the workshop possible.

June 2022                                                                                          Kyungmin Bae

# Organization

## Program Committee

| | |
|---|---|
| Erika Abraham | RWTH Aachen University, Germany |
| María Alpuente | Universitat Politècnica de València, Spain |
| Kyungmin Bae (Chair) | Pohang University of Science and Technology, South Korea |
| Roberto Bruni | Università di Pisa, Italy |
| Francisco Durán | University of Málaga, Spain |
| Santiago Escobar | Universitat Politècnica de València, Spain |
| Maribel Fernandez | King's College London, UK |
| Mark Hills | East Carolina University, USA |
| Nao Hirokawa | JAIST, Japan |
| Alexander Knapp | Universität Augsburg, Germany |
| Temur Kutsia | Johannes Kepler University Linz, Austria |
| Alberto Lluch Lafuente | Technical University of Denmark, Denmark |
| Dorel Lucanu | Alexandru Ioan Cuza University, Romania |
| Salvador Lucas | Universitat Politècnica de València, Spain |
| Narciso Martí-Oliet | Universidad Complutense de Madrid, Spain |
| José Meseguer | University of Illinois at Urbana-Champaign, USA |
| Aart Middeldorp | University of Innsbruck, Austria |
| Vivek Nigam | Federal University of Paraíba, Brazil |
| Kazuhiro Ogata | JAIST, Japan |
| Peter Csaba Ölveczky | University of Oslo, Norway |
| Adrián Riesco | Universidad Complutense de Madrid, Spain |
| Christophe Ringeissen | Inria, France |
| Camilo Rocha | Pontificia Universidad Javeriana Cali, Colombia |
| Vlad Rusu | Inria, France |
| Traian Florin Serbanuta | University of Bucharest, Romania |
| Carolyn Talcott | SRI International, USA |

## Additional Reviewers

Aoto, Takahito
Chiang, James
Fernet, Laouen
Sapiña, Julia

# Contents

**Tool Papers**

# Invited Papers

# From Static to Dynamic Analysis and Allocation of Resources for BPMN Processes

Francisco Durán[1], Yliès Falcone[2], Camilo Rocha[3], Gwen Salaün[2(✉)],
and Ahang Zuo[2]

[1] ITIS Software, University of Málaga, Málaga, Spain
[2] Univ. Grenoble Alpes, CNRS, Grenoble INP, Inria, LIG, 38000 Grenoble, France
gwen.salaun@inria.fr
[3] Pontificia Universidad Javeriana, Cali, Colombia

**Abstract.** Business process optimisation is a strategic activity in organisations because of its potential to increase profit margins and reduce operational costs. One of the main challenges in this context is concerned with the problem of optimising the allocation and sharing of resources. In this work, processes are described using the BPMN notation extended with an explicit description of execution time and resources associated with tasks, and can be concurrently executed multiple times. First, a simulation-based approach for computing certain metrics of interest, such as average execution time or resource usage, is presented. This approach applies off-line and is static in the sense that the number of resources does not evolve over the time of the simulation. In a second step, an alternative approach is presented, which works online, thus requiring the instrumentation of an existing platform for retrieving information of interest during the processes' execution. This second approach is dynamic because the number of resource replicas is updated over the time of the execution. This paper aims at stressing pros and cons of both approaches, and at showing how they complement each other.

## 1 Introduction

Business process optimisation is a strategic activity in organisations because of its potential to increase profit margins and reduce operational costs. Optimisation is, however, a difficult task to be achieved manually since several parameters should be taken into account (e.g., execution times, resources, costs, etc.). One of the main challenges in this context is concerned with the problem of optimising the allocation and sharing of resources. Resource usage is crucial because it directly impacts the time it takes to execute a process. Moreover, by associating a certain cost to each resource, the total cost of executing a process a certain number of times can be computed. Optimising resource usage reduces the process execution time and the costs associated with its execution.

In this work, we assume that a description of a business process is given using the BPMN [24] workflow-based modelling language. BPMN has been standardised by the International Organization for Standardization (ISO). It was first

K. Bae (Ed.): WRLA 2022, LNCS 13252, pp. 3–21, 2022.
https://doi.org/10.1007/978-3-031-12441-9_1

published in 2013 and since then it has become the *de facto* notation for developing business processes. The BPMN language defines the set of tasks involved in a process and the order in which they should be executed. Beyond this description of the model, the time it takes to execute each task is also needed, as well as an explicit description of the resources required for executing each task. This extended model is precise enough for modelling both behavioural and quantitative aspects of processes.

This paper presents two ways to analyse BPMN processes with time and resources. Both techniques assume that the process is executed multiple times and multiple concurrent executions of a process compete for the shared resources. Such multiple executions correspond to realistic scenarios where a process is not executed once but several times (one execution per client or user for instance). Furthermore, both approaches also aim at computing some metrics of interest such as average execution times, resource usage, and costs. The first approach applies to design models, without the need for an implementation of the system running on real resources. To compute the previously mentioned metrics, offline simulation techniques are used, assuming that the allocation of resources is static (i.e., no update of the number of resources during the simulation). This first approach relies on a specification of a subset of BPMN in rewriting logic [21]. This specification is executable in Maude [6], and the computation of metrics is achieved by using Maude's rewriting tools.

The second approach applies at runtime or online, and works by instrumenting an existing platform for executable BPMN (Activiti [2] in this work). In this case, access to a database is used for storing information related to the execution of the process. This information is particularly useful for computing the process execution time, resource usage, and costs. This approach is also dynamic in the sense that the number of replicas for each resource is not defined once and for all, but can be updated by using the metrics computed during the process execution. In particular, a strategy that relies on the resource usage values for dynamically updating the number of replicas of each resource is presented.

The static approach applies to a model of the process, and additional information is required such as the probability to execute exclusive branches. This approach is useful for processes under development, or for potential changes that need to be evaluated before being applied. The approach can help for instance to simulate several scenarios and decide whether the number of required resources needs to be adjusted before the deployment of the process in production. On the other hand, the dynamic approach accepts as input an executable BPMN process and provides strategies to update resources at execution time thus allowing a certain stabilisation of the computed metrics over time (such as execution times and resource usage). However, this dynamic change does not apply in all contexts since it is not systematically possible to dynamically update the number of any kind of resources (such as human beings).

The organisation of the rest of this paper is as follows. Section 2 introduces the BPMN notation used in this work. Section 3 overviews the static approach for analysing resource usage. Section 4 surveys the main ideas of the dynamic approach for the allocation of resources. Section 5 presents existing works on this topic. Section 6 concludes by comparing both approaches.

## 2   BPMN with Time and Resources

BPMN 2.0 (BPMN, as a shorthand, in the rest of this paper) was published as an ISO/IEC standard [17] in 2013 and is nowadays extensively used for modelling and developing business processes. In this paper, for the sake of simplicity, we focus on activity diagrams including the BPMN constructs related to control-flow modelling and behavioural aspects. Beyond those constructs, execution time and resources are also associated with tasks, and probabilities are specified for exclusive and inclusive split gateways. Figure 1 summarises some of the BPMN constructs used in this work, with a focus on how time and resources are associated with flows and tasks.

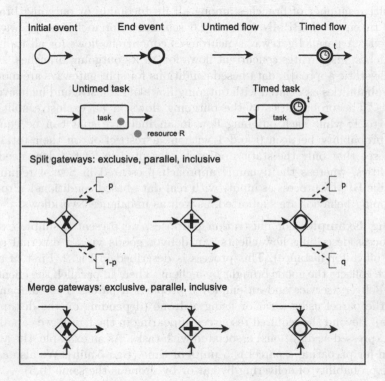

**Fig. 1.** Extended BPMN Syntax

Specifically, the node types *event*, *task*, and *gateway*, and the edge type *sequence flow* are considered. Start and end events are used, respectively, to initialise and terminate processes. A task represents an atomic activity that has exactly one incoming and one outgoing flow. A sequence flow describes two nodes executed one after the other in a specific execution order. A task and a flow may have a duration or delay. The timing information associated with tasks and flows is described as a literal value (a non-negative real number, possibly 0). Resources are explicitly defined at the task level. A task that requires resources can include, as part of its specification, the name of the required resources. Thus, a task is specified with an amount of time (its duration), and information on its required resources. Then, once the resources required by a task are acquired, the task is going to execute for the defined duration.

Gateways are used to control the divergence and convergence of the execution flow. Three types of gateways are considered for the static analysis: *exclusive*, *inclusive*, and *parallel*. Gateways with one incoming branch and multiple outgoing branches are called *splits*, e.g., split inclusive gateway. Gateways with one outgoing branch and multiple incoming branches are called *merges*, e.g., merge parallel gateway. An exclusive gateway chooses one out of a set of mutually exclusive alternative incoming or outgoing branches. For an inclusive gateway, any positive number of branches among all its incoming or outgoing branches may be taken (both BPMN 1.0 and 2.0 semantics for inclusive gateways are supported). A parallel gateway synchronises concurrent flows for all its incoming branches, and creates concurrent flows for all its outgoing branches.

In the static approach, data-based conditions for split gateways are modelled using probabilities associated with outgoing flows of exclusive and inclusive split gateways. The probabilities of the outgoing flows in an exclusive split must sum up to 1, while each outgoing flow in an inclusive split can be equipped with a probability between 0 and 1 without a restriction on their total sum. We will see that only the static approach presented in Sect. 3 does need such probabilities, whereas the dynamic approach presented in Sect. 4 requires an executable BPMN process as input (with real data-based conditions). Processes with looping behavior are supported, as well as unbalanced workflows.

**Running Example.** For illustration purposes, we present a simple example of a process describing how clients can deliver goods via an external service (a mail office for instance). This process is described in Fig. 2. First of all, an employee collects the goods brought by a client. Then, in parallel, the client pays for the delivery service and an employee prepares a parcel. The company can deliver the parcel using a car or using a drone (depending on the distance for example). Beyond the required resources appearing in the figure, we can also see times (expressed as durations) associated with tasks. As an example, the average duration for preparing a parcel is 5 units of time (e.g., 5 min). We also assume that the probability of delivering by car or by drone is the same (0.5).

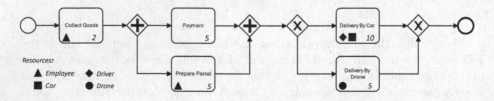

**Fig. 2.** Goods Delivery Process

# 3  Static Quantitative Analysis

In this section, we summarise the approach presented in [10] for analysing BPMN processes with resources. More precisely, we first introduce the specification of BPMN (syntax and semantics) in Maude's rewriting logic. Second, we present the quantitative properties of interest focusing on timing and resource-based properties. To compute these properties, we leverage Maude's rewriting capabilities to simulate and extract analysis results on a given BPMN process.

## 3.1  Process Description

In the Maude specification of BPMN, a process is represented as an object with sets of flows and nodes as attributes. Nodes can be of five different types: start, end, task, split, or merge. The representation of each of these types of elements includes the necessary information. A task node involves an identifier, a description, two flow identifiers (input and output), a stochastic function or a value modelling its duration (0 if there is no duration), and a set of resources required for its execution. A split node includes a node identifier, a gateway type (exclusive, inclusive, or parallel), an input flow identifier, and a set of output flow identifiers. A merge node includes a node identifier, a gateway type, a set of input flow identifiers, and an output flow identifier. The representation of a flow includes a probability distribution function corresponding to the probability of executing that flow (1 by default).

## 3.2  Execution Semantics

The operational semantics of BPMN is defined using a rewrite theory, with rewrite rules modeling how tokens evolve through a process. This rewrite theory is executable, which allows us to simulate BPMN processes. In this specification, each action is modeled as a rewrite rule. For instance, when a token arrives at a parallel split gateway, the token corresponding to the incoming flow is removed, and one token is added for each outgoing flow. Technically, rewrite rules operate on systems composed of a process object and a *simulation* object.

**Simulation Object.** While the process object introduced in Sect. 3.1 represents the BPMN process and does not change during an execution, the simulation object keeps information on the execution of the process. It stores a collection of tokens (in a scheduler, see below), a global time (gtime), and a set of resources. It also keeps track of the quantities being measured during the analysis of a process. Figure 3 presents the structure of the Simulation object.

*Tokens.* Tokens are used to represent the evolution of the workflow under execution. When a process instance is triggered, a token is added to the start node. The tokens move through nodes and flows of the process. When a token meets a split gateway (e.g., parallel gateway), several tokens are generated on outgoing flows, depending on the type of split gateway. On the contrary, when multiple tokens meet a merge gateway (e.g., inclusive gateway), they are merged into a

```
< s : Simulation | tokens : ...,          ---- scheduler
gtime : ...,            ---- global time
resources : ...,       ---- resource set
process-execs : ...., ---- execution times
sync-times : ...,      ---- synchronisation times
task-times : ...,      ---- task execution times
... >
```

**Fig. 3.** Representation of the Simulation Object

single token depending on the type of merge gateway. A token is represented as a term token(TId, Id, T). Since several executions may happen simultaneously, each execution has a unique identifier, and tokens are identified by the execution instance TId they belong to, and the flow or node Id they are attached to. The expression T represents a timer, of sort Time, modelling a delay on the token. Once this timer becomes 0, the token may be consumed.

*Scheduling.* Tokens are stored in a *scheduler* implemented as a priority queue, so that they are kept according to their due time. However, even with its timer set to 0, the token at the front of this queue may be not enough to fire some action. Consider, for example, a task that requires some resource that is not available or a parallel merge for which some incoming flow is not yet active. To avoid blocking situations, the scheduler is provided with a shifting mechanism, which moves the first active token to the front of the scheduler in case the current head cannot fire the corresponding action. This scheduler is similar to those used in typical discrete event simulations.

*Resources.* Each resource is represented with an identifier, the number of available replicas (initially the total number), the total amount of time this resource has been in use, and the intervals of time during which any replica of this resource was used. These two last parameters are stored during the simulation, and are particularly useful for analysis purposes. When a task requires several resources, it atomically uses all of them at once, or waits for them to become available.

*Workloads.* Simulation-based analysis techniques are typically parameterized by the workload that represents the way a system is used. They define the rate at which new instances of a given process are executed. Currently, closed workloads can be handled by specifying the number of executions and the rate at which executions are started, that is, their inter-arrival time. The inter-arrival time is specified as a stochastic expression.

**Rewrite Rules for BPMN Constructs.** Rewriting rules represent how tokens evolve through the process and how nodes are executed, thus defining the execution semantics of BPMN. Each action supported by the system is modelled as a rewrite rule. These rules are overviewed in the rest of this section to gather an intuition on the formal semantics (see [8] for the complete specification).

*Start/End Events.* Figure 4 depicts the rule for the start event. When there is a token in the execution TId in the start node NId with delay 0 (note the token at the front of the scheduler in the Simulation object in line 5), then this rule

```
1   crl [startProc] :
2      < PId : Process | nodes : (start(NId, FId), Nodes),
3                        flows : (flow(FId, SE), Flows),
4                        Atts >
5      < SId : Simulation | tokens : (token(TId, NId, 0) Tks), ... Atts1 >
6      < CId : Counter | counter : N >
7   => < PId : Process | nodes : (start(NId, FId), Nodes),
8                        flows : (flow(FId, SE), Flows),
9                        Atts >
10     < SId : Simulation | tokens : insert(Tks, token(TId, FId, T')), ... Atts1 >
11     < CId : Counter | counter : N' >
12  if {T', N'} := eval(SE, N) .
```

**Fig. 4.** Start Event Processing

generates a new token on the outgoing flow of the selected node to initiate the
execution of a process instance (line 10). The insert function puts this token in the
scheduler and the eval function evaluates the stochastic expression SE specifying
the delay of the outgoing flow FId to be assigned to the new token. Details on
the initialisation of time stamps and recorded times for the initiated execution
have been replaced by ellipses. A termination rule, associated to stop events,
consumes tokens when they arrive at those events.

*Tasks.* A task execution is modelled with two rules. The first rule, the initTask
rule shown in Fig. 5, represents the task initiation, which is applied when a token
with zero time is available for the incoming flow (line 5). If all the resources
required by this task are available, which is checked with the allResourcesAvailable function (line 8), then a new token is generated with the task identifier
and the task duration (line 12). Otherwise, the scheduler's token shifting mechanism is invoked (line 20). If available, all required resources are removed from
the set of resources, and the time those resources have been in use is updated
(grabResources&updateTime function, line 18). Since all auxiliary functions in the
right-hand side of the initTask rule are defined equationally, the checking and
grabbing of resources are performed atomically, without introducing any blocking issues. Note also that rules update the information on execution times, task
durations, etc. (see, e.g., the update of the task-tstamps attribute, lines 13–16).
This information is important for analysis purposes, as it will be seen in Sect. 3.3.
    A second rule, which models task completion, is triggered when there is a
token for that task with zero time. In that case, the token is consumed, a new
one is generated for the outgoing flow, and all resources are released.

*Exclusive Gateways.* There are two rules for the exclusive gateways, namely, one
for the split and one for the merge. The rule for the split applies when a token
with zero time is available on its incoming flow. A uniformly sampled probability
distribution is used to choose the branch to be executed. The newly created
token is assigned with its run-to-completion time generated by evaluating the
stochastic expression associated with the chosen outgoing flow—this is actually
the case every time a new token is added for a flow. The exclusive merge gateway
is triggered when one of its incoming flows has a token with zero time. In that
case, a new token is generated, assigned to the outgoing flow, and added to the
scheduler.

```
1   rl [initTask] :
2     < PId : Process |
3         nodes : (task(NId, TaskName, FId1, FId2, SE, RIds, SEI), Nodes), Atts >
4     < SId : Simulation |
5         tokens : (token(TId, FId1, 0) Tks),
6         task-tstamps : TTSs, gtime : T, resources : Rs, Atts1 >
7     < CId : Counter | counter : N >
8  => if allResourcesAvailable(RIds, Rs)
9     then < PId : Process |
10              nodes : (task(NId, TaskName, FId1, FId2, SE, RIds, SEI), Nodes), Atts >
11          < SId : Simulation |
12              tokens : insert(Tks, token(TId, NId, time(eval(SE, N)))),
13              task-tstamps : if TTSs[TId][NId] == undefined
14                              then insert(TId, insert(NId, T, TTSs[TId]), TTSs)
15                              else TTSs
16                              fi,            ---- for loops, stamps get overwritten
17              gtime : T,
18              resources : grabResources&updateTime(RIds, Rs, time(eval(SE, N)), T), Atts1 >
19          < CId : Counter | counter : int(eval(SE, N)) >
20     else ...                               ---- if necessary, the scheduler is updated
21     fi .
```

**Fig. 5.** Task Initiation Rule

*Parallel Gateways.* The parallel split gateway rule is triggered when a token with zero time corresponding to the input flow is available. If so, the token is consumed and one token is added to each of its outgoing flows. The merge rule for the parallel gateway is executed when there is a token with zero time for each incoming branch. In that case, these tokens are removed and a new token is generated for the outgoing flow. In the merge rule, synchronisation times are also updated.

*Inclusive Gateways.* The split rule applies when a token with zero time is available at the incoming flow. Since all outgoing branches are equipped with probabilities, a function in charge of computing the subset of branches to be triggered is invoked. For each one of the selected branches, a new token is added to the scheduler. Regarding merge gateways, both BPMN 1.0 and 2.0 semantics are supported in this research. In BPMN 2.0, merge inclusive gateways behave like exclusive ones. The 1.0 version of the semantics is more involved [5], since the merge rule for the inclusive gateway is executed when all the expected tokens are available with zero time. This requires a global analysis. To check whether all expected tokens have arrived, a backward traversal that explores the process upstream and checks whether there are tokens on their way to that merge is performed. In both cases, once the merge gateway is triggered, the incoming tokens are removed, a new token is added to the scheduler for the outgoing flow, and simulation information is updated with synchronisation times.

*Loops and Unbalanced Workflows.* The modelling of the BPMN execution semantics using tokens and their circulation through the process structure supports intricate constructs such as loops and unbalanced workflows. As far as looping behaviour is concerned, a token may circulate back to an already visited flow without any additional treatment. Similarly, tokens can advance through flows that are part of balanced or unbalanced gateways, independently of their structure.

## 3.3 Properties

Several kinds of properties or metrics can be computed, particularly timing and resource-based properties. These properties are meaningful when executing multiple instances of a process that compete for the shared resources. As for *timing properties*, the approach presented in this paper allows the computation of average execution times (AET) of a process, its variance (Var), and the average synchronisation time (AST) for merge gateways, representing the time that elapse from the arrival of the first token through one of its incoming flows to its activation. Synchronisation times make sense only for parallel and BPMN 1.0 inclusive gateways, since there is no waiting nor synchronisation time for the other gateways.

As far as *resource-based properties* are concerned, which is the main focus in this work, the following properties are computed:

- The global time of usage of all instances of each resource $R$ ($GTU_R$). E.g., when executing 10 instances of a process $P$, with an AET of 42, it is possible that the two instances of a resource $A$ are used for 56 time units and the three instances of resource $B$ for 60 time units.
- The expression $GTU_R^1$ denotes the average GTU of resource $R$ (i.e., the GTU per instance of resource $R$). Thus, although in the previous example $GTU_B$ is greater than $GTU_A$, $GTU_A^1$ is 28 and $GTU_B^1$ is 20.
- The average usage percentage $UP_R$ for a resource $R$ over the global execution time. E.g., continuing with the running example, on average, an instance of the resource $A$ is used 24% of the global execution time when executing 200 instances of a process $P$.

To compute these metrics, Maude rewriting capabilities are used to simulate and extract analysis results on a given BPMN process. The simulation object presented in Sect. 3.2 is used to accumulate information of synchronisation times, task durations, and resource usages. At the end of all executions, these results are used for computing the expected average times and resource usage percentages. Since the analysed processes are assumed syntactically correct and processes that may lead to non-terminating analysis are not considered (e.g., loops without end events), the verification process always terminates. Indeed, all splits are probabilistic, and time duration and probabilities assigned to the branches respect specific assumptions (e.g., all probabilities are between 0 and 1, they sum up to 1 in exclusive branches, and times are positive).

Last but not least, if one can associate a cost (in euros for example) to each kind of resource, we can compute the total cost of the simulation by using the collected data on execution times and resource usage. We can even go farther by computing the optimal allocation of resources. This is achieved by expressing this computation as a multi-objective optimisation problem since we may not want to reduce costs but also to reduce execution time for example. The solution to this optimisation problem is computed by using heuristic-based search algorithms such as gradient descent [26].

### 3.4    Example

Let us illustrate this approach with the running example presented in Sect. 2. Table 1 shows a few experiments consisting of 1,000 tokens with an inter-arrival time computed with an exponential probability distribution with 2 as the parameter. For each row, there is a variation in the input in terms of the number of replicas of the different resources. As a result, the table gives the total execution time for executing 1,000 times the process, the average execution time, and the total cost (assuming a cost per hour of 40, 30, 35, and 25 euros for each resource, respectively).

**Table 1.** Experimental Results for the Delivery Process

| Resources | | | | | | | | Total execution time | Average execution time | Total cost |
|---|---|---|---|---|---|---|---|---|---|---|
| Employee | | Car | | Driver | | Drone | | | | |
| Inst. | Usage % | Inst. | Usage % | Inst | Usage % | Inst. | Usage % | | | |
| 1 | 99.11 | 1 | 72.63 | 1 | 72.63 | 1 | 34.48 | 7 063.00 | 2 794.26 | 918 190.00 |
| 2 | 99.18 | 2 | 69.85 | 2 | 69.85 | 2 | 35.92 | 3 528.92 | 904.02 | 917 519.74 |
| 3 | 81.42 | 2 | 90.37 | 2 | 90.37 | 1 | 84.09 | 2 865.94 | 463.75 | 788 133.61 |
| 4 | 85.24 | 3 | .84.42 | 3 | 84.42 | 2 | 58.45 | 2 053.10 | 131.00 | 831 508.77 |
| 4 | 86.75 | 4 | 57.88 | 4 | 57.88 | 4 | 33.04 | 2 017.26 | 100.40 | 1 048 976.52 |

First of all, we can observe a clear correlation between the number of resources and the execution time/costs. The more resources, the shorter it takes to execute once the process (or all processes), but the more resources, the higher cost. Secondly, we can see that the critical resource is the employee since whatever is the number of replicas, this resource is always very busy (active more than 80% of his time). In contrast, drones are less busy except if there is a single drone and several replicas for the other resources. Finally, if we assume that we both want to reduce the average execution time and the total cost with an equal weight (0.5 and 0.5), the optimal resource allocation is 4, 3, 3, and 2 (before last row in Table 1).

## 4    Dynamic Quantitative Analysis

In this section, we will show how an existing platform (Activiti [2] in this work) can be instrumented to extract the required information from its database and compute properties periodically during the process execution. We will also show how we can develop dynamic resource allocation strategies for varying the number of resource replicas at runtime and thus impact the results of these properties. Note that in this section, we do not have any restrictions on the BPMN syntax, we just need BPMN processes to be executable. Moreover, there is no need to have probabilities associated to split exclusive and inclusive gateways, since we have real data-based conditions.

## 4.1   Instrumentation

In this section, we use Activiti as BPMN platform. Activiti is an open-source workflow engine written in Java that can execute business processes described in BPMN 2.0. We first require monitoring techniques [4,13] for BPMN processes at runtime. These techniques are useful because a process is usually not executed only once. Instead, a process can be executed multiple times. Each execution of the process is called an instance. An instance of the process can be in one of the following states: *initial* means that the instance is ready to start (one token in the start event), *running* means that the instance is currently executing and is not yet completed, *completed* means that all tokens have reached end events. Tokens are used to define the behaviour of a process. Similarly to the static approach, an identifier is used to characterise a specific instance of process execution, and this identifier is thus associated to all nodes (e.g., tasks) executed by this instance.

Monitoring techniques for BPMN executed using Activiti mostly aim at analysing the information stored in a database, and extracting the information required for computing the properties of interest (such as AET and resource usage percentage). Figure 6 gives an overview of this data extraction. We first need to retrieve the information regarding task execution and completion. This is what we can see in Fig. 6 (top right, (a)). For each task, we also extract the corresponding process execution instance and the times of beginning and end. This information is useful for determining which resources were in use and for what amount of time. Second, we retrieve execution traces for each process instance as shown in Fig. 6 (bottom right, (b)). An execution trace corresponds to a list of tasks executed by this specific instance. The tasks are not stored with a specific order in the database. Therefore, we have to order these tasks by using time stamps, corresponding to the time at which each task is executed. These time stamps are computed by the process execution engine, which relies on a global clock. The execution trace corresponding to a specific instance can be computed only when the instance is in its *completed* state.

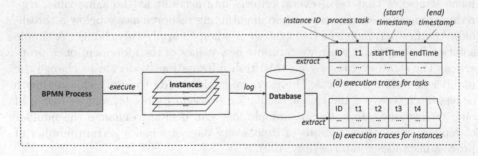

**Fig. 6.** Runtime Monitoring of Multiple Executions of a BPMN Process

## 4.2  Computation of Properties

Since new instances can execute at any time and possibly infinitely, the dynamic approach requires extracting data and computing properties on time and resources periodically. There are several possible strategies to choose the period. It can be based on a fixed amount of time (e.g., every 10 min) or it can apply when a certain number of process instances have been completed. These two strategies can also be combined, e.g., we get data whenever 100 instances have been completed or every hour if after one hour less than 100 instances have been completed. The choice of one of these strategies may have a different impact on the actual results. Note that the choice of this strategy is a parameter of the approach. In the rest of this section, we rely on a time-based strategy.

When the period completes, the data extraction is triggered. Then, we extract the required information from these data to compute the properties presented in Sect. 3.3 on execution times and resource usage. As an example, to compute the resource usage percentage per resource replica, we analyse the tasks executed during the last period of time. For these tasks, we look at the resources associated with each task and sum up the durations each resource was active during that period. Then, we divide this total time by the number of replica and compute a percentage out of these numbers by using the time of activity for each replica out of the time of the period.

As we will see below, the results are represented using curves that show the different property values (e.g., average execution time) along time.

## 4.3  Dynamic Resource Allocation

Several strategies can be defined for dynamically changing the number of replicas for each resource. These strategies rely on the metrics computed before and thus can vary in their choice and implementation. For instance, one strategy can aim at reducing the average execution time whereas another one may maintain the resource usage under a certain level, e.g., under 90%. We could also implement strategies that take several criteria into account at the same time, e.g., reduce process execution time while maintaining resource usage below a threshold. Another parameter of the strategy is when to apply this change. A simple solution is to apply it when we compute new values of the aforementioned properties. The strategy can rely on this fresh information to decide to change the number of resource replicas. However, we could decide to apply changes more or less often to avoid the classic oscillation problem (add one, remove one, add one, remove one, etc.). As an example, one can decide to change the number of replicas every three periods of time, every day, or when a certain number of process instances complete (e.g., 100).

For illustration purposes, we will present an example of strategy in the rest of this section. This strategy focuses on one specific property, namely the percentage of resource usage per replica. The strategy aims at maintaining this percentage within a certain interval, for instance, [70%, 90%]. After completion of a period of time, all properties are computed and the strategy then checks if the usage percentage for each resource is still included in this interval. If, for a given resource, this percentage goes above the highest value (e.g., 90% in our example), one replica of that resource is added. If this percentage goes below the lowest value (e.g., 70% in our example), one replica of that resource is removed. Note that we choose in this strategy to follow the same period of time as the one used for the computation of properties.

### 4.4   Example

Let us focus again on the goods delivery example introduced in Sect. 2. The only difference in terms of the input BPMN process is that here we do not need to make explicit the probabilities of executing the split exclusive gateway. This decision is taken based on internal data belonging to the (executable) BPMN process. We use the same workload as in Sect. 3.4, that is, 1000 tokens with $\exp(2)$ as inter-arrival time. There are additional parameters that are required for the dynamic approach. We use as initial allocation of resource one replica for each resource type. The targeted interval for resource usage is $[70\%, 90\%]$. The period for updating the metrics is fixed to 10 units of time whereas the strategy for dynamic resource update applies every 60 units of time.

In the rest of this section, we will show three different figures to give different insights on the results of the multiple process execution. Figure 7 describes the evolution of the number of replicas for each kind of resource. The employee is particularly important because every execution of the process requires an employee to collect goods and prepare parcels whereas the other resources are not systematically used for every process execution. One can see that this execution requires 2 or 3 employees to work properly. Cars and drivers take more time than drones to deliver goods (10 units of times for cars and 5 units for drones), therefore more replicas are required for allowing the delivery by car with driver.

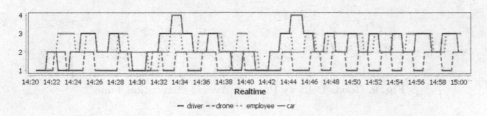

**Fig. 7.** Goods Delivery Process: Evolution of the Number of Replicas

Figure 8 focuses on the usage percentage per replica for each type of resource. It is worth reminding that the strategy used for these experiments aim at maintaining the percentage in the interval $[70\%, 90\%]$. We can see that from the beginning the usage percentage for employees is higher than 90% thus explaining why several replicas of employees were added at the beginning in Fig. 7. After the addition of these replicas for employee, the percentage remains lower.

The usage percentage for drones is the lower of all resources. We can observe important variations in all these percentages because we use a short period for computing these numbers (10 units of time) and because the use of an exclusive gateway for the delivery induces variations between the use of drones or cars.

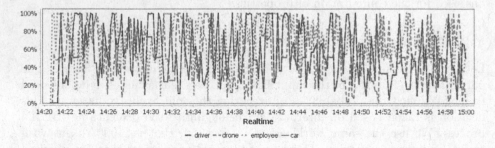

**Fig. 8.** Goods Delivery Process: Resource Usage

Figure 9 shows the evolution of the average execution time. The curve shows that this time tends to increase at the beginning, but at some point stabilises (since new executions occur on a periodic basis) and remains around 40 units of time. We can see peaks at some points of the execution corresponding to an increase in the number of delivery by car, which takes more time than drones. This increase in time can be correlated with the addition in Fig. 7 of additional replicas of cars and drivers.

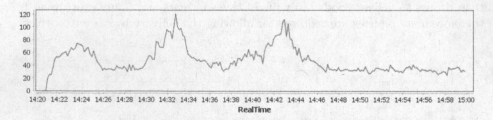

**Fig. 9.** Goods Delivery Process: Average Execution Time

## 5   Related Work

Several works on the analysis and provisioning of resources can be found in the literature. Schömig and Rau [27] use coloured stochastic Petri nets to specify and analyse business processes in the presence of dynamic routing, simultaneous resource allocation, forking/joining of process-control threads, and priority-based queuing. In their work, each resource is equipped with properties grouped in a role defining if the resource is eligible to perform a certain activity. Li *et al.* [20] introduce *multidimensional workflow nets* to model and analyse resource availability and workload. Oliveira *et al.* [23] use generalised stochastic Petri nets for correctness verification and performance evaluation of business processes. In their work, an activity can be associated with multiple roles and the completion of an activity can use a portion of the resources available for a role. They also propose metrics for evaluating process performance such as: the minimum number of resources needed for a role in order to complete a process, the expected number of activity instances when completing a process under the assumption of sufficient resources, and the expected activity response time. Colored Petri Nets are used in [22] for understanding how bounded resources can impact the behaviour of a process. They introduce the notion of "flexible resource allocation" as a way to assign resources associated with a given role based on priorities. In their approach, alternative strategies are used to better allocate a fixed number of available resources. Havur *et al.* [15] study the problem of resource allocation in business processes management systems where constraints can be assigned to resources (e.g., time of availability) and have dependencies. Their technique is based on the answer set programming formalism and is capable of deriving optimal schedules. Sperl *et al.* [28] describe a stochastic method for quantifying resource utilisation relative to structural properties of processes and past executions.

In [29], a solution is presented to optimise resource allocation by focusing on the structure of the process, and more precisely on dependencies between resources and tasks. The approach then proposes a solution to adapt the structure of the business process to better fit the resources available in the enterprise. The authors in [7] focus on the specification and verification of concurrently running processes, operating in time-critical scenarios and having assigned a limited amount of resources. The authors propose to use a fragment of first-order logic to capture process fragments along the timeline and to combine them in a sound model, by observing constraints defined on both activity durations and resource availability. In [25], a contribution to the field of business process simulation is made by providing a new simulation engine, which supports advanced resource specificities such as queuing mechanisms, resource dependencies, or simulation parameters. A conceptual model supports these features and a prototype implementation of this conceptual model are proposed. Incorporating these features also allows for more accurate simulation of the processes and obtaining more relevant performance metrics. Finally, [16] presents a framework to integrate optimised resource allocation in business processes by adding a new component called *resource manager*. It is responsible for maintaining all relevant information concerning the availability of resources and for allocating resources to a process instance. The process designer can specify resource requirements within the business process model through dedicated resource-allocation activities.

There are many tools supporting the design and management of business processes (e.g., Activiti, Bonita, Camunda, or Signavio), of which a subset supports the analysis and optimisation of processes. For instance, this is the case of

Signavio [1], which packs tools such as the Signavio Process Intelligence for process optimisation. It automatically mines process models from currently running systems and monitors those processes with the purpose of collecting data that enables end-users to make decisions for process improvement. The proposal here takes a different approach since the idea is to compare the possibility to make the decision at design time or at runtime, with static or dynamic allocation of resources.

This work is part of a long term project with the goal of developing different tools for the analysis of BPMN processes. [18,19] present an approach transforming BPMN into the input language of the CADP model checker, thus allowing the automated verification of functional properties and the comparison of BPMN processes. In [12], basic BPMN processes were specified. This work provides operations for the estimation of execution times, and uses model-checking techniques to verify reachability problems and LTL properties. In [9], a model similar to the current one was proposed and was used for stochastic analysis using the statistical model checker PVeStA [3]. In [10], Maude is used to model and analyse the resource allocation of business processes. In that work, optimal allocation is presented as a multi-objective optimisation problem, where response time and resource usage are minimised. [11] proposes an automatic analysis technique to evaluate and compare the execution time and resource occupancy of a business process relative to a workload and a provisioning strategy. Four different strategies were implemented and compared from an experimental perspective. [14] presents an approach to perform probabilistic model checking of multiple executions of a BPMN process (including time and resources) at runtime.

# 6    Concluding Remarks

In this paper, the focus is on business processes developed using the BPMN notation extended with a description of time and resources. Processes are executed several times and those multiple instances compete for the shared resources. In this context, several metrics can be computed, such as average execution time or resource usage percentage. These metrics are helpful to optimise processes by, for instance, increasing the usage of resources or reducing the average execution time. Two different options to compute these metrics have been presented. The first approach relies on off-line simulation techniques and assumes that the allocation of resources is static (same number of resources). The second approach applies at runtime, which requires the instrumentation of an existing platform for executing BPMN processes. This latter approach is dynamic and the number of replicas can be updated for each resource during execution to adapt to a change in the resource usage. Both approaches are fully automated and have been applied to realistic processes.

The static approach is useful for a process that is under development and thus can be refined before being effectively deployed. This approach thus allows users to better understand the process and improve it in the early stage of its development. The static approach does not permit adjusting the resources to the workload, but still corresponds to realistic scenarios. This is the case, for instance, when the number of resources cannot be changed with simple or quick fixes. Complementarily, the dynamic approach adjusts at runtime the number of resources, resulting in a more stable resource usage in terms of occupancy percentages. However, this dynamic change is not always possible since there are some specific kinds of resources (such as human beings) that cannot be immediately or automatically updated. Another difference of the dynamic approach is that it applies to any executable BPMN (no restriction at the syntactic level), whereas the static approach works for a subset of BPMN and also requires probabilities for split exclusive and inclusive gateways.

The main perspective of this work is to investigate how AI techniques could help to develop new allocation strategies based on prediction analytics. More precisely, such techniques could be used to predict the resource usage in the short future and the strategy would rely on these values in order to anticipate the change in the number of resource replicas.

**Acknowledgements.** This work was supported by the Région Auvergne-Rhône-Alpes within the *"Pack Ambition Recherche"* programme. The first author was partially supported by projects UMA18-FEDERJA-180 (J. Andalucía/FEDER) and PGC2018-094905-B-I00 (Spanish MINECO/FEDER). The third author was in part supported by the ECOS-NORD project FACTS (C19M03).

# References

1. Signavio (2019). https://www.signavio.com
2. Activiti: Open source business automation. Accessed Dec 2021
3. AlTurki, M., Meseguer, J.: PVESTA: a parallel statistical model checking and quantitative analysis tool. In: Corradini, A., Klin, B., Cîrstea, C. (eds.) CALCO 2011. LNCS, vol. 6859, pp. 386–392. Springer, Heidelberg (2011). https://doi.org/10.1007/978-3-642-22944-2_28
4. Bartocci, E., Falcone, Y. (eds.): Lectures on Runtime Verification. LNCS, vol. 10457. Springer, Cham (2018). https://doi.org/10.1007/978-3-319-75632-5
5. Christiansen, D.R., Carbone, M., Hildebrandt, T.: Formal semantics and implementation of BPMN 2.0 inclusive gateways. In: Bravetti, M., Bultan, T. (eds.) WS-FM 2010. LNCS, vol. 6551, pp. 146–160. Springer, Heidelberg (2011). https://doi.org/10.1007/978-3-642-19589-1_10
6. Clavel, M., Durán, F., Eker, S., Lincoln, P., Martí-Oliet, N., Meseguer, J., Talcott, C.: All About Maude - A High-Performance Logical Framework. LNCS, vol. 4350. Springer, Heidelberg (2007). https://doi.org/10.1007/978-3-540-71999-1
7. Combi, C., Sala, P., Zerbato, F.: A logical formalization of time-critical processes with resources. In: Weske, M., Montali, M., Weber, I., vom Brocke, J. (eds.) BPM 2018. LNBIP, vol. 329, pp. 20–36. Springer, Cham (2018). https://doi.org/10.1007/978-3-319-98651-7_2
8. Durán, F., Rocha, C., Salaün, G.: A Note on Resource Allocation Analysis of BPMN Processes (2018). http://maude.lcc.uma.es/BPMN-R
9. Durán, F., Rocha, C., Salaün, G.: Stochastic analysis of BPMN with time in rewriting logic. Sci. Comput. Program. **168**, 1–17 (2018)

10. Durán, F., Rocha, C., Salaün, G.: A rewriting logic approach to resource allocation analysis in business process models. Sci. Comput. Program. **183** (2019)
11. Durán, F., Rocha, C., Salaün, G.: Resource provisioning strategies for BPMN processes: specification and analysis using Maude. J. Log. Algebraic Methods Program. **123**, 100711 (2021)
12. Durán, F., Salaün, G.: Verifying timed BPMN processes using Maude. In: Jacquet, J.-M., Massink, M. (eds.) COORDINATION 2017. LNCS, vol. 10319, pp. 219–236. Springer, Cham (2017). https://doi.org/10.1007/978-3-319-59746-1_12
13. Falcone, Y., Krstić, S., Reger, G., Traytel, D.: A taxonomy for classifying runtime verification tools. Int. J. Softw. Tools Technol. Transfer **23**(2), 255–284 (2021). https://doi.org/10.1007/s10009-021-00609-z
14. Falcone, Y., Salaün, G., Zuo, A.: Probabilistic model checking of BPMN processes at runtime. In: ter Beek, M.H., Monahan, R. (eds.) IFM 2022. LNCS, vol. 13274, pp. 191–208. Springer, Heidelberg (2022). https://doi.org/10.1007/978-3-031-07727-2_11
15. Havur, G., Cabanillas, C., Mendling, J., Polleres, A.: Resource allocation with dependencies in business process management systems. In: La Rosa, M., Loos, P., Pastor, O. (eds.) BPM 2016. LNBIP, vol. 260, pp. 3–19. Springer, Cham (2016). https://doi.org/10.1007/978-3-319-45468-9_1
16. Ihde, S., Pufahl, L., Lin, M.-B., Goel, A., Weske, M.: Optimized resource allocations in business process models. In: Hildebrandt, T., van Dongen, B.F., Röglinger, M., Mendling, J. (eds.) BPM 2019. LNBIP, vol. 360, pp. 55–71. Springer, Cham (2019). https://doi.org/10.1007/978-3-030-26643-1_4
17. ISO/IEC. International Standard 19510, Information technology - Business Process Model and Notation (2013)
18. Krishna, A., Poizat, P., Salaün, G.: VBPMN: automated verification of BPMN processes (tool paper). In: Polikarpova, N., Schneider, S. (eds.) IFM 2017. LNCS, vol. 10510, pp. 323–331. Springer, Cham (2017). https://doi.org/10.1007/978-3-319-66845-1_21
19. Krishna, A., Poizat, P., Salaün, G.: Checking business process evolution. Sci. Comput. Program. **170**, 1–26 (2019)
20. Li, J., Fan, Y., Zhou, M.: Performance modeling and analysis of workflow. IEEE Trans. Syst. Man Cybern. **34**(2), 229–242 (2004)
21. Meseguer, J.: Conditional rewriting logic as a unified model of concurrency. Theor. Comput. Sci. **96**(1), 73–155 (1992)
22. Netjes, N., van der Aalst, W., Reijers, H.: Analysis of resource-constrained processes with colored Petri Nets. In: Proceedings of CPN. DAIMI, vol. 576, pp. 251–266 (2005)
23. Oliveira, C., Lima, R., Reijers, H., Ribeiro, J.: Quantitative analysis of resource-constrained business processes. Trans. Syst. Man Cybern. **42**(3), 669–684 (2012)
24. OMG. Business Process Model and Notation (BPMN) - Version 2.0, January 2011
25. Peters, S.P.F., Dijkman, R.M., Grefen, P.W.P.J.: Advanced simulation of resource constructs in business process models. In: Weske, M., Montali, M., Weber, I., vom Brocke, J. (eds.) BPM 2018. LNBIP, vol. 329, pp. 159–175. Springer, Cham (2018). https://doi.org/10.1007/978-3-319-98651-7_10
26. Polyak, B.: Introduction to Optimization. Translations Series in Mathematics and Engineering. Optimization Software Inc. (1987)
27. Schömig, A.K., Rau, H.: A Petri Net Approach for the Performance Analysis of Business Processes. Technical Report 116, Universität Würzburg, Würzburg, Germany, May 1995

28. Sperl, S., Havur, G., Steyskal, S., Cabanillas, C., Polleres, A., Haselböck, A.: Resource utilization prediction in decision-intensive business processes. In: Proceedings of SIMPDA, CEUR Workshop Proceedings, pp. 128–141 (2017)
29. Xu, J., Liu, C., Zhao, X.: Resource allocation vs. business process improvement: how they impact on each other. In: Dumas, M., Reichert, M., Shan, M.-C. (eds.) BPM 2008. LNCS, vol. 5240, pp. 228–243. Springer, Heidelberg (2008). https://doi.org/10.1007/978-3-540-85758-7_18

# Rewriting Privacy

Sebastian Mödersheim[(⊠)]

DTU Compute, Richard Petersens Plads, Building 321,
2800 Kongens Lyngby, Denmark
samo@dtu.dk

**Abstract.** This invited paper extends on the invited talk of the same title held at WRLA 2022. It highlights, summarizes and connects the research works on $(\alpha, \beta)$-privacy, an approach to the verification of privacy properties of security protocols. While the de-facto standard is to express privacy as the trace equivalence of two processes, $(\alpha, \beta)$-privacy goes a radically different way to formulate privacy a reachability problem, where every state is characterized by two formulae $\alpha$ and $\beta$. $\alpha$ formalizes all the information that has been deliberately given to the intruder. $\beta$ formalizes what the intruder actually has found out by observing messages, interacting with other agents, and the knowledge of the protocol. $(\alpha, \beta)$-privacy means that in no reachable state $\beta$ allows to derive more than $\alpha$. We describe research papers that define $(\alpha, \beta)$-privacy for a fixed state; the application to vote secrecy and receipt-freeness; and finally a rewriting-based definition of $(\alpha, \beta)$-privacy for a distributed system.

## 1 Introduction

Privacy is important: you may not feel free to read any book you want, if your choice of books can be observed by others. In the same way, the privacy of voting is essential for the democracy. Another example is the vast variety of cards that can communicate with card readers: an attacker (passive or even active) who can link several uses of the same card is able to perform mass surveillance. In fact, vote privacy and unlinkability for RFID protocols are prime examples of privacy properties of security protocols.

The de-facto standard is to express privacy as the *trace equivalence* of two processes (see [8] for a survey): intuitively, from any interaction of sending and receiving messages, the intruder cannot tell which of the two processes they are interacting with (we use the gender-neutral *they* for the intruder). Unlinkability in RFID protocols, for instance, can be specified as the equivalence between the scenario where any number of tags perform one session each, and the scenario where the same tag performs every session. Vote secrecy in a voting protocol

This paper is based on [10,13,14,17], and I would like to thank my co-authors Luca Viganò, Sébastien Gondron, and Laouen Fernet for a great collaboration as well as Santiago Escobar for helpful comments. This work has been supported by the EU H2020-SU-ICT-03-2018 Project No. 830929 CyberSec4Europe (cybersec4europe.eu).

K. Bae (Ed.): WRLA 2022, LNCS 13252, pp. 22–41, 2022.
https://doi.org/10.1007/978-3-031-12441-9_2

can for instance be specified by two scenarios that differ from each other only by swapping the votes of two honest voters.

This way of specifying privacy goals is quite technical and not very declarative. For instance, one may wonder whether slightly different scenarios could be distinguishable to the intruder and indeed break what we intuitively wanted to achieve. This is an even bigger problem in more complex properties like receipt-freeness of voting protocols, i.e., the goal that a bribed[1] voter cannot prove to the intruder how they voted.

$(\alpha, \beta)$-privacy provides an alternative way of specifying privacy goals that is in many cases more declarative. Given a fixed state of the system, we specify two formulae $\alpha$ and $\beta$ in Herbrand logic [12], a variant of first-order logic, as explained below. $\alpha$ formalizes all the information that has been deliberately released, i.e., that the intruder may know. $\beta$ formalizes what the intruder actually has found out by observing messages, interacting with other agents, and their knowledge of the protocol. Our privacy goal is, roughly speaking, that all relevant derivations the intruder can make from $\beta$ are already entailed by $\alpha$.

For instance, one possible formalization of unlinkability in $(\alpha, \beta)$-privacy has in every state a formula $\alpha$ consisting of conjuncts $T_i \in$ Tags. Here, Tags is the set of RFID tags, and each $T_i$ is a distinct free variable of $\alpha$ representing the tag that performed the $i$-th transaction. Thus, $\alpha$ simply specifies that the intruder must not find out more about the tags than the trivial fact that they are tags, in particular the intruder must not be able to tell whether $T_i = T_j$ for any $i \neq j$.

Furthermore, in a possible formalization of vote privacy, $\alpha$ in each state has just a free variable $v_i$ for every cast vote and the information that $v_i \in \{0,1\}$ (if it is a binary vote). In the state after the voting has finished and ballots have been tallied, $\alpha$ additionally tells the sum of the $v_i$. This specifies that the intruder must not learn more about the votes than their sum—the published election result. In fact, it turns out that this is in some sense equivalent to the vote swap formulation mentioned above. The more declarative formulation can thus provide a justification of an existing notion. In contrast to vote swap, the $(\alpha, \beta)$-privacy approach works also for other voting systems where voters can give a list of preferences for instance. Finally in $(\alpha, \beta)$-privacy, receipt-freeness can be specified as the *same* goal as vote privacy, just giving as part of $\beta$ more information to the intruder about a voter they bribed.

For what concerns $\beta$, the most common case will be that it contains a list of messages that the intruder has observed. While such a message may contain an encrypted vote $v_i$ that the intruder cannot see, they may know the structure of that message. This gives rise to the concept of a *message-analysis problem*: can the intruder perform any experiment on their knowledge that would rule out a model of $\alpha$, e.g., showing that two particular votes must be the same. We show that sometimes this allows for very intuitive manual proofs of privacy when there is a simple construction to extend an arbitrary model of $\alpha$ to a model of $\beta$.

---

[1] There may be a variety of reasons that a voter may try to prove how they voted, e.g., peer pressure, for simplicity we just say "bribed" in all cases.

Up till this point, we are only looking at a single state. This leaves aside many questions of the interaction between the intruder and honest agents. We give a model of honest agents as sets of *transactions*, where a transaction models a small process that is atomic (and contains no repetitions). A transaction may contain a condition and consequently go into different branches. This gives rise to a generalization of the message-analysis problem, since the intruder may not a priori know which branch was taken, so there may be several viable candidates for the structure of a given message. Vice-versa, if the intruder can establish which structure the message actually has, they may learn the value of the condition and thus possibly something about the variables in $\alpha$. This generalization of the $(\alpha, \beta)$-privacy approach to transition systems is defined using a set of rewrite rules that symbolically evaluate a given transaction and contrast it with the observations of the intruder. For example, if in one branch no message is sent, but the intruder observes a message, then the respective branch is also excluded.

There are two reasons why this paper is called "Rewriting Privacy": First, it provides an alternative approach that is more declarative than existing ones; it can even serve as a justification for the more technical formulation of privacy goals in previous approaches. This declarativity can also be beneficial for automation, since we can treat privacy as a reachability problem (rather than a bi-similarity problem) without restrictions like those that come with diff-equivalence (cf. Sect. 5). Second, the definition of $(\alpha, \beta)$-privacy is very closely related to term rewriting, as it is based on term algebraic models and Herbrand logic. Also state transition involve a symbolic execution by the intruder, which is formalized using rewrite rules.

This paper highlights, summarizes, and connects material from the following publications and is organized in sections as follows:

- Section two introduces $(\alpha, \beta)$-privacy for a fixed state and the message analysis problem. This is based on the article "Mödersheim, S., Viganò, L.: Alpha-beta privacy. ACM Trans. Priv. Secur. 22(1), 1–35 (2019)" [17].
- Section three discusses the use in voting privacy and receipt freeness. This is based on the paper "Gondron, S., Mödersheim, S.: Formalizing and Proving Privacy Properties of Voting Protocols Using Alpha-Beta Privacy. In: ESORICS 2019. LNCS 11735 (2019)" [13].
- Section four gives the definition of a transition system based on rewrite rules. This is based on the technical report "Gondron, S., Mödersheim, S., Viganò, L.: Privacy as reachability (extended version). Tech. rep., DTU (2021)" [14] available at https://people.compute.dtu.dk/samo/abg.pdf.
- Section five gives conclusions and outlook.

## 2    Alpha-Beta Privacy for a Fixed State

The inspiration to $(\alpha, \beta)$-privacy came from zero-knowledge proofs, i.e., small protocols between a prover and a verifier that should convince the verifier of a certain statement, e.g., "Alice is over 18". This statement being proved, is definitely revealed to the verifier. The zero-knowledge property means that the

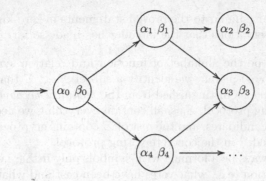

**Fig. 1.** Illustration of the state space.

verifier (or others) should not be able to learn anything else from the zero-knowledge proof—besides the proved statement. For instance they should not be able to determine whether or not "Alice is over 65". However, they can make inferences from what was proved (and from whatever they already know), for instance "Alice is over 15".

$(\alpha, \beta)$-privacy is an approach rooted in logic: we want to model an intruder who can reason about their observations such as received messages and combine it with their knowledge of the protocol. To that end, we use as a logical framework *Herbrand logic* [12]: we model the intruder knowledge as formulae in that logic, and we consider what the intruder can logically deduce from this knowledge.

*Herbrand Logic* is a variant of First-Order Logic FOL. The difference to FOL is that instead of an arbitrary universe, we take the Herbrand universe—the set of terms generated by the function symbols. For instance, using as function symbols the constant $z$ and the unary function $s$, we get a universe isomorphic to the natural numbers. Herbrand logic is thus very expressive (it can formalize arithmetic), but the actual reason for using it is to be close to the term rewriting-based models of messages we have: this allows us to represent cryptographic operations with ("free") function symbols as standard in most black-box cryptography models. As it is standard, we define the semantics as a model relation $\mathcal{I} \models \phi$ between a formula $\phi$ and an interpretation $\mathcal{I}$ (mapping variables to the universe, and $n$-ary relation symbols to a set of $n$-tuples of the universe).

Since relation symbols in Herbrand logic are interpreted, we can use them to simulate also interpreted function symbols [12]. We thus can introduce interpreted functions as syntactic sugar, and for distinction with normal uninterpreted functions, we write those functions with square brackets like $concr[l]$.

## 2.1 The State Space

The general idea is that we model a world of reachable states and each reachable states contains a pair $\alpha_i$ and $\beta_i$ of Herbrand logic formulae that model the intruder knowledge, as sketched in Fig. 1. Here, $\alpha_i$ is the information that has been released so far (similar to the proved statements in zero-knowledge proofs) and $\beta_i$ are the observations that the intruder has made so far.

*Alphabets.* Let $\Sigma$ be the alphabet of function and relation symbols used in a specification. In $(\alpha, \beta)$-privacy we identify a subset $\Sigma_0 \subset \Sigma$ that represents the *high-level vocabulary* as distinguished from the *technical vocabulary* $\Sigma \setminus \Sigma_0$. For instance, in a voting protocol, $\Sigma_0$ shall contain everything we need to talk about votes, voters, and candidates, and the rest of $\Sigma$ contains cryptographic operators and everything needed in the concrete voting protocol.

Now $\alpha$ shall always be a formula with symbols only in $\Sigma_0$, i.e., talking about high-level information (e.g., what votes have been cast and what the sum of the votes is), while $\beta$ can be over they entire $\Sigma$. We can then define what $(\alpha, \beta)$-privacy means:

**Definition 1 ($(\alpha, \beta)$-privacy).** *Given a formula $\alpha$ over $\Sigma_0$ and a formula $\beta$ over $\Sigma$. We say $(\alpha, \beta)$-privacy holds iff every model of $\alpha$ is consistent with $\beta$.*

Intuitively that means that from $\beta$ the intruder does not learn anything (except "technical" stuff in $\Sigma \setminus \Sigma_0$) that is not implied by $\alpha$ already.

## 2.2 Example

To model a binary voting protocol, we may use an interpreted function $x[\cdot] \in \Sigma_0$ to map voters (for simplicity say 1, 2, 3...) to their votes in $\{0, 1\}$. A reachable state in the protocol could be that three votes $x[1], x[2], x[3]$ have been cast and the result is published; let the concrete result be *one vote for yes*. The formula $\alpha$ in this state is then:

$$\alpha \equiv x[1], x[2], x[3] \in \{0, 1\} \wedge x[1] + x[2] + x[3] = 1$$

The models of $\alpha$ are thus:

| Model | $x[1]$ | $x[2]$ | $x[3]$ |
|-------|------|------|------|
| $\mathcal{I}_1$ | 1 | 0 | 0 |
| $\mathcal{I}_2$ | 0 | 1 | 0 |
| $\mathcal{I}_3$ | 0 | 0 | 1 |

These three models represent the three possible worlds that the intruder can imagine in which $\alpha$ is true. One of the models is the *reality*, and the intruder usually does not know which one. Indeed privacy now means that the intruder *cannot exclude* any of the models by reasoning in $\beta$.

For $\beta$ we need to specialize to a particular voting protocol and we use the old and simple but very instructive example of FOO'92 [11]. Here, the cast votes are published on a bulletin board, and as a first approximation let us forget about the cryptography and model that the raw votes are published in some permutation, which we model as an interpreted function $\pi[\cdot] \in \Sigma \setminus \Sigma_0$. Let us have as the $\beta$ in the reached state the following bulletin board:

$$\beta \equiv x[\pi[1]] = 1 \wedge x[\pi[2]] = 0 \wedge x[\pi[3]] = 0 \ \wedge \pi \text{ is a permutation on } \{1, 2, 3\}$$

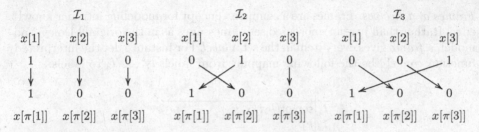

**Fig. 2.** For each model of $\alpha$ there is a suitable interpretation of the permutation $\pi$, indicated by the arrows.

Now it is easy to see that $(\alpha, \beta)$-privacy holds in this state: any model of $\alpha$ can be extended to a model of $\beta$ using a suitable interpretation of $\pi$ as depicted by the arrows in Fig. 2.

In fact, this is an example of how easy a privacy proof in $(\alpha, \beta)$-privacy can sometimes be: we just need to find a simple argument why every model of $\alpha$ is compatible with $\beta$, i.e., how the intruder could explain all their observations for every model of $\alpha$. Let us conclude the example with an additional twist: suppose for some security bug the intruder can actually derive that voters 1 and 2 have voted differently (let us we write $\doteq$ for equality in Herbrand logic formulae):

$$\beta \models x[1] \not\doteq x[2]$$

This would in fact mean a violation of $(\alpha, \beta)$-privacy, because model $\mathcal{I}_3$ above cannot be extended to a model of $\beta$. Note that the intruder still cannot tell whether $\mathcal{I}_1$ or $\mathcal{I}_2$ is the reality, but they have found out strictly more than we have allowed by $\alpha$.

## 2.3   Message Analysis Problems

We want to model algebraic properties of cryptographic operators like

$$\mathsf{dscrypt}(\mathsf{scrypt}(k, m), m) = m \text{ and } \mathsf{vscrypt}(k, \mathsf{scrypt}(k, m)) = \mathsf{true}$$

expressing that symmetric encryption (scrypt) and decryption (dscrypt) cancel each other out, and that one can tell if the decryption was correct (vscrypt). To model this, we simply take as the Herbrand universe the *quotient algebra* of the term algebra by the algebraic identities. Thus, two terms are equal iff that is a consequence of the given algebraic properties. For instance, let $h$ be another function symbol modeling a hash function. Then $h(m)$ and $h(m')$ are interpreted differently unless $m = m'$. Thus hash functions in our model (and in most black-box cryptography models) are collision-free (even though they are not in reality). We will use generally $h$ for public one-way functions, and we may in some examples use it as a unary and in some as a binary function.

*Frames and Recipes.* Frames are a common concept for modeling intruder knowledge. Rather than just an unordered set of messages as in the original Dolev-Yao model, a *frame* gives every item in the set a *label.* For instance, let the interpreted function $concr[\cdot]$ be the following mapping from labels $l_1, \ldots, l_3$ to labels:

| | concr |
|---|---|
| $l_1$ | $\mathsf{scrypt}(k_1, r_1, k_2)$ |
| $l_2$ | $k_1$ |
| $l_3$ | $\mathsf{scrypt}(h(k_1, k_2), r_2, \mathsf{secret})$ |

The use of labels allows us to speak about *recipes* which are terms built from only labels and cryptographic operators. When we apply a frame to a recipe, we simply use the frame like a substitution that substitutes each label of the recipe with the corresponding term of the frame, for instance:

$$concr[\mathsf{dscrypt}(l_2, l_1)] = k_2$$
$$concr[\mathsf{dscrypt}(h(l_2, \mathsf{dscrypt}(l_2, l_1)), l_3)] = \mathsf{secret}$$

It is straightforward to define the behavior of such a frame $concr[\cdot]$ by a formula in Herbrand logic.

Another standard concept is that of *static equivalence* which says whether the intruder is unable to distinguish between two frames: they cannot make an *experiment* by building two recipes $r_1$ and $r_2$ that yield the same term in one frame, and a different term in the other frame:

**Definition 2 (Static Equivalence).** *Two frames $concr_1$ and $concr_2$ are called statically equivalent, written $concr_1 \sim concr_2$, iff for all recipes $r_1, r_2$:*

$$concr_1[r_1] = concr_1[r_2] \text{ iff } concr_2[r_1] = concr_2[r_2]$$

Also this notion is straightforward to formalize in Herbrand logic.

*Structural Knowledge.* In contrast to trace equivalence approaches, $(\alpha, \beta)$-privacy has only one "reality" in each state with *one* intruder knowledge (the messages actually observed), and thus never has two alternative "intruder knowledges" to compare with static equivalence. However, we model, and this is one of the crucial ideas of $(\alpha, \beta)$-privacy, that the intruder knows something about the structure of the messages in their knowledge. This structural knowledge can be characterized similarly to the concrete intruder knowledge $concr[\cdot]$ as another frame, called $struct[\cdot]$ that has the same labels but the terms may contain variables.

For instance, consider the following state of a hypothetical protocol, and have $\alpha \equiv x \in \{a, b, c\}$ (where $a, b, c$ are constants the intruder knows initially) and the intruder has received the message $h(a)$. The intruder cannot directly see $a$, but knows that the protocol message is $h(x)$ for whatever $x$ is:

|       | struct | concr |
|-------|--------|-------|
| $l_1$ | $a$    | $a$   |
| $l_2$ | $b$    | $b$   |
| $l_3$ | $c$    | $c$   |
| $l_4$ | $h(x)$ | $h(a)$|

In fact, we shall always have that *concr* is an instance of *struct*, namely filling in for every variable the value that it really has (here $a$ for $x$). More formally, let $\mathcal{I}$ be the model of $\alpha$ that represents what really happened, then $concr = \mathcal{I}(struct)$.

Since $h$ is a public function, the intruder can now make an experiment: namely whether $h(l_1) = l_4$. This succeeds, and the intruder thus learns that $x = a$. This corresponds of course to a guessing attack of the intruder. The simple way to formalize this and similar reasoning in general is to include in $\beta$ the information *concr* $\sim$ *struct*: since $concr = \mathcal{I}(struct)$, the intruder can rule out any model $\mathcal{I}'$ that allows them to distinguish between *concr* and *struct*. Note that by making *concr* $\sim$ *struct* part of $\beta$, we do *not* state that these two frames are statically equivalent (in general they are not, due to the variables) but rather that *under every model* of $\beta$ they are statically equivalent. In the above example we have the experiment $h(l_1)$ vs $l_4$ which gives the same in *struct* iff $x = a$ and thus excludes the other models of $\alpha$.

If we modify the example by including an unknown nonce $n$ into the hash:

|       | struct    | concr    |
|-------|-----------|----------|
| $l_1$ | $a$       | $a$      |
| $l_2$ | $b$       | $b$      |
| $l_3$ | $c$       | $c$      |
| $l_4$ | $h(n,x)$  | $h(n,a)$ |

then $(\alpha, \beta)$-privacy actually holds: there is no experiment that allows one to exclude any of the models of $\alpha$. Roughly speaking, this is because the intruder has no way to re-construct the term of $l_4$, not knowing the nonce.

Suppose now there are several runs of the protocol, and we have

$$\alpha \equiv x_1 \in \{a, b, c\} \ \wedge \ x_2 \in \{a, b, c\}$$

and the protocol (unwisely) uses the same nonce $n$ twice:

$\beta \equiv$

|       | struct      | concr     |
|-------|-------------|-----------|
| $l_1$ | $a$         | $a$       |
| $l_2$ | $b$         | $b$       |
| $l_3$ | $c$         | $c$       |
| $l_4$ | $h(n, x_1)$ | $h(n, a)$ |
| $l_5$ | $h(n, x_2)$ | $h(n, b)$ |

The simple experiment to compare $l_4$ and $l_5$ reveals whether $x_1 = x_2$, violating $(\alpha, \beta)$-privacy. Using two distinct nonces instead prevents the privacy breach, roughly speaking, because the intruder cannot re-construct the messages of $l_4$ or $l_5$ (not knowing the nonces) and also comparing them does not help (they are already different because of the distinct nonces).

Thus, we define the message analysis problem as follows:

**Definition 3 (Message Analysis Problem).** *Given a formula $\alpha$, a model $\mathcal{I} \models \alpha$ (the reality), and a frame struct such that all variables in struct occur freely in $\alpha$. Let concr $= \mathcal{I}(struct)$ be the concrete frame resulting from instantiating struct with the real values, then we call the message analysis problem induced by $\alpha$, $\mathcal{I}$ and struct the following formula $\beta$:*

$$\beta = \alpha \wedge \phi_{concr} \wedge \phi_{struct} \wedge concr \sim struct$$

*where $\phi_{concr}$ and $\phi_{struct}$ are formulae formalizing the two said frames.*

## 3  Applications in Vote Privacy

We now illustrate how $(\alpha, \beta)$-privacy can be practically used for verifying a simple voting protocol. This also illustrates some general properties of $(\alpha, \beta)$-privacy like stability under background knowledge. As the concrete protocol we use the already mentioned protocol FOO'92. Our precise formalization is found in [13] and it this was inspired by the first formalization in [15].

For this protocol we need the following operators and properties:

$$\mathsf{getMsg}(\mathsf{sign}(\mathsf{priv}(K), M)) = M \qquad \mathsf{vsign}(\mathsf{pub}(K), \mathsf{sign}(\mathsf{priv}(K), M)) = \mathsf{true}$$
$$\mathsf{open}(R, \mathsf{commit}(R, M)) = M \qquad \mathsf{vcommit}(R, \mathsf{commit}(R, M)) = \mathsf{true}$$

Here all operators except priv are public operators, i.e., the intruder can use them in recipes. The first line describes electronic signatures: in our model one can retrieve the text of the signature by getMsg (without knowing any keys) and verify the signature knowing the corresponding public key. The second line describes commitments: the first argument of commit is a random secret $R$ and the second argument the message $M$ being committed on. Until the author of the commitment releases the secret $R$, nobody can read the message $M$. It is a commitment in the sense that when $R$ is released, everybody can check it is a commitment with $R$ to message $M$, so a dishonest agent cannot change what they committed to.

The messages that will be on the bulletin board in FOO'92 have the form:

$$\mathsf{sign}(\mathsf{priv}(a), \mathsf{commit}(r[\pi[j]], x[\pi[j]]))$$

where $\mathsf{priv}(a)$ is the private key of the administrator,[2] and the commitment is by some voter $\pi[j]$ (i.e., the voter whose vote is published in position $j$ on the board), containing a random secret $r[\pi[j]]$ and the vote $x[\pi[j]]$ of that voter.

---

[2] In the first part of the protocol (that we omit here for simplicity) these signatures are issued by the administrator using a blinding signature scheme to prevent that the administrator can see the votes.

The reason for the commitment schemes is that a dishonest bulletin board cannot block votes it does not like, but voters cannot change their vote once it is on the board. In the final phase, the commitments are opened by the voters: they identify their position $\pi[j]$ on the board and send $r[\pi[j]]$ which is then stored along with the corresponding signature.

In $(\alpha, \beta)$-privacy, we have of course the structural knowledge containing messages of the above form, and the concrete knowledge where $r[\cdot]$, $\pi[\cdot]$, and $x[\cdot]$ are instantiated with concrete values. For instance, again with three voters of whom one has voted 1, we have the following frames:[3]

|       | struct | concr |
|-------|--------|-------|
| $l_0$ | $\mathsf{pub}(a)$ | $\mathsf{pub}(a)$ |
| $l_1$ | $\mathsf{sign}(\mathsf{priv}(a), \mathsf{commit}(r[\pi[1]], x[\pi[1]]))$ $r[\pi[1]]$ | $\mathsf{sign}(\mathsf{priv}(a), \mathsf{commit}(17, 1))$ $17$ |
| $l_2$ | $\mathsf{sign}(\mathsf{priv}(a), \mathsf{commit}(r[\pi[2]], x[\pi[2]]))$ $r[\pi[2]]$ | $\mathsf{sign}(\mathsf{priv}(a), \mathsf{commit}(42, 0))$ $42$ |
| $l_3$ | $\mathsf{sign}(\mathsf{priv}(a), \mathsf{commit}(r[\pi[3]], x[\pi[3]]))$ $r[\pi[3]]$ | $\mathsf{sign}(\mathsf{priv}(a), \mathsf{commit}(24, 0))$ $24$ |

To show that $(\alpha, \beta)$-privacy is preserved in this state, we can show that an arbitrary model of $\alpha$ can be extended to a model of $\beta$, i.e., such that struct $\sim$ concr. This is fairly easy, following the idea of Fig. 2 discussed earlier: let $\mathcal{I}_0$ be the model of $\alpha$ that is the reality (thus $\mathcal{I}_0(x[i])$ is $i$'s true vote), let $\pi_0$ be the real permutation used by the bulletin board, and let finally $\mathcal{I}$ by any model of $\alpha$. Since the sum of the votes must be the same in $\mathcal{I}$ and $\mathcal{I}_0$, there must be a permutation $\psi$ such that $\mathcal{I}_0(x[i]) = \mathcal{I}(x[\psi(i)])$. Now interpret $\pi[\cdot]$ as follows:

$$\mathcal{I}(\pi) := \psi \circ \pi_0$$

We write here := to denote that we extend the model $\mathcal{I}$. Thus we have $\mathcal{I}(x[\pi[j]]) = \mathcal{I}(x[\psi[\pi_0[j]]]) = \mathcal{I}_0(x[\pi_0[j]])$ and with a suitable interpretation for $r[\cdot]$ we arrive at a model $\mathcal{I}$ where $\mathcal{I}(concr) = \mathcal{I}(struct)$; i.e., in $\mathcal{I}$, the frames concr and struct are identical and thus of course also statically equivalent. Since we can do this for any model $\mathcal{I}$ of $\alpha$, $(\alpha, \beta)$-privacy holds.

This demonstrates that also the proof of $(\alpha, \beta)$-privacy can be rather declarative and suitable for manual reasoning. There are many cases where the proof is a bit more difficult: for instance in voting with preferences where the models of $\alpha$ are not permutations of the votes, the proof is more difficult, and we cannot in general arrive at constructions where every model $\mathcal{I} \models \alpha$ can be extended so that $\mathcal{I} \models concr \doteq struct$ but only $\mathcal{I} \models concr \sim struct$, making the proofs a bit more difficult.

## 3.1  Receipt Freeness

A related problem to vote secrecy is receipt freeness: we do not want that a voter can *prove* how they voted, because this could make the elections vulnerable to

---

[3] For more intuitive reading we have displayed some concrete numbers for the random values $r[\cdot]$, but in the actual model they are symbolic unguessable constants.

pressure, extortion, and bribery. Voters may of course tell others how they voted, but it must always be possible for them to *lie* about how they voted.

The interesting notion here is the *inability to prove* how one has voted. The intruder may pressure an agent to reveal all their secrets like private keys and all the session data from the voting protocol, but they might lie about it. The question is thus, if the messages that the intruder receives from a bribed voter can act as a proof how they voted.

Thanks to the logical approach of $(\alpha, \beta)$-privacy, this question can directly be specified in $\beta$: consider a voter $Dan$ (say $Dan = 1$ is the voter number 1) who is being bribed by the intruder and we simply model also the structural and concrete knowledge of $Dan$ as frames $struct_{Dan}$ and $concr_{Dan}$. Let us pick labels for these frames that are initially disjoint from those of the intruder knowledge $struct$ and $concr$. We now model that $Dan$ tells the intruder some *story* (be it true or not) about what is in his knowledge. Let us look again at the FOO'92 example from before and consider that $Dan$'s vote is the second on the board, then we may have:

| | $struct$ | $concr$ | $struct_{Dan}$ | $concr_{Dan}$ |
|---|---|---|---|---|
| $l_0$ | ... | ... | | |
| $l_1$ | ... | ... | | |
| $l_2$ | ... | ... | | |
| $l_3$ | ... | ... | | |
| $d_0$ | $\mathsf{priv}(Dan)$ | $\mathsf{priv}(Dan)$ | $\mathsf{priv}(Dan)$ | $\mathsf{priv}(Dan)$ |
| $d_1$ | $x[Dan]$ | 1 | $x[Dan]$ | 0 |
| $d_2$ | $r[Dan]$ | 17 | $r[Dan]$ | 42 |
| $d_3$ | $\pi^{-1}[Dan]$ | 1 | $\pi^{-1}[Dan]$ | 2 |
| $d_4$ | ... | ... | ... | ... |

The extension of the intruder knowledge by the story of $Dan$ is as follows: the intruder knows what the structural knowledge of $Dan$ is, because that is inherent in the protocol; thus this part (highlighted in blue) is determined to be the true structural knowledge of $Dan$. The concrete knowledge is what $Dan$ can lie about: highlighted in red we have the messages that $Dan$ may have told the intruder. For these red messages $Dan$ can choose any messages he can construct from his actual knowledge.

So in the example, $Dan$ has told the truth about his private key (note that both the symbolic and the concrete message are identical as there is nothing to interpret). He cannot really lie about that, because the intruder knows everybody's public key (since the function $\mathsf{pub}$ is public): Even though we have no algebraic properties that directly relate public and private keys, the properties of signatures allow us to tell that: the intruder can sign an arbitrary message with whatever $Dan$ tries to pass on as his private key and check if that signature verifies with $Dan$'s public key.

About the other things $Dan$ has lied here: he claims to have voted 1 (i.e. supposedly $x[Dan]$), and that his vote (supposedly $\pi^{-1}[Dan]$) is the first item on the bulletin board. He also takes as his commitment secret the value 17 that he found in that entry on the board. (We have abbreviated with ... in *Dan*'s knowledge that he also knows the entire bulletin board; for these items he must of course tell the truth, but the intruder may not even ask, as this is public knowledge).

The clou is now that for every model $\mathcal{I} \models \alpha$, *Dan* can make such a story (i.e., the messages highlighted in red in the above example) that is consistent with the rest of the knowledge of the intruder, i.e., $\mathcal{I}$ can be extended such that *concr* $\sim$ *struct* holds.

More generally, we can thus describe receipt freeness as follows:

**Definition 4.** *Given a message analysis problem with a formula $\alpha$, with the structural and concrete knowledge of the intruder struct and concr using labels $l_1, \ldots, l_n$, as well as the structural and concrete knowledge struct$_{Dan}$ and concr$_{Dan}$ of an agent Dan with labels $d_1, \ldots, d_k$.*

*Then* receipt freeness *(with respect to this message analysis problem and Dan's knowledge) is obtained by extending $\beta$ with the following* liar's axiom:

$$\phi_{lie}(struct, concr, struct_{Dan}, concr_{Dan}) \equiv$$
$$struct[d_1] \doteq struct_{Dan}[d_1] \wedge \ldots \wedge struct[d_k] \doteq struct_{Dan}[d_k]$$
$$\wedge \exists s_1, \ldots, s_k. \Big(concr[d_1] \doteq concr_{Dan}[s_1] \wedge \ldots \wedge concr[d_k] \doteq concr_{Dan}[s_k]$$
$$\wedge gen_{Dan}(s_1) \wedge \ldots \wedge gen_{Dan}(s_k)\Big)$$

*where gen$_{Dan}(s)$ formalizes that s is a recipe over labels $d_1, \ldots, d_k$, i.e., the concr$_{Dan}[s_i]$ are messages Dan can construct.*

This means that for every model $\mathcal{I} \models \alpha$ and every label $d_i$, *Dan* can generate a story (the *concr*$[s_i]$) so that the extension of *struct* by *struct*$_{Dan}$ and the extension of *concr* by this story still is consistent with *concr* $\sim$ *struct*.

The idea is thus: receipt freeness means that bribed agents like *Dan* can consistently lie about their knowledge for every model of $\alpha$ and therefore nothing they can say allows the intruder to logically rule out any model of $\alpha$. That is in our opinion a very declarative way to express receipt freeness, i.e., the inability to prove what one has done.

Observe that the definition of receipt freeness is independent of the formula $\alpha$ or the details of the protocol: receipt freeness of a private choice of an agent simply means that, for every model of $\alpha$, the agent can consistently lie about their knowledge. This makes the model also applicable to other areas where we want to prevent pressuring or bribery of decisions, e.g., in medical prescriptions.

Unfortunately, FOO'92 satisfies receipt freeness only in its final state; in the intermediate states where the signed commitments are published but not yet opened, each agent is in the unique position to reveal their own secret to the intruder, and thus there is not a consistent story for every model. However, the resulting $\beta$ formula itself is always consistent: since an agent can say the truth ($s_i = d_i$), there is always one model of $\alpha$ for which a consistent story exists.

## 3.2  Stability Under Background Knowledge

Consider the following situation: a person from a large city moves to a small village. At the next election, there is one vote for a party that has never received a vote in this village before. It is quite likely that the person moving in was the one who cast this vote. But can this actually be consider a breach of voting privacy? Apparently this works in any voting system where the number of votes for each party (or candidate) is revealed, no matter how this is implemented. In fact, we argue that this is *background knowledge* that one may have about voters, e.g., having surveys how people tend to vote in different areas.

A question that arises from this is: could such background knowledge together with observations be used to further degrade the privacy of voters? More formally, consider a state with formulae $\alpha$ and $\beta$ such that $(\alpha, \beta)$-privacy holds. Suppose the intruder has some background knowledge $\alpha_0$ (over the high-level alphabet $\Sigma_0$). Then, they clearly know $\alpha \wedge \alpha_0 \wedge \beta$, i.e., the intruder can make inferences from all of these three sub-formulae combined. Does this possibly violate privacy in a way which knowing just $\alpha$ and $\beta$ just would not? Of course, $\alpha_0$ may allow the intruder to exclude some models of $\alpha$, but that is just combining the information $\alpha$ that we deliberately gave them and the background knowledge $\alpha_0$ that we cannot "take away" from them. The question is thus: can we be sure that the intruder cannot derive even more than $\alpha \wedge \alpha_0$ in this case? The following result (from [17]) shows that we can indeed be sure:

**Theorem 1 (Stability under Background Knowledge).** *Given a state with $(\alpha, \beta)$ as above such that $(\alpha, \beta)$-privacy holds, and let $\alpha_0$ be a $\Sigma_0$ formula. Then also $(\alpha \wedge \alpha_0, \beta \wedge \alpha_0)$-privacy holds.*

Thus, once $(\alpha, \beta)$-privacy is established, we do not need to worry that any background knowledge could cause breaches beyond that background knowledge and what we deliberately reveal.

## 4  Privacy as Reachability

So far, we have looked at one state at a time. This does not tell much how the intruder *interacts* with other principals, and in fact, what honest agents send often depends on what they received before and checks that they have made. Thus, their reaction to a particular input can be revealing to the intruder.

Several popular approaches formulate protocols using rewrite rules, in particular OFMC [2], Maude-NPA [9], and Tamarin [16]. The basic idea is that a state is a multi-set of facts which can be easily induced by an operator " $\cdot$ " with the algebraic properties of being associative, commutative, and having a unit. The rewrite rules can then have the form

$$\text{AgentState}(t_1, \ldots, t_n) \cdot \text{Net}(t_{in}) \Rightarrow \text{AgentState}(t'_1, \ldots, t'_n) \cdot \text{Net}(t_{out})$$

meaning that they can be applied in a state that contains a subset of facts matching the left-hand side (modeling an honest agent in a particular local state characterized by terms $t_1, \ldots, t_n$) and an available input message $t_{in}$ on the network. These facts are then replaced by the right-hand side under the match, thereby updating the agent state to terms $t'_1, \ldots, t'_n$ and replacing the received message $t_{in}$ with the reply $t_{out}$. An advantage of rewrite rules is that we have directly a notion of an *atomic* transaction that includes agents receiving a message, checking conditions of their current state, and if successful, updating their state, and sending a reply; no other transitions can be made concurrently.

Another popular approach, e.g., used in the tool ProVerif [3], is based instead on the applied $\pi$-calculus. This has the advantage that one does not need to reason about agents "saving" their local state after each received message and that we can directly formulate conditional behavior and repetition.

For $(\alpha, \beta)$-privacy we have chosen to define a notion of transaction that combines some advantages of both: we have an atomic unit that consists of receiving, checking, updating, and replying, where the checks may have conditionals branches for both the positive and negative case. A transaction is a process following the grammar of $\mathcal{P}_l$:

$$
\begin{aligned}
\mathcal{P}_l ::= &\ \star\, x \in D.\mathcal{P}_l & \qquad \mathcal{P}_r ::= &\ \mathsf{snd}(t).\mathcal{P}_r \\
\mid &\ \mathsf{rcv}(x).\mathcal{P}_l & \mid &\ \mathsf{cell}(s) := t.\mathcal{P}_r \\
\mid &\ x := \mathsf{cell}(s).\mathcal{P}_l & \mid &\ \star\, \phi.\mathcal{P}_r \\
\mid &\ \text{if } \phi \text{ then } \mathcal{P}_l \text{ else } \mathcal{P}_l & \mid &\ 0 \\
\mid &\ \nu \overline{N}.\mathcal{P}_r
\end{aligned}
$$

$\mathcal{P}_l$ represents everything that in terms of rewriting rules would be on the left-hand side, namely receiving $\mathsf{rcv}(x)$, reading the current state $x := \mathsf{cell}(s)$, and checking the condition of if $\phi$ then $\mathcal{P}_l$ else $\mathcal{P}_l$. Dually, $\mathcal{P}_r$ is everything that would be on the right-hand side of a rewrite rule, namely sending $\mathsf{snd}(t)$ and modifying the state $\mathsf{cell}(s) := t$. Note that memory cells here can model also a session-independent state of an agent like a database.

A new concept is $\star\, x \in D$, which means to non-deterministically pick a value $x$ from the finite domain $D$.[4] This is indeed where privacy comes into play: for instance a binary vote may be modeled as $\star\, x \in \{0, 1\}$. Moreover, we always release such information to the intruder, i.e., we augment $\alpha$ by the conjunct $x \in D$ upon this state transition. On the right-hand side, we similarly have $\star\, \phi$ meaning that when reaching this, the process declassifies the formula $\phi$ (which thus also gets added to $\alpha$ upon this state transition). Similar to process calculi, $\nu \overline{N}$ represents introduction of fresh constants, and 0 the finished process.

We have one restriction though: we assume that in each path from root to leaves through the if-then-else tree of a transaction, we encounter the same $\mathsf{rcv}(x)$ steps and the same $\star\, x \in D$ steps. This is not a big restriction as the number of messages received and the number of private choices should normally not depend on any conditions.

---

[4] The full grammar in [14] is conceptually a bit richer: we may have also a non-deterministic choice of variables that are not directly relevant for privacy, as well as random choices with a given distribution. Both we have left out for simplicity here.

## 4.1   Example: AF2

As an example, let us use the second protocol from Abadi and Fournet [1] (and our model in [17]). It is a handshake between agents designed to reveal as little as possible about participants names and who they are willing to talk to.

Let us have a finite set Agent (there does not need to be a bound on it, but finiteness means we can in principle enumerate) and let us have a binary relation talk on Agent with a fixed (but again arbitrary) initial interpretation. Here talk($a, b$) for two agents $a$ and $b$ means that $a$ is willing to talk to $b$. The start of the protocol is then the following transaction:

$$\star\ A \in \text{Agent.} \star\ B \in \text{Agent.}$$
$$\text{if talk}(A, B) \text{ then } \nu NA.\text{snd(crypt(pub}(B), NA, A))$$
$$\text{else } 0.$$

This non-deterministically chooses two agents $A$ and $B$, and if $A$ indeed wants to talk to $B$, it sends its name encrypted with the public key of $B$. Here $NA$ is a randomization of the encryption (to prevent guessing attacks when public key and content is known). This is characterized by the following properties:

$$\text{dcrypt(priv}(X), \text{crypt(pub}(X), R, M)) = M$$
$$\text{rand(priv}(X), \text{crypt(pub}(X), R, M)) = R$$
$$\text{vcrypt(priv}(X), \text{crypt(pub}(X), R, M)) = \text{true}$$

where all function symbols are public except priv. Thus, knowing the private key, one can obtain the message $M$, the randomness $R$, and check that it is a valid encryption.

As syntactic sugar, let us write let $x = e\ P$ for replacing in $P$ every occurrence of $x$ by $e$. The transaction of $B$ receiving and answering such messages is now:

$$\star\ B \in \text{Agent.rcv}(M).\nu NB.$$
$$\text{if vcrypt(priv}(B), M) = \text{true then}$$
$$\quad \text{let } NA = \text{rand(priv}(B), M)$$
$$\quad \text{let } A = \text{dcrypt(priv}(B), M)$$
$$\quad \text{if talk}(B, A) \text{ then}$$
$$\qquad \text{snd(crypt(pub}(A), NB, NA)) \qquad (a)$$
$$\quad \text{else snd}(NB) \qquad\qquad\qquad\qquad\quad (b)$$
$$\text{else snd}(NB) \qquad\qquad\qquad\qquad\qquad\ \ (c)$$

Thus $B$ is any agent who receives some message $M$, checks if the message is encrypted, and if so extracts the sender name $A$ and randomness $NA$ from it. If $B$ is willing to talk to $A$, then it now sends the acknowledgement which consists of some new randomness $NB$ and sending back the randomness $NA$, encrypted for $A$. In all other cases, i.e., if $B$ cannot decrypt or does not want to talk to $A$, the answer is just the random $NB$.

These two transactions together ensure that $A$ and $B$ can make a handshake, but the intruder observing it cannot find out the name of any of them, since we never release anything but the information that the involved $A$ and $B$ are agents, and also the intruder never knows who is willing to talk to whom.

This privacy property does not hold if the intruder is also an agent $i \in$ Agent knowing their own key pair $\mathsf{pub}(i)$ and $\mathsf{priv}(i)$: if the intruder sends a message to a particular agent $b$ and receives an answer that can be decrypted with $\mathsf{priv}(i)$ then the intruder learns that the responder $B$ was indeed $b$ and that $\mathsf{talk}(b, i)$ holds. Of course, the intruder should in this case indeed learn this—the best protocol could not prevent that. So we can simply specify that we allow this release of information by replacing line (a) with the following expression:

if $A \doteq i$ then $\star$ $\mathsf{talk}(B, A) \wedge B = \mathcal{I}(B)$. $\mathsf{snd}(\mathsf{crypt}(\mathsf{pub}(A), NB, NA))$
else $\mathsf{snd}(\mathsf{crypt}(\mathsf{pub}(A), NB, NA))$

Here $\mathcal{I}$ stands for the reality (the model of $\alpha$ that is truly the case) and thus $\mathcal{I}(B)$ means the real value of $B$. Therefore, this says: in case the intruder is the sender of $M$, and all checks are satisfied, then the intruder learns exactly that ($B$ is the person they contacted, and $B$ is willing to talk to them).

However it still has an attack: in the negative cases when one of the checks failed, the intruder learns that one of the two conditions was not satisfied. Also this is actually reasonable, so let us also release this information in these cases, namely replacing the line labeled (b) by the following:

if $A \doteq i$ then $\star$ $\neg\mathsf{talk}(B, A) \vee B \neq \mathcal{I}(B)$. $\mathsf{snd}(NB)$
else $\mathsf{snd}(NB)$

A similar correction is more tricky for the last case (labeled (c)). In the case that this message was not even an encryption of the expected form, the intruder should not learn anything from this. However, if the intruder has addressed the input message to some different agent $C \neq \mathcal{I}(B)$ and using $i$ as the sender name, then they should be allowed to learn from the response that either $C$ was not the agent who received the message or is not talking to $i$. Thus instead of line (c) we have:

if $\mathsf{vcrypt}(\mathsf{priv}(C), M) = \mathsf{true} \wedge \mathsf{dcrypt}(\mathsf{priv}(C), M) = i$ then
    $\star$ $B \neq C \vee \neg\mathsf{talk}(B, i)$ .$\mathsf{snd}(NB)$
else $\mathsf{snd}(NB)$

Finally, a similar change must be done to the first transaction: if $B \doteq i$, the intruder learns the name of $A$ and that $\mathsf{talk}(A, i)$. This is of course similar to the previous two modifications.

With all these modifications, the protocol is safe, and thus the intruder can only learn what is really unavoidable: when the intruder sends under their real name a message to some honest agent and gets a decipherable response, then the intruder learns the name of that agent and the fact that they are prepared to talk. Otherwise either it is another agent than addressed or one that is not willing to talk to the intruder.

In general, $(\alpha, \beta)$-privacy thus allows us to start with a specification that has a very strict privacy goal, by releasing as few as possible information to the intruder, and then checking if the protocol is indeed so strong to achieve this privacy goal. If not, one can deliberately make the choice to relax the privacy goal by releasing further information that are unavoidable – or in some case to actually strengthen the protocol.

## 4.2  Semantics

So far we have just given the syntax of our transaction formalism and relied on the reader's intuition how these are actually performed. It is straightforward to define a "simple" operational semantics that just checks how the protocol executes without worrying to define what the intruder might actually learn, i.e., $\alpha$ and $\beta$ of each state. To see that this defining of the intruder knowledge is tricky, consider the second transaction of the above example where both $A \neq i$ and $B \neq i$. In this case, the intruder cannot tell which of the conditions evaluates to true in this transaction, and thus the intruder cannot know the *structure* of the outgoing message for sure, i.e., whether it is a nonce or an encryption.

Thus, in contrast to the *struct* introduced in the message analysis problem, we now have to deal with a number of $struct_i$ and corresponding conditions $\phi_i$: for every label $l$, the intruder knows a concrete message $concr[l]$ (the message actually received) but for the structure they know only that it is $struct_i[l]$ if $\phi_i$ holds. What we can ensure though is that exactly one of the $\phi_i$ is true, i.e., these conditions are pairwise disjoint and cover all cases.

We thus define a *symbolic evaluation* of the transaction; this can be thought of as an evaluation performed by the intruder to figure out (as far as possible) what is actually happening. The state of this symbolic evaluation is characterized as a set of *possibilities* where each possibility has the form $(P, \phi, struct_i)$ where $P$ is the remainder of the transaction in the possibilities. Also, exactly one of these possibilities is marked as the real one, i.e., what really happens in this state—this will in general not be known by the intruder.

The symbolic evaluation of a condition is as follows:

$$\{(\text{if } \psi \text{ then } \mathcal{P}_1 \text{ else } \mathcal{P}_2, \phi, struct)\} \cup \mathcal{P} \implies$$
$$\{(\mathcal{P}_1, \phi \wedge \psi, struct), (\mathcal{P}_2, \phi \wedge \neg\psi, struct)\} \cup \mathcal{P}$$

thus splitting the first possibility into two depending on the two possible outcomes. The remaining transaction is thus either the process of the then branch or the else branch. There are similar rules for evaluating memory cell access, the choice of new variables and release of conditions in $\alpha$; for the complete set see [14]. All these rules are performed as *normalization* rules: i.e., we apply them to the current symbolic evaluation state until no more rules can be applied. In the resulting set of possibilities, each process starts with either snd, rcv, or 0.

For evaluating these normalized processes, we have another set of rewrite rules that apply depending on what is the first step in the possibility that is marked as the real one:

– If it is rcv($x$).$P_i$, then the restriction on the occurrences of rcv steps in transactions implies that all other possibilities also start with a rcv($x$) step. Now the intruder can choose an arbitrary recipe $r$ (with labels from the current frame). In the $i$th possibility, the resulting message is thus $struct_i[r]$, and we can symbolically simulate that this message is received by replacing each rcv($x$).$P_i$ with $P_i[x \mapsto struct_i[r]]$.

– If it is snd($t$).$P$, then we can discard all possibilities that do not currently send, since the intruder can observe whether a message is sent. Thus, we choose a new label and add to each $struct_i$ the respective message $t$ for $l$.

– If it is 0, then no other possibility can have a rcv step due to our restriction on their occurrence. We can also discard all possibilities that are sending (because it is observable that there are no more outgoing messages). Thus the symbolic evaluation is done since all remaining processes are 0.

From this evaluation we construct the $\beta$ of the resulting state, because all the final possibilities have the form $(0, struct_i, \phi_i)$, and we have one $concr$ (the actual messages derived). We can now formulate this as

$$concr \sim struct \wedge \bigwedge \phi_i \implies struct = struct_i$$

Recall that, by construction, exactly one of the $\phi_i$ is true.

## 5    Conclusions and Outlook

As a conclusion, let us briefly review what challenges $(\alpha, \beta)$-privacy poses for automated or semi-automated verification methods. The message analysis problem defined in Sect. 2 is certainly the most basic problem as it arises for every reachable state even for only a passive intruder. A first work by Fernet and Mödersheim [10] that gives a decision procedure for a standard set of operators.

In Sect. 4 we have shown that in case of an active intruder and conditional branching we arrive at a generalization of this problem, namely where we have several $struct_i$ (attached to a condition $\phi_i$) depending on which branches the transaction may have taken. We may call this the *multi message-analysis problem*. There is a rather naïve decision procedure: generate for each model of $\mathcal{I}$ with $\mathcal{I} \models \phi_i$ the corresponding $concr_{\mathcal{I}} = \mathcal{I}(struct_i)$ and check that all resulting $concr_{\mathcal{I}}$ are statically equivalent. This creates an exponential blowup in general, and we are currently working on a more clever procedure that avoids this by a symbolic representation.

A second challenge arises from the fact that at each rcv($x$) of a process, the intruder should choose a recipe $r$ over the labels of the current state. This choice is infinite and, even under reasonable restrictions, very large. The common approach is the *lazy intruder* or constraint-based approach (see [2,5] for instance), where we stop searching for solutions whenever the constraint for this intruder is to generate an arbitrary message. Our ongoing work of integrating this technique with multi message-analysis problems promises a decision procedure for a bounded number of transactions.

For the unbounded case, it is common to abstract from the single "sessions" by an over-approximation. Unfortunately, this often produces false positives when not restricting the problem significantly. An approach using type-systems by Cortier et al. [7] can tame this problem in many instances, and we plan to investigate if similar type systems could also be helpful for $(\alpha, \beta)$-privacy with unbounded sessions.

To compare with the de-facto standard of trace equivalence, we have shown in [17] that $(\alpha, \beta)$-privacy has at least comparable expressive power, while also being often more declarative. Tools for deciding trace equivalence (or minor restrictions of it) for bounded sessions exist, for instance DEEPSEC [6]. For unbounded sessions, the typical approach is to consider the restriction to diff-equivalence, which does not compare arbitrary processes, but rather a bi-process that has for each message a *left* and *right* variant [4]. This side-steps the problem of the different possibilities that arise form the conditionals (i.e., what corresponds to the multi-message analysis problem in $(\alpha, \beta)$-privacy) since diff-equivalence requires that the left and right variant either both satisfy the condition or both do not. This allows for efficient unbounded session verification in tools like ProVerif [3], Maude NPA [9], and Tamarin [16], but limits the number of protocols that can be considered. We plan to investigate here if a type system similar to [7] could provide a better compromise between feasibility and expressiveness.

# References

1. Abadi, M., Fournet, C.: Private authentication. Theor. Comput. Sci. **322**(3), 427–476 (2004)
2. Basin, D., Mödersheim, S., Viganò, L.: OFMC: a symbolic model checker for security protocols. Int. J. Inf. Secur. **4**(3), 181–208 (2004). https://doi.org/10.1007/s10207-004-0055-7
3. Blanchet, B.: An efficient cryptographic protocol verifier based on prolog rules. In: CSFW 2001, pp. 82–96. IEEE Computer Society (2001)
4. Blanchet, B., Abadi, M., Fournet, C.: Automated verification of selected equivalences for security protocols. In: LICS 2005 (2005)
5. Cheval, V., Comon-Lundh, H., Delaune, S.: A procedure for deciding symbolic equivalence between sets of constraint systems. Inf. Comput. **255**, 94–125 (2017)
6. Cheval, V., Kremer, S., Rakotonirina, I.: DEEPSEC: deciding equivalence properties in security protocols - theory and practice. In: S&P 2018 (2018)
7. Cortier, V., Grimm, N., Lallemand, J., Maffei, M.: A type system for privacy properties. In: CCS 2017 (2017)
8. Delaune, S., Hirschi, L.: A survey of symbolic methods for establishing equivalence-based properties in cryptographic protocols. J. Log. Algebraic Methods Program. **87**, 127–144 (2017)
9. Escobar, S., Meadows, C.A., Meseguer, J.: A rewriting-based inference system for the NRL protocol analyzer and its meta-logical properties. Theor. Comput. Sci. **367**(1–2), 162–202 (2006)
10. Fernet, L., Mödersheim, S.: Deciding a fragment of $(\alpha, \beta)$-privacy. In: Roman, R., Zhou, J. (eds.) STM 2021. LNCS, vol. 13075, pp. 122–142. Springer, Cham (2021). https://doi.org/10.1007/978-3-030-91859-0_7

11. Fujioka, A., Okamoto, T., Ohta, K.: A practical secret voting scheme for large scale elections. In: Seberry, J., Zheng, Y. (eds.) AUSCRYPT 1992. LNCS, vol. 718, pp. 244–251. Springer, Heidelberg (1993). https://doi.org/10.1007/3-540-57220-1_66
12. Genesereth, M., Hinrichs, T.: Herbrand logic. Technical report. LG-2006-02, Stanford University (2006)
13. Gondron, S., Mödersheim, S.: Formalizing and proving privacy properties of voting protocols using alpha-beta privacy. In: Sako, K., Schneider, S., Ryan, P.Y.A. (eds.) ESORICS 2019. LNCS, vol. 11735, pp. 535–555. Springer, Cham (2019). https://doi.org/10.1007/978-3-030-29959-0_26
14. Gondron, S., Mödersheim, S., Viganò, L.: Privacy as reachability (extended version). Technical report, DTU (2021). https://people.compute.dtu.dk/samo/abg.pdf
15. Kremer, S., Ryan, M.: Analysis of an electronic voting protocol in the applied Pi calculus. In: Sagiv, M. (ed.) ESOP 2005. LNCS, vol. 3444, pp. 186–200. Springer, Heidelberg (2005). https://doi.org/10.1007/978-3-540-31987-0_14
16. Meier, S., Schmidt, B., Cremers, C., Basin, D.: The TAMARIN prover for the symbolic analysis of security protocols. In: Sharygina, N., Veith, H. (eds.) CAV 2013. LNCS, vol. 8044, pp. 696–701. Springer, Heidelberg (2013). https://doi.org/10.1007/978-3-642-39799-8_48
17. Mödersheim, S., Viganò, L.: Alpha-beta privacy. ACM Trans. Priv. Secur. **22**(1), 1–35 (2019)

# Invited Tutorials and Experience Report

# Canonical Narrowing with Irreducibility and SMT Constraints as a Generic Symbolic Protocol Analysis Method

Raúl López-Rueda$^{(\boxtimes)}$ and Santiago Escobar

VRAIN, Universitat Politècnica de València, Valencia, Spain
{rloprue,sescobar}@upv.es

**Abstract.** Nowadays, formal cryptographic protocol analysis relies on symbolic techniques such as narrowing and equational unification, e.g. Maude-NPA, Tamarin or AKISS crypto tools. In previous works, we developed a new narrowing strategy, called canonical narrowing, which manages to reduce the state explosion problem by introducing irreducibility constraints. In this paper, we extend canonical narrowing to handle conditional rules with SMT constraints. We demonstrate the viability of this method with the Brands and Chaum protocol using time and location information described as SMT constraints on the real numbers.

**Keywords:** Canonical narrowing · SMT solver · Maude · Security protocols · Brands and Chaum

## 1 Introduction

Formal protocol analysis allows to determine whether an attacker can cause a protocol to fail any of its security objectives. One of the ways to perform this type of analysis is through the use of symbolic techniques, such as narrowing. There are tools for protocol analysis, like Maude-NPA [8], that use narrowing together with equational unification as a basis. These techniques are efficiently supported by the Maude language, and are also used in other protocol analysis tools such as Tamarin [15] or AKISS [4]. In our works [10,14], we already developed a new narrowing algorithm, called canonical narrowing, which manages to reduce the state explosion problem by introducing irreducibility constraints.

In a large number of protocols, the use of laws of physics that use real numbers to represent distances, time, or coordinates is essential. The formal analysis of this type of protocols can be done using either an explicit model with physical information, or by using an abstract model without physical information, e.g.,

This work has been partially supported by the EC H2020-EU grant agreement No. 952215 (TAILOR), by the grant RTI2018-094403-B-C32 funded by MCIN/AEI/10.13039/501100011033 and ERDF "A way of making Europe", by the grant PROMETEO/2019/098 funded by Generalitat Valenciana, and by the grant PCI2020-120708-2 funded by MICIN/AEI/10.13039/501100011033 and by the European Union NextGenerationEU/PRTR.

K. Bae (Ed.): WRLA 2022, LNCS 13252, pp. 45–64, 2022.
https://doi.org/10.1007/978-3-031-12441-9_3

untimed, and showing it is sound and complete with respect to a model with physical information. The former is more intuitive for the user, but the latter is often chosen because not all cryptographic protocol analysis tools support reasoning about, e.g., time or space. SMT solvers allow precisely the use of explicit models with physical information, translating the physical laws into SMT constraints. In order to analyze these models using narrowing algorithms, there is a need to extend them so that they are capable of handling these restrictions. One way to do it is by having narrowing to handle conditional rules, as in [19], in which each of the constraints will be collected at runtime. In the following example, we show one of the protocols that use laws of physics. This protocol really goes beyond existing narrowing approaches such as [10,14,19], since two cryptographic primitives are combined, exclusive-or over a set of nonces and a commitment scheme, apart of time and location represented as real numbers, requiring both irreducibility and SMT constraints.

*Example 1.* The Brands-Chaum protocol [3] specifies communication between a verifier V and a prover P. P needs to authenticate itself to V, and also needs to prove that it is within a distance "d" of it. A typical interaction between the prover and the verifier is as follows, where $N_A$ denotes a nonce generated by $A$, $S_A$ denotes a secret generated by $A$, $X; Y$ denotes concatenation of two messages $X$ and $Y$, $commit(N, S)$ denotes commitment of secret $S$ with a nonce $N$, $open(N, S, C)$ denotes opening a commitment $C$ using the nonce $N$ and checking whether it carries the secret $S$, $\oplus$ is the exclusive-or operator, and $sign(A, M)$ denotes $A$ signing message $M$.

$P \rightarrow V : commit(N_P, S_P)$

//The prover sends his name and a commitment

$V \rightarrow P : N_V$

//The verifier sends a nonce and records the time when this message was sent

$P \rightarrow V : N_P \oplus N_V$

//The verifier checks the answer message arrives within two times a fixed distance

$P \rightarrow V : S_P$

//The prover sends the committed secret and the verifier opens the commitment

$P \rightarrow V : sign_P(N_V; N_P \oplus N_V)$

//The prover signs the two rapid exchange messages

We assume the participants are located at an arbitrary given topology (participants do not move from their assigned locations) with distance constraints, where travelled time and coordinates are represented by a real number. We assumed coordinates $P_x$, $P_y$, $P_z$ for each participant $P$.

The previous informal Alice&Bob notation was naturally extended to include time in [1] and to include both time and location in [2]. First, we add the time when a message was sent or received as a subindex $P_{t_1} \rightarrow V_{t_2}$. Second, the sending and receiving times of a message differ by the distance between them just by adding some location constraints

$$\lfloor d(A, B) \rfloor := (d(A, B) \geq 0 \wedge d(A, B)^2 = (A_x - B_x)^2 + (A_y - B_y)^2 + (A_z - B_z)^2)$$

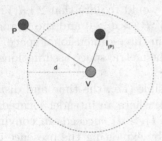

**Fig. 1.** Mafia Attack

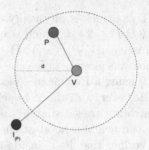

**Fig. 2.** Hijacking Attack

Third, the distance bounding constraint of the verifier is represented as an arbitrary distance $d$. Time and space constraints are written using quantifier-free formulas in real arithmetic. For convenience, we allow both $2 * x = x + x$ and the monus function $x \dot{-} y = if\ y < x\ then\ x - y\ else\ 0$ as definitional extensions.

*Example 2* (Cont'd Example 1). In the following time and space sequence of actions, a vertical bar differentiates between the process and corresponding constraints associated to the metric space. The following action sequence differs from [1] only on the terms $\lfloor d(P, V) \rfloor$.

$$
\begin{array}{ll}
P_{t_1} \to V_{t'_1} : commit(N_P, S_P) & |\ t'_1 = t_1 + d(P, V) \land \lfloor d(P, V) \rfloor \\
V_{t_2} \to P_{t'_2} : N_V & |\ t'_2 = t_2 + d(P, V) \land t_2 \geq t'_1 \land \lfloor d(P, V) \rfloor \\
P_{t_3} \to V_{t'_3} : N_P \oplus N_V & |\ t'_3 = t_3 + d(P, V) \land t_3 \geq t'_2 \land \lfloor d(P, V) \rfloor \\
\quad\quad V: t'_3 \dot{-} t_2 \leq 2 * d & \\
P_{t_4} \to V_{t'_4} : S_P & |\ t'_4 = t_4 + d(P, V) \land t_4 \geq t_3 \land \lfloor d(P, V) \rfloor \\
\quad\quad V: open(N_P, S_P, commit(N_P, S_P)) & \\
P_{t_5} \to V_{t'_5} : sign_P(N_V; N_P \oplus N_V) & |\ t'_5 = t_5 + d(P, V) \land t_5 \geq t_4 \land \lfloor d(P, V) \rfloor
\end{array}
$$

The Brands-Chaum protocol is designed to defend against mafia frauds, where an honest prover is outside the neighborhood of the verifier (i.e., $d(P, V) > d$) but an intruder is inside (i.e., $d(I, V) \leq d$), pretending to be the honest prover as depicted in Fig. 1. The following is an example of an *attempted* mafia fraud, in which the intruder simply forwards messages back and forth between the prover and the verifier. We write $I(P)$ to denote an intruder pretending to be an honest prover $P$.

$$
\begin{array}{ll}
P_{t_1} \to I_{t_2} \quad : commit(N_P, S_P) & |\ t_2 = t_1 + d(P, I) \land \lfloor d(P, I) \rfloor \\
I(P)_{t_2} \to V_{t_3} \quad : commit(N_P, S_P) & |\ t_3 = t_2 + d(V, I) \land \lfloor d(V, I) \rfloor \\
V_{t_3} \to I(P)_{t_4} : N_V & |\ t_4 = t_3 + d(V, I) \land \lfloor d(V, I) \rfloor \\
I_{t_4} \to P_{t_5} \quad : N_V & |\ t_5 = t_4 + d(P, I) \land \lfloor d(P, I) \rfloor \\
P_{t_5} \to I_{t_6} \quad : N_P \oplus N_V & |\ t_6 = t_5 + d(P, I) \land \lfloor d(P, I) \rfloor \\
I(P)_{t_6} \to V_{t_7} \quad : N_P \oplus N_V & |\ t_7 = t_6 + d(V, I) \land \lfloor d(V, I) \rfloor \\
\quad\quad V \quad\quad : t_7 \dot{-} t_3 \leq 2 * d & \\
P_{t_8} \to I_{t_9} \quad : S_P & |\ t_9 = t_8 + d(P, I) \land t_8 \geq t_5 \land \lfloor d(P, I) \rfloor \\
I(P)_{t_{10}} \to V_{t_{11}} : S_P & |\ t_{11} = t_{10} + d(V, I) \land t_{11} \geq t_7 \land \lfloor d(V, I) \rfloor \\
I(P)_{t_{12}} \to V_{t_{13}} : sign_P(N_V; N_P \oplus N_V)|\ t_{13} = t_{12} + d(V, I) \land t_{13} \geq t_{11} \land \lfloor d(V, I) \rfloor
\end{array}
$$

This attack is physically unfeasible, since it would require that $2 * d(V, I) + 2 * d(P, I) \leq 2 * d$, which is unsatisfiable by $d(V, P) > d > 0$ and the triangular inequality $d(V, P) \leq d(V, I) + d(P, I)$, satisfied in three-dimensional space. This attack was already unfeasible in [1] using only the metric space assumptions and in [2] using a Euclidean space.

However, a distance hijacking attack is possible (i.e., the time and distance constraints are satisfiable), as depicted in Fig. 2, where an intruder located outside the neighborhood of the verifier (i.e., $d(V, I) > d$) succeeds in convincing the verifier that he is inside the neighborhood by exploiting the presence of an honest prover in the neighborhood (i.e., $d(V, P) \leq d$) to achieve his goal. The following is an example of a *successful* distance hijacking, in which the intruder listens to the exchanged messages between the prover and the verifier but builds the last message.

$$
\begin{aligned}
P_{t_1} \rightarrow V_{t_2} \quad &: commit(N_P, S_P) & &| \; t_2 = t_1 + d(P, V) \wedge \lfloor d(P, V) \rfloor \\
V_{t_2} \rightarrow P_{t_3}, I_{t'_3} &: N_V & &| \; t_3 = t_2 + d(P, V) \wedge \lfloor d(P, V) \rfloor \\
& & &| \; t'_3 = t_2 + d(I, V) \wedge \lfloor d(V, I) \rfloor \\
P_{t_3} \rightarrow V_{t_4}, I_{t'_4} &: N_P \oplus N_V & &| \; t_4 = t_3 + d(P, V) \wedge \lfloor d(P, V) \rfloor \\
& & &| \; t'_4 = t_3 + d(I, P) \wedge \lfloor d(I, P) \rfloor \\
V \quad\quad &: t_4 \dot{-} t_2 \leq 2 * d & & \\
P_{t_5} \rightarrow V_{t_6} \quad &: S_P & &| \; t_6 = t_5 + d(P, V) \wedge \lfloor d(P, V) \rfloor \\
& & &| \; t_5 \geq t_3 \wedge t_6 \geq t_4 \\
I(P)_{t_7} \rightarrow V_{t_8} &: sign_I(N_V; N_P \oplus N_V) & &| \; t_8 = t_7 + d(I, V) \wedge \lfloor d(I, V) \rfloor \\
& & &| \; t_7 \geq t'_4 \wedge t_8 \geq t_6
\end{aligned}
$$

This attack was feasible in [1] using the metric space assumptions, and it was also possible in three-dimensional space in [2].

In Sect. 2, we provide some preliminaries. In Sect. 3, we introduce our new canonical narrowing with irreducibility and SMT constraints. In Sect. 4, we describe our implementation. In Sect. 5, we present some experiments using the Brands and Chaum protocol that prove its viability. In Sect. 6, we conclude and give some future work.

## 2  Preliminaries

We follow the classical notation and terminology from [21] for term rewriting, and from [16,19] for rewriting logic and order-sorted notions.

We assume an order-sorted signature $\Sigma$ with a poset of sorts $(S, \leq)$. The poset $(\mathsf{S}, \leq)$ of sorts for $\Sigma$ is partitioned into equivalence classes, called *connected components*, by the equivalence relation $(\leq \cup \geq)^+$. We assume that each connected component $[\mathsf{s}]$ has a *top element* under $\leq$, denoted $\top_{[\mathsf{s}]}$ and called the *top sort* of $[\mathsf{s}]$. This involves no real loss of generality, since if $[\mathsf{s}]$ lacks a top sort, it can be easily added.

We assume an S-sorted family $\mathcal{X} = \{\mathcal{X}_\mathsf{s}\}_{\mathsf{s} \in \mathsf{S}}$ of disjoint variable sets with each $\mathcal{X}_\mathsf{s}$ countably infinite. $\mathcal{T}_\Sigma(\mathcal{X})_\mathsf{s}$ is the set of terms of sort s, and $\mathcal{T}_{\Sigma,\mathsf{s}}$ is the set of ground terms of sort s. We write $\mathcal{T}_\Sigma(\mathcal{X})$ and $\mathcal{T}_\Sigma$ for the corresponding order-sorted term algebras. Given a term $t$, $Var(t)$ denotes the set of variables in $t$.

A *substitution* $\sigma \in \mathcal{S}ubst(\Sigma, \mathcal{X})$ is a sorted mapping from a finite subset of $\mathcal{X}$ to $\mathcal{T}_\Sigma(\mathcal{X})$. Substitutions are written as $\sigma = \{X_1 \mapsto t_1, \ldots, X_n \mapsto t_n\}$ where the domain of $\sigma$ is $Dom(\sigma) = \{X_1, \ldots, X_n\}$ and the set of variables introduced by terms $t_1, \ldots, t_n$ is written $Ran(\sigma)$. The identity substitution is $id$. Substitutions are homomorphically extended to $\mathcal{T}_\Sigma(\mathcal{X})$. The application of substitution $\sigma$ to a term $t$ is denoted by $t\sigma$ or $\sigma(t)$.

A $\Sigma$-*equation* is an unoriented pair $t = t'$, where $t, t' \in \mathcal{T}_\Sigma(\mathcal{X})_s$ for some sort $s \in S$. Given $\Sigma$ and a set $E$ of $\Sigma$-equations, order-sorted equational logic induces a congruence relation $=_E$ on terms $t, t' \in \mathcal{T}_\Sigma(\mathcal{X})$ (see [17]). Throughout this paper we assume that $\mathcal{T}_{\Sigma,s} \neq \emptyset$ for every sort s, because this affords a simpler deduction system. We write $\mathcal{T}_{\Sigma/E}(\mathcal{X})$ and $\mathcal{T}_{\Sigma/E}$ for the corresponding order-sorted term algebras modulo the congruence closure $=_E$, denoting the equivalence class of a term $t \in \mathcal{T}_\Sigma(\mathcal{X})$ as $[t]_E \in \mathcal{T}_{\Sigma/E}(\mathcal{X})$.

The first-order language of equational $\Sigma$-formulas is defined as: $\Sigma$-equations $t = t'$ as basic atoms, conjunction $\wedge$ of formulas, disjunction $\vee$ of formulas, negation $\neg$ of a formula, universal quantification $\forall$ of a variable $x{:}s$ in a formula, and existential quantification $\exists$ of a variable $x{:}s$ in a formula. A formula is quantifier-free (QF) if it does not contain any quantifier. Given a $\Sigma$-algebra $A$, a formula $\varphi$, and an assignment $\alpha \in X \mapsto A$ for the free variables $X$ in $\varphi$, $A, \alpha \models \varphi$ denotes that $\varphi$ is satisfied and $A \models \varphi$ holds if $\forall \alpha : A, \alpha \models \varphi$.

An *equational theory* $(\Sigma, E)$ is a pair with $\Sigma$ an order-sorted signature and $E$ a set of $\Sigma$-equations. An equational theory $(\Sigma, E)$ is *regular* if for each $t = t'$ in $E$, we have $Var(t) = Var(t')$. An equational theory $(\Sigma, E)$ is *linear* if for each $t = t'$ in $E$, each variable occurs only once in $t$ and in $t'$. An equational theory $(\Sigma, E)$ is *sort-preserving* if for each $t = t'$ in $E$, each sort s, and each substitution $\sigma$, we have $t\sigma \in \mathcal{T}_\Sigma(\mathcal{X})_s$ iff $t'\sigma \in \mathcal{T}_\Sigma(\mathcal{X})_s$. An equational theory $(\Sigma, E)$ is *defined using top sorts* if for each equation $t = t'$ in $E$, all variables in $Var(t)$ and $Var(t')$ have a top sort. Given two equational theories $G = (\Sigma, E)$ and $T = (\Delta, \Gamma)$, we say $T$ is the background theory of $E$ iff $\Sigma \subseteq \Delta$ and for each ground $\Sigma$-formula $\varphi$, $\mathcal{T}_{\Sigma/E} \models \varphi \iff T \models \varphi$.

An $E$-*unifier* for a $\Sigma$-equation $t = t'$ is a substitution $\sigma$ such that $t\sigma =_E t'\sigma$. For $Var(t) \cup Var(t') \subseteq W$, a set of substitutions $CSU_E^W(t = t')$ is said to be a *complete* set of unifiers for the equality $t = t'$ modulo $E$ away from $W$ iff: (i) each $\sigma \in CSU_E^W(t = t')$ is an $E$-unifier of $t = t'$; (ii) for any $E$-unifier $\rho$ of $t = t'$ there is a $\sigma \in CSU_E^W(t = t')$ such that $\sigma|_W \sqsupseteq_E \rho|_W$ (i.e., there is a substitution $\eta$ such that $(\sigma\eta)|_W =_E \rho|_W$); and (iii) for all $\sigma \in CSU_E^W(t = t')$, $Dom(\sigma) \subseteq (Var(t) \cup Var(t'))$ and $Ran(\sigma) \cap W = \emptyset$.

A *conditional rewrite rule* is an oriented pair $l \to r$ if $\varphi$, where $l \notin \mathcal{X}$, $\varphi$ is a QF $\Sigma$-formula, and $l, r \in \mathcal{T}_\Sigma(\mathcal{X})_s$ for some sort $s \in S$. An unconditional rewrite rule is written $l \to r$. A *conditional order-sorted rewrite theory* is a triple $(\Sigma, E, R, T)$ with $\Sigma$ an order-sorted signature, $E$ a set of $\Sigma$-equations, $T$ is a background theory, and $R$ a set of conditional rewrite rules. The set $R$ of rules is *sort-decreasing* if for each $t \to t'$ (or $t \to t'$ if $\varphi$) in $R$, each $\mathsf{s} \in \mathsf{S}$, and each substitution $\sigma$, $t'\sigma \in \mathcal{T}_\Sigma(\mathcal{X})_\mathsf{s}$ implies $t\sigma \in \mathcal{T}_\Sigma(\mathcal{X})_\mathsf{s}$.

The rewriting relation on $\mathcal{T}_\Sigma(\mathcal{X})$, written $t \to_R t'$ or $t \to_{p,R} t'$ holds between $t$ and $t'$ iff there exist a $p \in Pos_\Sigma(t)$, $l \to r$ if $\varphi \in R$ and a substitution $\sigma$, such that $T \models \varphi\sigma$, $t|_p = l\sigma$, and $t' = t[r\sigma]_p$. The relation $\to_{R/E}$ on $\mathcal{T}_\Sigma(\mathcal{X})$ is $=_E; \to_R; =_E$. The transitive (resp. transitive and reflexive) closure of $\to_{R/E}$ is denoted $\to_{R/E}^+$ (resp. $\to_{R/E}^*$). A term $t$ is called $\to_{R/E}$-irreducible (or just $R/E$-irreducible) if there is no term $t'$ such that $t \to_{R/E} t'$. For $\to_{R/E}$ confluent and terminating, the irreducible version of a term $t$ is denoted by $t\!\downarrow_{R/E}$.

A relation $\to_{R,E}$ on $\mathcal{T}_\Sigma(\mathcal{X})$ is defined as: $t \to_{p,R,E} t'$ (or just $t \to_{R,E} t'$) iff there are a position $p \in Pos_\Sigma(t)$, a rule $l \to r$ if $\varphi$ in $R$, and a substitution $\sigma$ such that $T \models \varphi\sigma$, $t|_p =_E l\sigma$ and $t' = t[r\sigma]_p$. Reducibility of $\to_{R/E}$ is undecidable in general since $E$-congruence classes can be arbitrarily large. Therefore, $R/E$-rewriting is usually implemented [13] by $R, E$-rewriting under some conditions on $R$ and $E$ such as confluence, termination, and coherence.

We call $(\Sigma, B, E)$ a *decomposition* of an order-sorted equational theory $(\Sigma, E \cup B)$ if $B$ is regular, linear, sort-preserving, defined using top sorts, and has a finitary and complete unification algorithm, which implies that $B$-matching is decidable, and the equations $E$ oriented into rewrite rules $\overrightarrow{E}$ are *convergent*, i.e., confluent, terminating, and strictly coherent [18] modulo $B$, and sort-decreasing.

Given a decomposition $(\Sigma, B, E)$ of an equational theory, $(t', \theta)$ is an $E, B$-*variant* [6,11] (or just a variant) of term $t$ if $t\theta\!\downarrow_{E,B} =_E t'$ and $\theta\!\downarrow_{E,B} =_E \theta$. A *complete set of $E, B$-variants* [11] (up to renaming) of a term $t$ is a subset, denoted by $[\![t]\!]_{E,B}$, of the set of all $E, B$-variants of $t$ such that, for each $E, B$-variant $(t', \sigma)$ of $t$, there is an $E, B$-variant $(t'', \theta) \in [\![t]\!]_{E,B}$ such that $(t'', \theta) \sqsupseteq_{E,B} (t', \sigma)$, i.e., there is a substitution $\rho$ such that $t' =_B t''\rho$ and $\sigma|_{Var(t)} =_B (\theta\rho)|_{Var(t)}$. A decomposition $(\Sigma, B, E)$ has the *finite variant property* (FVP) [11] (also called a *finite variant decomposition*) iff for each $\Sigma$-term $t$, a complete set $[\![t]\!]_{E,B}$ of its most general variants is finite.

In what follows, the set $G$ of equations will in practice be $G = E \uplus B$ and will have a decomposition $(\Sigma, B, E)$.

**Definition 1 (Reachability goal).** *Given an order-sorted rewrite theory* $(\Sigma, G, R, T)$, *a reachability goal is defined as a pair* $t \xrightarrow{?}^*_{R/G} t'$, *where* $t, t' \in \mathcal{T}_\Sigma(\mathcal{X})_\mathsf{s}$. *It is abbreviated as* $t \xrightarrow{?}^* t'$ *when the theory is clear from the context;* $t$ *is the* source *of the goal and* $t'$ *is the* target. *A substitution* $\sigma$ *is a $R/G$-solution of the reachability goal (or just a solution for short) iff there is a sequence* $\sigma(t) \to_{R/G} \sigma(u_1) \to_{R/G} \cdots \to_{R/G} \sigma(u_{k-1}) \to_{R/G} \sigma(t')$.

*A set $\Gamma$ of substitutions is said to be a complete set of solutions of* $t \xrightarrow{?}^*_{R/G} t'$ *iff (i) every substitution* $\sigma \in \Gamma$ *is a solution of* $t \xrightarrow{?}^*_{R/G} t'$, *and (ii) for any solution* $\rho$ *of* $t \xrightarrow{?}^*_{R/G} t'$, *there is a substitution* $\sigma \in \Gamma$ *more general than* $\rho$ *modulo $G$, i.e.,* $\sigma|_{Var(t) \cup Var(t')} \sqsupseteq_G \rho|_{Var(t) \cup Var(t')}$.

This provides a tool-independent semantic framework for symbolic reachability analysis of protocols under algebraic properties. Note that we have removed the condition $Var(\varphi) \cup Var(r) \subseteq Var(l)$ for rewrite rules $l \to r$ if $\varphi \in R$ and thus a solution of a reachability goal must be applied to all terms in the rewrite sequence. If the terms $t$ and $t'$ in a goal $t \overset{?}{\to}{}^*_{T/G} t'$ are ground and rules have no extra variables in their right-hand sides, then goal solving becomes a standard rewriting reachability problem. However, since we allow terms $t, t'$ with variables, we need a mechanism more general than standard rewriting to find solutions of reachability goals. *Narrowing* with $R$ modulo $G$ generalizes rewriting by performing *unification* at non-variable positions instead of the usual matching modulo $G$. Soundness and completeness of narrowing for solving reachability goals are proved in [13,20] for unconditional rules $R$ modulo an equational theory $G$ and in [19] for conditional rules $R$ modulo an equational theory $G$, both with the restriction of considering only order-sorted *topmost* rewrite theories, i.e., rewrite theories were all the rewrite steps happen at the top of the term.

# 3    Canonical Narrowing with Irreducibility and SMT Constraints

This section extends the canonical narrowing strategy of [10] with SMT constraints.

When $(\Sigma, E \cup B)$ has a decomposition as $(\Sigma, B, E)$, then the initial algebra $\mathcal{T}_{\Sigma/E\cup B}$ is isomorphic to the canonical term algebra $\mathcal{C}_{\Sigma/E\cup B} = (C_{\Sigma/E\cup B}, \to_{R/E\cup B})$, where $C_{\Sigma/E\cup B} = \{C_{\Sigma/E\cup B,\mathsf{s}}\}_{\mathsf{s}\in S}$ and $C_{\Sigma/E\cup B,\mathsf{s}} = \{[t{\downarrow}_{\overrightarrow{E},B}]_B \in T_{\Sigma/B} \mid t{\downarrow}_{\overrightarrow{E},B} \in T_{\Sigma,\mathsf{s}}\}$ and where for each $f \in \Sigma$, $f_{\mathcal{C}_{\Sigma/E\cup B}}([t_1]_B, \ldots, [t_n]_B) = [f(t_1, \ldots, t_n){\downarrow}_{\overrightarrow{E},B}]_B$.

We have an isomorphism of initial algebras $\mathcal{T}_{\Sigma/E\cup B} \cong \mathcal{C}_{\Sigma/E\cup B}$. Likewise, we have an isomorphism of free $(\Sigma, E \cup B)$-algebras $\mathcal{T}_{\Sigma/E\cup B}(\mathcal{X}) \cong \mathcal{C}_{\Sigma/E\cup B}(\mathcal{X})$, where $\mathcal{C}_{\Sigma/E\cup B}(\mathcal{X}) = (C_{\Sigma/E\cup B}(\mathcal{X}), \to_{R/E\cup B})$ and

$$C_{\Sigma/E\cup B,\mathsf{s}}(\mathcal{X}) = \{[t{\downarrow}_{\overrightarrow{E},B}]_B \in T_{\Sigma/B}(\mathcal{X}) \mid t{\downarrow}_{\overrightarrow{E},B} \in T_{\Sigma}(\mathcal{X})_{\mathsf{s}}\}.$$

The key point of canonical rewriting is that we can simulate rewritings $[t]_{E\cup B} \to_{R/E\cup B} [t']_{E\cup B}$ by corresponding rewritings $[t{\downarrow}_{\overrightarrow{E},B}]_B \to_{R/E\cup B} [t'{\downarrow}_{\overrightarrow{E},B}]_B$ and make rewriting decidable when $(\Sigma, B, \overrightarrow{E})$ is FVP.

**Definition 2 (Canonical Rewriting).** *Let $\mathcal{R} = (\Sigma, E \cup B, R, T)$ be a topmost order-sorted rewrite theory such that $(\Sigma, E \cup B)$ has an FVP decomposition $(\Sigma, B, E)$. Let $\mathcal{C}^{\circ}_{\Sigma/E\cup B}(\mathcal{X})_{\mathsf{State}} = \bigcup \mathcal{C}_{\Sigma/E\cup B}(\mathcal{X})_{\mathsf{State}}$, i.e., $\mathcal{C}^{\circ}_{\Sigma/E\cup B}(\mathcal{X})_{\mathsf{State}} = \{t{\downarrow}_{\overrightarrow{E},B} \mid t{\downarrow}_{\overrightarrow{E},B} \in T_{\Sigma}(\mathcal{X})_{\mathsf{State}}\}$, so that $\mathcal{C}^{\circ}_{\Sigma/E\cup B}(\mathcal{X})_{\mathsf{State}} \subseteq T_{\Sigma}(\mathcal{X})_{\mathsf{State}}$. We then define the $\to_{R/E,B}$ canonical rewrite relation with rules $R$ modulo $E \cup B$ as the following binary relation $\to_{R/E,B} \subseteq \mathcal{C}^{\circ}_{\Sigma/E\cup B}(\mathcal{X})_{\mathsf{State}} \times \mathcal{C}^{\circ}_{\Sigma/E\cup B}(\mathcal{X})_{\mathsf{State}}$, where $t \to_{R/E,B} t'$ iff $\exists l \to r$ if $\varphi \in R$ and $\exists \theta$ with $Dom(\theta) \subseteq Var(l) \cup Var(r) \cup Var(\varphi)$ and $\theta = \theta{\downarrow}_{\overrightarrow{E},B}$ such that: (i) $T \models \varphi\theta$, (ii) $(l\theta){\downarrow}_{\overrightarrow{E},B} =_{E\cup B} t$, and (iii) $t' =_B (r\theta){\downarrow}_{\overrightarrow{E},B}$.*

The claim that $\rightarrow_{R/E,B}$ exactly captures/bisimulates the $\rightarrow_{R/E\cup B}$ rewrite relation is justified by the following result.

**Theorem 1.** *For each* $t, t' \in T_\Sigma(\mathcal{X})_{\text{State}}$, $t \rightarrow_{R/E\cup B} t'$ *iff* $t\downarrow_{\overrightarrow{E},B} \rightarrow_{R/E,B}$ $t'\downarrow_{\overrightarrow{E},B}$.

A term $t(x_1{:}s_1, \ldots, x_n{:}s_n)$ can be viewed as a symbolic, effective method to describe a (typically infinite) set of terms, namely the set

$$\lceil t(x_1{:}s_1, \ldots, x_n{:}s_n)\rceil = \{t(u_1, \ldots, u_n) \mid u_i \in T_\Sigma(\mathcal{X})_{s_i}\} = \{t\theta \mid \theta \in \mathcal{S}ubst(\Sigma, \mathcal{X})\}.$$

We think as $t$ as a *pattern*, which symbolically describes all its *instances* (including non-ground). However, since $(\Sigma, B, E)$ is a decomposition of an equational theory $(\Sigma, E \cup B)$, we can consider only normalized instances of $t$

$$\lceil t\rceil_{\overrightarrow{E},B} = \{(t\theta)\downarrow_{\overrightarrow{E},B} \mid \theta \in \mathcal{S}ubst(\Sigma, \mathcal{X})\}$$

However, since we are interested in terms that may satisfy some irreducibility and SMT conditions, we can obtain a more expressive symbolic pattern language where patterns are *constrained by both irreducibility and SMT constraints*. That is, we consider constrained patterns of the form $\langle t, \Pi, \varphi \rangle$ where $\Pi$ is a finite set of normalized terms and $\varphi$ is a QF $\Sigma$-formula. Then we can define:

$$\lceil \langle t, (u_1, \ldots, u_k), \varphi \rangle \rceil_{\overrightarrow{E},B} = \{(t\theta)\downarrow_{\overrightarrow{E},B} \mid \theta \in \mathcal{S}ubst(\Sigma, \mathcal{X}), T \models \varphi\theta,$$

$$u_1\theta, \ldots, u_k\theta \text{ are } \overrightarrow{E}, B\text{-normalized}\}.$$

The canonical narrowing relation $\rightsquigarrow_{R/E,B}$ includes irreducibility constraints only for the left-hand sides of the rules and SMT constraints only from the conditional part of the rules.

**Definition 3 (Canonical Narrowing).** *Given a topmost order-sorted rewrite theory* $(\Sigma, E \cup B, R, T)$ *such that* $(\Sigma, B, E)$ *is a decomposition of* $(\Sigma, E \cup B)$, *the* canonical narrowing relation with irreducibility constraints *holds between* $\langle t, \Pi, \varphi \rangle$ *and* $\langle t', \Pi', \varphi' \rangle$, *denoted*

$$\langle t, \Pi, \varphi \rangle \rightsquigarrow_{\alpha, R/E,B} \langle t', \Pi', \varphi' \rangle$$

*iff there exists* $l \rightarrow r$ *if* $\varphi'' \in R$, *which we always assume renamed, so that* $Var(\langle t, \Pi, \varphi \rangle) \cap (Var(r) \cup Var(l) \cup Var(\varphi'')) = \emptyset$, *and a unifier* $\alpha \in CSU_{E\cup B}^W(t = l)$, *where* $W = Var(\langle t, \Pi, \varphi \rangle) \cup Var(r) \cup Var(l) \cup Var(\varphi'')$, *and*

1. $\langle t', \Pi', \varphi' \rangle = \langle r\alpha, \Pi\alpha \cup \{(l\alpha)\downarrow_{\overrightarrow{E},B}\}, \varphi\alpha \wedge \varphi''\alpha \rangle$,
2. $\Pi\alpha \cup \{(l\alpha)\downarrow_{\overrightarrow{E},B}\}$ *are* $\overrightarrow{E}, B$-*irreducible, and*
3. $\varphi'$ *is satisfiable, i.e.,* $\exists \alpha'$ *s.t.* $T \models \varphi'\alpha'$.

Note that we do not require a narrowing step to compute $CSU_{E\cup B}(t=l)$ anymore, we perform regular equational unification but impose an irreducibility constraint on the normal form of the instantiated left-hand side, which can be handled in Maude by using asymmetric unification [7], i.e., equational unification is done with irreducibility constraints.

Irreducibility constraints are computed by using the normalized left-hand side of the rules that are used in the narrowing steps. SMT constraints are simply added to the third component and check for satisfiability. Note that we assume that satisfiability of QF $\Sigma$-formulas is decidable, indeed for a subsignature $\Sigma_0 \subseteq \Sigma$ associated to the background theory $T$. Maude is using the CVC4 SMT solver for satisfiability.

Each trace will carry a different set of irreducibility and SMT constraints, although some of the conditions are shared by having common predecessor nodes. In each new narrowing step, the list of irreducibility constraints computed previously in that sequence must be taken into account, so that if it is necessary to reduce one of the terms appearing in the list to compute a new step, it will be discarded. Similary, the SMT formula carried along the sequence must be taken into account, so that if it becomes unsatisfiable after one narrowing step, it will be discarded.

In this way, we eliminate redundancy as well as branches of the reachability tree, which will be less and less wide than the tree resulting from using standard narrowing. In some cases, we will even get infinite reachability trees to become finite, ensuring termination.

The key completeness property about this relation is the following.

**Lemma 1 (Lifting Lemma).** *Given $\langle t, \Pi, \varphi \rangle$, a $\overrightarrow{E}, B$-normalized substitution $\theta$, and terms $u, v \in \mathcal{C}^\circ_{\Sigma/E,B}(\mathcal{X})$ such that $u = (t\theta)\!\downarrow_{\overrightarrow{E},B}$, $T \models \varphi\theta$, and $\Pi\theta$ are $\overrightarrow{E}, B$-normalized and $u \rightarrow_{R/E,B} v$, there is a canonical narrowing step with irreducibility and SMT constraints*

$$\langle t, \Pi, \varphi \rangle \rightsquigarrow_{\alpha, R/E,B} \langle r\alpha, \Pi\alpha \cup \{(l\alpha)\!\downarrow_{\overrightarrow{E},B}\}, \varphi' \rangle$$

*and a $\overrightarrow{E}, B$-normalized substitution $\gamma$ such that*

$$\langle t, \Pi, \varphi \rangle \qquad \rightsquigarrow_{\alpha, R/E,B} \langle r\alpha, \Pi\alpha \cup \{(l\alpha)\!\downarrow_{\overrightarrow{E},B}\}, \varphi' \rangle$$
$$\downarrow_\theta \qquad\qquad\qquad\qquad \downarrow_\gamma$$
$$\lceil \langle t, \Pi, \varphi \rangle \rceil_{\overrightarrow{E},B} \rightarrow_{R/E,B} \lceil \langle r\alpha, \Pi\alpha \cup \{(l\alpha)\!\downarrow_{\overrightarrow{E},B}\}, \varphi' \rangle \rceil_{\overrightarrow{E},B}$$

*(i)* $\theta =_B (\alpha\gamma)|_{Var(\langle t, \Pi, \varphi \rangle)}$,
*(ii)* $(r\alpha\gamma)\!\downarrow_{\overrightarrow{E},B} =_B v$,
*(iii)* $\Pi\alpha\gamma \cup \{((l\alpha)\!\downarrow_{\overrightarrow{E},B})\gamma\}$ *are $\overrightarrow{E}, B$-normalized,*
*(iii)* $T \models \varphi'\gamma$.

Note that this shows that $v \in \lceil \langle r\alpha, \Pi\alpha \cup \{(l\alpha)\!\downarrow_{\overrightarrow{E},B}\}, \varphi' \rangle \rceil_{\overrightarrow{E},B}$.

# 4    Implementation

To implement SMT constraint handling in the narrowing algorithm, we have used our implementation of standard/canonical narrowing [14] as a starting point. To do this, we use the features of the Maude meta-level, thus creating an extension of the previous meta-level command.

## 4.1    Our Previous Narrowing Command

The meta-level command we use as a starting point already allows us to choose between several narrowing algorithms to use. First of all, it allows to invoke the standard narrowing algorithm, with a behavior similar to the standard narrowing built-in in Maude. It also allows the canonical narrowing algorithm [14] to be invoked, in which irreducibility constraints are used to reduce the width of the computed reachability tree. To control the algorithm used along with other parameters, such as the maximum depth of the tree or the maximum number of solutions to search for, the command uses ten arguments:

```
narrowing(Module, Term, SearchArrow, Term, AlgorithmOptionSet, VariantOptionSet, TermList, Qid,
          Bound, Bound)
```

In the implementation of that command, we already prepared an adequate infrastructure to allow future extensions. Several data structures and substructures were defined to represent the reachability tree, its nodes, and the solutions found. Additionally, we divided the implementation into three main parts, which correspond to the main steps of the algorithm at a theoretical level: (i) the generation of nodes (terms) in the reachability tree, (ii) the attempt to unify each new term with the target term, and (iii) the computation of solutions in case the unification is successful. Those main parts are further broken down into highly distinguishable subparts, making it easy to make extensions or modifications to some parts without having to change the rest of the implementation.

## 4.2    Using Conditional Rules in Narrowing

To manage SMT constraints, our approach has been to use Maude's conditional rules to add them as a condition in each of the narrowing steps. The problem that arises is that the Maude narrowing mechanisms are not capable of processing the conditional rules. The way to fix this is to transform those conditional rules into normal rules, in which the new left-hand side of the new rules will contain both the left-hand side of the conditional rules and the SMT constraints. An operator should separate both parts, so that later the original term can be distinguished from the SMT restrictions.

We have implemented a module that is responsible for carrying out the process of transformation of conditional rules. This module defines two operators:

```
op transformMod : Module -> Module .
op transformRls : RuleSet -> RuleSet .
```

The first receives a module, theory, module with strategy or theory with strategy. In either case, a new operator is added to the set of operators of the module or theory, which will be used to separate the terms from the SMT constraints in the transformed rules. It is also necessary to add the import of the Maude META-TERM module to the converted module, so that it is capable of processing the addition of this new operator. Finally, this operator calls the other defined operator, using as an argument the set of rules of the module to be transformed. For example, the equation used to transform a module without strategy would be the following:

```
eq transformMod(mod ModId is Imports sorts Sorts . Subsorts Ops Membs Eqs Rls1 endm)
   = mod ModId is Imports (protecting 'META-TERM .)
     sorts Sorts . Subsorts
     (Ops (op '_>>_ : 'Boolean 'State -> 'State [ctor poly (0 2)] .))
     Membs Eqs transformRls(Rls1) endm .
```

The second operator, therefore, receives a set of rules, and is in charge of iterate through it looking for conditional rules. Each time a conditional rule is found, it is transformed into a new unconditional rule, in which the condition is added to the left-hand side using the >> operator defined above. The equations used to do this are as follows:

```
eq transformRls(Rls1 (crl Lhs => Rhs if (SmtConst = BooleanValue) [Attrs].) Rls2)
   = transformRls(Rls1 Rls2) (rl Lhs => '_>>_[SmtConst,Rhs] [Attrs narrowing] .) .
eq transformRls(Rls1) = Rls1 [owise] .
```

Thus, if we have a conditional rule of the form crl Lhs => Rhs if (SmtConst = BooleanValue) [Attrs], it will be automatically transformed into an unconditional rule of the form rl Lhs => (SmtConst >> Rhs) [Attrs narrowing], where Lhs and Rhs are variables of Universal type (that is, they can be instantiated as any sort), SmtConst is a variable that represents the SMT constraints, and BooleanValue is a Boolean variable expected to be true, used only to be able to encode SMT constraints in the conditions of the rules. The new form of the rule after transforming it will allow us later to make the >> operator disappear and separate the term from the SMT constraints. This is explained in detail in the following section.

## 4.3 Extension to Handle SMT Constraints

Once we have prepared the module transformation to convert all the conditional rules into unconditional ones, we can extend the previous command so that it processes the SMT terms that will be generated with the new rules. This extension has been done without making changes at the user level, except for the addition of possible values to one of the existing arguments, as well as a new argument that allows to indicate initial SMT constraints:

```
narrowing(Module, Term, SearchArrow, Term, AlgorithmOptionSet, VariantOptionSet, TermList,
          Term, Qid, Bound, Bound)
```

Until now, the fifth argument, of type AlgorithmOptionSet, only accepted the standard and canonical values, used to indicate the type of narrowing algorithm to use. Now, it also accepts combinations of those two values with the smt and noCheck values, although the second is a limitation of the first, so it cannot appear without it. By using the smt value, the transformation of the conditional rules will be performed in the module used as rewrite theory if necessary. Subsequently, the SMT constraints will be processed during the execution of the algorithm to check if they are satisfiable at each node of the reachability tree. If it is also accompanied by the value noCheck, only the transformation of the rules will be carried out, ignoring the satisfiability of the SMT constraints.

The most relevant changes to the algorithm occur before trying to unify the term of a new generated node with the target term, since the satisfiability of the SMT constraints for that node will have to be checked first. Until reaching that step, not many modifications are needed, since the narrowing steps will be given using the rules in a usual way, because the transformation of conditional rules will have been previously carried out just at the beginning of the algorithm, if the user indicates that SMT constraints are being used. Furthermore, we need to modify the previously used data structures. Now the main structure must save the initial SMT constraints indicated by the user. It will also be necessary for each of the nodes to contain a list of the SMT constraints carried so far. We have stored that list at each node in a {Term, Bool} pair, where the second value of the pair indicates the satisfiability of the constraints found in the first value. Two new operators are introduced in the algorithm that run after the generation of a new node and renaming of its variables, although they will only be used if the user indicates that SMT restrictions must be processed:

```
op evaluateSMT : UserArguments TreeInfo SolutionList -> NarrowingInfo .
op checkSat : UserArguments TreeInfo SolutionList -> NarrowingInfo .
```

The evaluateSMT operator performs the separation of the SMT constraints from the new term generated with one of the transformed rules. In turn, it joins these restrictions with the list of restrictions carried so far, which will come from the predecessor nodes to the current one and from the initial restrictions indicated by the user. Additionally, it launches to evaluate all those restrictions, to know if they are satisfiable or not. To do this, we rely on Maude's SMT interface, which is available in the meta-level. Specifically, we use the metaCheck [5, §16], which receives the module to use and the term to evaluate, returning a value of type Bool. If the result is true, the constraints are satisfiable. Otherwise, false is returned:

```
op metaCheck : Module Term ~> Bool [special (...)] .
```

Note that in case the user has indicated, in addition to the smt value as an argument, the noCheck value, the evaluateSMT operator will only separate the SMT constraints from the term, ignoring the rest of the process, since we are not interested in checking the satisfiability, but in being able to process the constraints of the initial conditional rules.

The checkSat operator is responsible for processing the result obtained when executing the metaCheck function. If the restrictions are satisfiable, the next execution step should be the attempt to unify the term of the node with the objective term, to check if it corresponds to one or more solutions of the reachability problem. If the constraints are not satisfiable, then it will not make sense to perform the unification step, since we will not consider the term of the node as valid. We therefore return to the step of generating new nodes, marking the current node as invalid, so that it is not taken into account later, since we do not want to generate the possible child nodes of this node either.

### 4.4   Variable Consistency

As we explained in our previous work [14] on which we based this algorithm, the way Maude generates the fresh variables may lead to clashes. For this reason, the fresh variables that are generated in each narrowing step must be renamed using an internal counter, and using the $ symbol as an identifier. Since the variables in the SMT constraints are related to those used in the terms, as well as to the variables in the previously processed SMT constraints, there is a consistency problem with this renaming. That is why in each narrowing step, we now have to apply variable substitutions to the SMT constraints so that there is no such loss of consistency. Specifically, at each narrowing step, the computed substitution that must be applied to the term of the previous node to take that step must be applied to the new node's SMT constraint. The substitution must also be applied to the SMT constraint list carried along the node branch. In turn, this list will already come with the variables renamed in the previous steps, so consistency builds up. Note that the initial SMT restrictions indicated by the user will also have to be renamed. This is not a problem, since these constraints are also automatically added to the list of constraints of each node, so it can be renamed at the same time as the rest.

## 5   Experiments

For the experiments, we have considered the Brands and Chaum protocol of Example 2 in two forms: its version with only time, published in [1], and its version with time and space, published in [2]. In both, the use of SMT restrictions is necessary, which in our case are codified with conditional rules. As explain in Sect. 4, these conditional rules will be processed to transform them into unconditional rules, in order to correctly obtain the SMT constraints at each narrowing step.

All the files used to define the new narrowing algorithm, as well as the experiments that we will see next and their results, can be found at the following link: https://github.com/ralorueda/smt-narrowing.

### 5.1   Handling SMT Constraints

We rely on the generic rewrite theory for protocol specification, inspired on the strand spaces [12] used by Maude-NPA [8], used in our previous work on canonical narrowing [14], but with some modifications that adapt it to include

SMT constraints on the real numbers, inspired on the constraints used in [1,2]. It is a module that allows us to specify a state, made up of sets of strands and the intruder knowledge, which represents the communication channel. With it we can represent the protocols in a generic way, adding the corresponding equational theories for each of them. Later, when coding the narrowing calls, we will specify the exact strands of each protocol.

In the original module, we had two transition rules. One of them processes the sent messages, and the other the received messages:

```
var IK : IntruderKnowledge .   var SS : StrandSet .   var M : Msg .   vars L1 L2 : SMsgList .

rl [receive-msg] : { (SS & [ ( L1 , -(M)) | L2 ]) { (inI(M) , IK) } } =>
                   { (SS & [ L1 | (-(M) , L2) ]) { (inI(M) , IK) } } [narrowing] .

rl [send-msg] : { (SS & [ (L1 , +(M)) | L2 ]) { (inI(M) , IK) } } =>
                { (SS & [ L1 | (+(M) , L2) ]) { (nI(M) , IK) } } [narrowing] .
```

It can be seen in each of them how, for each set of strands, represented in square brackets, there is a list to the left of the operator | and one to the right. The first contains the messages to be processed, while the second contains the processed messages. At each transition, a message (sent or received) is taken from the end of the list of messages to be processed and moved to the list of processed messages. In the event that it is a sent message, the correspondence of that message will also be modified in the communication channel or intruder knowledge.

To adapt the module to protocols using non-linear arithmetic constraints on the real numbers via satisfiability, we add a conditional rule that is responsible for processing a new type of data that can appear in the strands sets: constraints. Specifically in our case, SMT constraints (type `Boolean`), which will be represented in the channel between the messages with the operator {_}. We will therefore now have three rules. One of them is responsible for processing the messages sent, another the messages received, and another the restrictions that occur at any given time:

```
var IK : IntruderKnowledge .   var SS : StrandSet .   var SSR : StrandSetR .
var SSN : StrandSetN .   var M : Msg .   vars LeE2 : SMsgList-eE .
var LREe1 : SMsgListR-Ee .

crl [check-constraint] : { (SSR & [ LREe1 , {B:Boolean} | LeE2 ]) { IK } } =>
                        { (SSR & [ LREe1 | {B:Boolean} , LeE2 ]) { IK } }
                        if B:Boolean = true [nonexec] .
rl [receive-msg] : { (SSN & [ LREe1 , -(M) | LeE2 ]) { (inI(M) , IK) } } =>
                   { (SSN & [ LREe1 | -(M) , LeE2 ]) { (inI(M) , IK) } } [narrowing] .
rl [send-msg] : { (SS & [ LREe1 , +(M) | LeE2 ]) { (inI(M) , IK) } } =>
                { (SS & [ LREe1 | +(M) , LeE2 ]) { (nI(M) , IK) } } [narrowing] .
```

Note that in this case we use variables from different sorts, `SMsgListR-Ee` and `SMsgListR-eE`, rather than the ones we used in [14]. This is because we have created a rule hierarchy, mimicking some optimizations of the Maude-NPA [9], in such a way that a more defined processing order is followed, significantly reducing the computation time in the experiments. In this way, whenever there is a constraint at the end of the list of messages to be processed in a strand set, it will be processed first. If this is not the case, it will check if there is any received message at the end of the list of messages to be processed in a strand set, and

will be processed. If neither of these two cases occurs, then a sent message will
be processed.

## 5.2   Brands and Chaum with Time

The previous module allows us, in a generic way, to specify protocols that contain
SMT restrictions. To this must be added the specific equational theories of each
protocol. In our case, the first protocol used is Brands and Cham with time [1],
which can be seen as a simplified version of the protocol seen in Example 1, but
does not take into account the coordinates of the messages. Two cryptographic
primitives are combined: exclusive-or over a set of nonces and a commitment
scheme. Exclusive-or is defined with the following properties:

```
sort NNSet .
subsorts Nonce Secret < NNSet .

op null : -> NNSet .
op _*_ : NNSet NNSet -> NNSet [assoc comm] .
vars X Y : [NNSet] .

eq [idem] :      X * X = null      [variant] .
eq [idem-Coh] : X * X * Y = Y      [variant] .
eq [id] :        X * null = X      [variant] .
```

The commitment scheme allows a participant to commit to a chosen hidden value
at an early protocol stage and reveal it later. It is defined with the following
properties:

```
op commit : Nonce Secret -> NTMsg .
op open : Nonce Secret NTMsg -> [Boolean] .
eq open(N1:Nonce,Sr:Secret,commit(N1:Nonce,Sr:Secret)) = true [variant] .
```

The open function is defined only for the successful case. This implies the use of
the kind [Boolean] rather than the sort Boolean. We also use additional operators for this protocol, which allow us to define signing, message concatenation,
and the creation of nonces and secrets.

```
sorts Msg NTMsg TMsg .
sorts Name Honest Intruder Fresh Secret Nonce .
subsorts NNSet < NTMsg < Msg .
subsorts Nonce Secret < NNSet .
subsort Name < Msg .
subsort Honest Intruder < Name .

ops a b : -> Honest .
op i : -> Intruder .
ops ra1 rb1 rb2 : -> Fresh .
op n : Name Fresh -> Nonce .
op s : Name Fresh -> Secret .
op sign : Name NTMsg -> NTMsg .
op _;_ : NTMsg  NTMsg  -> NTMsg [gather (e E)] .
```

Additionally, we add several operators that will allow us to add metadata to
the messages. In them, the sending and receiving times of the messages will be
saved, as well as the identifier of the sender and the receiver.

```
sorts TimeInfo NameTime NameTimeSet .
subsort NameTime < NameTimeSet .
subsort TMsg < Msg .

op _@_ : NTMsg TimeInfo -> TMsg .
op _:_ : Name Real -> NameTime .
op mt : -> NameTimeSet .
op _#_ : NameTimeSet NameTimeSet -> NameTimeSet [assoc comm id: mt] .
op _->_ : NameTime NameTimeSet -> TimeInfo .
```

Note that times will be represented as real numbers, one of the data types manageable by Maude's SMT interface. The distance between two participants $A$ and $B$ is represented by a variable dab:Real.

The module defined with the previous sorts, operators and rules allows us to code the strands of the Brands and Chaum protocol of Example 1 only with time. This will be done in the call to the narrowing algorithm, with an initial state and a target state. In the initial state, the strand sets will contain a list of messages and constraints to be processed and a list of messages and constraints processed, which will be empty. In the target state, the lists will have been inverted, so that all the messages and restrictions to be processed become processed. Consider, for example, the strands of a prover and a verifier in a regular execution of the Brands and Chaum protocol with time. With our syntax, they would be specified in the initial state as follows:

```
--- Alice, verifier
([nilEe,
  -(Commit:NTMsg                            @ b : t1:Real -> a : t1':Real),
      {(t1':Real === t1:Real + dab:Real) and dab:Real > 0/1},
  +(n(a,ra1)                                @ a : t2:Real -> b : t2':Real),
  -((n(a,ra1) * NB:Nonce)                   @ b : t3:Real -> a : t3':Real),
      {(t3':Real === t3:Real + dab:Real) and dab:Real > 0/1 and t3:Real >= t2':Real},
      {(t3':Real - t2':Real) <= (2/1 * dab:Real) and dab:Real > 0/1},
  -(SB:Secret                               @ b : t4:Real -> a : t4':Real),
      {open(NB:Nonce,SB:Secret,Commit:NTMsg)},
      {(t4':Real === t4:Real + dab:Real) and dab:Real > 0/1 and t4:Real >= t3':Real},
  -(sign(b,(n(a,ra1) * NB:Nonce) ; n(a,ra1))  @ b : t5:Real -> a : t5':Real),
      {(t5':Real === t5:Real + dab:Real) and dab:Real > 0/1 and t5:Real >= t4':Real}
| nileE]
&
--- Bob, prover
[nilEe,
  +(commit(n(b,rb1),s(b,rb2))               @ b : t1:Real -> a : t1':Real),
  -(NA:Nonce                                @ a : t2:Real -> b : t2':Real),
      {(t2':Real === t2:Real + dab:Real) and dab:Real > 0/1 and t2:Real >= t1':Real},
  +((NA:Nonce * n(b,rb1))                   @ b : t3:Real -> a : t3':Real),
  +(s(b,rb2)                                @ b : t4:Real -> a : t4':Real),
  +(sign(b,(NA:Nonce * n(b,rb1)) ; NA:Nonce)  @ b : t5:Real -> a : t5':Real)
| nileE])
```

We can see how the prover, Bob, will first send a commit to the verifier. Afterwards, the verifier, Alice, sends her nonce to the prover. Subsequently, the prover will send the XOR of his nonce with the received one, and then sends the secret. The verifier will open it to confirm everything is okay. Finally, the prover will send the signs messages. An @ operator appears in each message, after which the sending and receiving times of the message are saved, as well as the identifier of the sender and receiver. We can also see how SMT constraints are introduced after each received message. In them, conditions to be met are specified regarding the delivery and reception times. Conditions to satisfy relative to distances are also specified. For example, in the SMT constraint that is introduced on the strands of the prover, it is specified that the arrival time of the received message must be equal to its departure time plus the distance between the prover and the verifier. It is also specified that this distance must be greater than zero, and that the sending time of the message must be equal to or greater than the time in which the previous message was received.

Using this syntax and coding methodology, we have defined three experiments in which we test a regular execution of the protocol, a mafia-like attack pattern, and a hijacking-like attack pattern. In regular execution, we get a solution, which is expected, since if the protocol is well defined, this execution should be possible. In the case of the mafia attack, a priori, a solution is also found, which translates into a possible vulnerability. However, adding the triangle inequality $(d(a,i) + d(b,i)) > d$ as the initial constraint, together with the constraint $d(V, P) > d > 0$, no solution is found. This is because, for consistency to exist in this execution, it is necessary that $2 * d(V, I) + 2 * d(P, I) \leq 2 * d$. As mentioned in Sect. 4, the initial SMT constraints can be written in one of the arguments of the narrowing command. However, it is possible to perform a hijacking attack, and that is why by specifying this pattern in one of the experiments, a solution is found. The attack occurs when an intruder located outside the neighborhood of the verifier (i.e., $d(V, I) > d$) succeeds in convincing the verifier that he is inside the neighborhood by exploiting the presence of an honest prover in the neighborhood (i.e., $d(V, P) \leq d$).

## 5.3  Brands and Chaum with Time and Space

The second protocol that we have used for the experiments is an extension of the previous one: Brands and Chaum with time and space, detailed at a theoretical level in Example 2. In this case, the coordinates related to the sending and receiving of each message appear in the metadata of the messages and in the restrictions, that is, the coordinates of the participants. To be able to write this, a slight modification of the previous protocol specification is enough, as well as the addition of a new operator:

```
sort CoordNameTime .
op _:_,_,_,_ : Name Real Real Real Real -> CoordNameTime .
op _->_ : CoordNameTime NameTimeSet -> TimeInfo .
```

Once the modification is done, it is possible to encode the new strands. For example, the strands for a verifier and a prover in a regular execution of the protocol would now be as follows:

```
--- Alice, verifier
[nilEe,
 -(Commit:NTMsg
      @ b : x1:Real,y1:Real,z1:Real,t1:Real -> a : t2:Real),
        {(t2:Real === t1:Real + dab1:Real) and (dab1:Real > 0/1) and
        ((dab1:Real * dab1:Real) === (((x1:Real - ax:Real) * (x1:Real - ax:Real)) +
        ((y1:Real - ay:Real) * (y1:Real - ay:Real))) +
        ((z1:Real - az:Real) * (z1:Real - az:Real)))},
```

```
+(n(a,ra1)
    @ a : ax:Real,ay:Real,az:Real,t2:Real -> b : t3:Real),
-((n(a,ra1) * NB:Nonce)
    @ b : x3:Real,y3:Real,z3:Real,t3:Real -> a : t4:Real),
        {(t4:Real === t3:Real + dab3:Real) and (dab3:Real > 0/1) and
        ((dab3:Real * dab3:Real) === (((x3:Real - ax:Real) * (x3:Real - ax:Real)) +
        ((y3:Real - ay:Real) * (y3:Real - ay:Real))) +
        ((z3:Real - az:Real) * (z3:Real - az:Real)))},
        {((t4:Real - t2:Real) <= (2/1 * d:Real)) and (d:Real > 0/1)},
-(SB:Secret
    @ b : x4:Real,y4:Real,z4:Real,t5:Real -> a : t6:Real),
        {open(NB:Nonce,SB:Secret,Commit:NTMsg)},
        {(t6:Real === t5:Real + dab4:Real) and (dab4:Real > 0/1) and
        ((dab4:Real * dab4:Real) === (((x4:Real - ax:Real) * (x4:Real - ax:Real)) +
        ((y4:Real - ay:Real) * (y4:Real - ay:Real))) +
        ((z4:Real - az:Real) * (z4:Real - az:Real)))},
-(sign(b,(n(a,ra1) * NB:Nonce) ; n(a,ra1))
    @ b : x5:Real,y5:Real,z5:Real,t7:Real -> a : t8:Real),
        {(t8:Real ===  t7:Real + dab5:Real) and (dab5:Real > 0/1) and
        ((dab5:Real * dab5:Real) === (((x5:Real - ax:Real) * (x5:Real - ax:Real)) +
        ((y5:Real - ay:Real) * (y5:Real - ay:Real))) +
        ((z5:Real - az:Real) * (z5:Real - az:Real)))}
| nileE]
&
--- Bob, prover
[nilEe,
  +(commit(n(b,rb1),s(b,rb2))
    @ b : bx:Real,by:Real,bz:Real,t1:Real -> a : t2:Real),
  -(NA:Nonce
    @ a : x2:Real,y2:Real,z2:Real,t2:Real -> b : t3:Real),
        {(t3:Real === t2:Real + dab2:Real) and (dab2:Real > 0/1) and
        ((dab2:Real * dab2:Real) === (((x2:Real - bx:Real) * (x2:Real - bx:Real)) +
        ((y2:Real - by:Real) * (y2:Real - by:Real))) +
        ((z2:Real - bz:Real) * (z2:Real - bz:Real)))},
  +((NA:Nonce * n(b,rb1))
    @ b : bx:Real,by:Real,bz:Real,t3:Real -> a : t4:Real),
  +(s(b,rb2)
    @ b : bx:Real,by:Real,bz:Real,t3:Real -> a : t6:Real),
  +(sign(b,(NA:Nonce * n(b,rb1)) ; NA:Nonce)
    @ b : bx:Real,by:Real,bz:Real,t3:Real -> a : t8:Real)
| nileE]
```

The exchange of messages is very similar to what we have seen before, but in this case the metadata is somewhat more complex, since the sending coordinates are attached to each sending time. In addition, the restrictions are also complicated, since in this case it will also be necessary to verify that the conditions required for those coordinates are satisfied at each moment. In fact, since the new constraints are non-linear arithmetic, Maude's SMT is not capable of processing them. In order to correctly execute the traces related to this protocol, we have used a version of Maude called Maude-NRA, which provides an SMT solver (CVC4) that is capable of processing this type of arithmetic.

Once more, we have performed experiments for this protocol with a regular execution, a mafia-like attack pattern, and a hijacking-like attack pattern. The results are similar to the previous ones, although more complex. Regular execution returns a solution, since it is possible to do it without problems. The hijacking attack is again possible as well, so a solution is again returned. Regarding the mafia attack, the same thing happens: a priori it is possible, but by adding the initial SMT restrictions necessary for the trace to be consistent, the attack is impossible. These restrictions are the same as before, but in this case some relative to coordinates are also added.

## 6    Conclusions and Future Work

The canonical narrowing strategy with irreducibility and SMT constraints opens the door to the use of narrowing to analyze protocols that use laws of physics, such as the Brands and Chaum protocol. It is a greatly generic methodology of symbolic reachability analysis that manages to prove the existence of traces of a protocol, giving greater flexibility when defining and specifying them. In this article we have presented an implementation of canonical narrowing capable of handling SMT constraints. This allows us to carry out symbolic analysis of two versions of the Brands and Chaum protocol. Maude-NPA already handled such protocols, as shown in [1,2], but in an ad-hoc way without the canonical narrowing presented here. We now have a new algorithm with a powerful theoretical framework behind it, which can be useful to both Maude-NPA and other symbolic protocol analysis tools. As future work, we expect to expand this canonical narrowing to more general cases, clearly increasing its power for protocol analysis.

## References

1. Aparicio-Sánchez, D., Escobar, S., Meadows, C., Meseguer, J., Sapiña, J.: Protocol analysis with time. In: Bhargavan, K., Oswald, E., Prabhakaran, M. (eds.) INDOCRYPT 2020. LNCS, vol. 12578, pp. 128–150. Springer, Cham (2020). https://doi.org/10.1007/978-3-030-65277-7_7
2. Aparicio-Sánchez, D., Escobar, S., Meadows, C., Meseguer, J., Sapiña, J.: Protocol analysis with time and space. In: Dougherty, D., Meseguer, J., Mödersheim, S.A., Rowe, P. (eds.) Protocols, Strands, and Logic. LNCS, vol. 13066, pp. 22–49. Springer, Cham (2021). https://doi.org/10.1007/978-3-030-91631-2_2
3. Brands, S., Chaum, D.: Distance-bounding protocols. In: Helleseth, T. (ed.) EUROCRYPT 1993. LNCS, vol. 765, pp. 344–359. Springer, Heidelberg (1994). https://doi.org/10.1007/3-540-48285-7_30
4. Chadha, R., Cheval, V., Ciobâcă, Ş., Kremer, S.: Automated verification of equivalence properties of cryptographic protocols. ACM Trans. Comput. Log. 17(4), 23:1–23:32 (2016)
5. Clavel, M., et al: Maude Manual (Version 3.2.1). Technical report, SRI International Computer Science Laboratory (2022). http://maude.cs.illinois.edu
6. Comon-Lundh, H., Delaune, S.: The finite variant property: how to get rid of some algebraic properties. In: Giesl, J. (ed.) RTA 2005. LNCS, vol. 3467, pp. 294–307. Springer, Heidelberg (2005). https://doi.org/10.1007/978-3-540-32033-3_22
7. Erbatur, S., et al.: Asymmetric unification: a new unification paradigm for cryptographic protocol analysis. In: Bonacina, M.P. (ed.) CADE 2013. LNCS (LNAI), vol. 7898, pp. 231–248. Springer, Heidelberg (2013). https://doi.org/10.1007/978-3-642-38574-2_16

8. Escobar, S., Meadows, C., Meseguer, J.: Maude-NPA: cryptographic protocol analysis modulo equational properties. In: Aldini, A., Barthe, G., Gorrieri, R. (eds.) FOSAD 2007-2009. LNCS, vol. 5705, pp. 1–50. Springer, Heidelberg (2009). https://doi.org/10.1007/978-3-642-03829-7_1

9. Escobar, S., Meadows, C.A., Meseguer, J., Santiago, S.: State space reduction in the Maude-NRL protocol analyzer. Inf. Comput. **238**, 157–186 (2014)

10. Escobar, S., Meseguer, J.: Canonical narrowing with irreducibility constraints as a symbolic protocol analysis method. In: Guttman, J.D., Landwehr, C.E., Meseguer, J., Pavlovic, D. (eds.) Foundations of Security, Protocols, and Equational Reasoning. LNCS, vol. 11565, pp. 15–38. Springer, Cham (2019). https://doi.org/10.1007/978-3-030-19052-1_4

11. Escobar, S., Sasse, R., Meseguer, J.: Folding variant narrowing and optimal variant termination. J. Log. Algebr. Program. **81**(7–8), 898–928 (2012)

12. Thayer Fabrega, F.J., Herzog, J., Guttman, J.: Strand spaces: what makes a security protocol correct? J. Comput. Secur. **7**, 191–230 (1999)

13. Jouannaud, J.-P., Kirchner, H.: Completion of a set of rules modulo a set of equations. SIAM J. Comput. **15**(4), 1155–1194 (1986)

14. López-Rueda, R., Escobar, S., Meseguer, J.: An efficient canonical narrowing implementation for protocol analysis. In: Bae, K. (ed.) WRLA 2022. LNCS, vol. 13252, pp. 151–170. Springer, Cham (2022). Held as a Satellite Event of ETAPS, Munich, Germany, 2–3 April 2022, Proceedings

15. Meier, S., Schmidt, B., Cremers, C., Basin, D.: The TAMARIN prover for the symbolic analysis of security protocols. In: Sharygina, N., Veith, H. (eds.) CAV 2013. LNCS, vol. 8044, pp. 696–701. Springer, Heidelberg (2013). https://doi.org/10.1007/978-3-642-39799-8_48

16. Meseguer, J.: Conditioned rewriting logic as a united model of concurrency. Theor. Comput. Sci. **96**(1), 73–155 (1992)

17. Meseguer, J.: Membership algebra as a logical framework for equational specification. In: Presicce, F.P. (ed.) WADT 1997. LNCS, vol. 1376, pp. 18–61. Springer, Heidelberg (1998). https://doi.org/10.1007/3-540-64299-4_26

18. Meseguer, J.: Strict coherence of conditional rewriting modulo axioms. Theor. Comput. Sci. **672**, 1–35 (2017)

19. Meseguer, J.: Generalized rewrite theories, coherence completion, and symbolic methods. J. Log. Algebraic Methods Program. **110**, 100483 (2020)

20. Meseguer, J., Thati, P.: Symbolic reachability analysis using narrowing and its application to verification of cryptographic protocols. High.-Order Symb. Comput. **20**(1–2), 123–160 (2007)

21. TeReSe (ed.): Term Rewriting Systems. Cambridge University Press, Cambridge (2003)

# An Overview of the Maude Strategy Language and its Applications

Rubén Rubio[✉][iD]

Universidad Complutense de Madrid, Madrid, Spain
rubenrub@ucm.es

**Abstract.** In the Maude specification language, the behavior of systems is modeled by nondeterministic rewrite rules, whose free application may not always be desirable. Hence, a strategy language has been introduced to control the application of rules at a high level, without the intricacies of metaprogramming. In this paper, we give an overview of the Maude strategy language, its applications, related verification tools, and extensions, illustrated with examples.

## 1 Introduction

Computation in rewriting logic [29,30] is the succession of independent rule applications in any positions within the terms. This flexibility is the cornerstone of its natural representation of nondeterminism and concurrency, but it is sometimes useful to restrict or guide the evolution of rewriting. For example, a theorem prover does not blindly apply its inference rules, and the local reactions of a chemical system may be modulated by the environment. Strategies are the traditional resource to express these concerns, but specifying them in Maude involved the not so easy task of using its reflective capabilities. This has changed in Maude 3 with the inclusion of an object-level strategy language to explicitly control the application of rules [15]. Several operators resembling the usual programming language constructs and regular expressions allow combining the basic instruction of rule application to program arbitrarily complex strategies, which can be compositionally defined in strategy modules. The language was originally designed in the mid-2000s by Narciso Martí-Oliet, José Meseguer, Alberto Verdejo, and Steven Eker [27] based on previous experience with *internal strategies* at the metalevel [12,14] and earlier strategy languages like ELAN [8], Stratego [10], and Tom [6]. Other similar strategy languages appeared later like ρLog [26] and Porgy [18]. The first prototype was available as a Full Maude extension and it was already given several applications [16,21,40–42]. Now, since Maude 3.0, the language is efficiently implemented in C++ as part of the official interpreter [15].

As well as an executable specification language, Maude is also a verification tool and systems modeled with strategies need also be verified. Together with Narciso Martí-Oliet, Isabel Pita, and Alberto Verdejo, we have extended the Maude LTL model checker to work with strategy-controlled specifications [38]

K. Bae (Ed.): WRLA 2022, LNCS 13252, pp. 65–84, 2022.
https://doi.org/10.1007/978-3-031-12441-9_4

and established connections with external model checkers for evaluating CTL, CTL*, and $\mu$-calculus properties [36]. More recently, we have also developed a probabilistic extension of the Maude strategy language whose specifications can be analyzed using probabilistic and statistical model-checking techniques [32].

This paper is based on an invited tutorial on the Maude strategy language given at WRLA 2022 and explains the language and the aforementioned related topics. Section 2 starts with an introduction to the strategy language, Sect. 3 illustrates it with some more examples and includes references for others, Sect. 4 reviews some related tools and extensions, and Sect. 5 concludes with some remarks for future developments. More information about the strategy language, examples, and its related tools is available at maude.ucm.es/strategies.

## 2    A Brief Introduction to the Maude Strategy Language

In this section, we give an introduction to the Maude strategy language through an example, without claiming to be exhaustive or systematic. For a comprehensive informal reference about the language, we suggest its dedicated chapter in the Maude manual [13, §10]. Formal semantics, both denotational [32] and operational [28,38], are also available [35].

Let us introduce the following system module WORDS as running example, where Words are defined as lists of Letters in the latin alphabet. Three rules, swap, remove, and append, are provided to manipulate words.

```
mod WORDS is
  sorts Letter Word .
  subsort Letter < Word .

  ops a b c d ··· z : -> Letter [ctor] .
  op nil : -> Word [ctor] . *** empty word
  op __ : Word Word -> Word [ctor assoc id: nil] .

  rl [swap]   : L W R => R W L .
  rl [remove] : L => nil .
  rl [append] : W => W L [nonexec] .
endm
```

The swap rule permutates two letters in a word, removes removes one, and append attaches a new letter L to the end of the word. This latter rule is marked nonexec(utable), since it includes an unbound variable in the right-hand side. However, we will be able to execute it with the strategy language.

This rewrite system is nonterminating due to the idempotent swap rule. In fact, for every word with at least two letters, Maude's rewrite command will loop.

```
Maude> rewrite i t .                    *** does not terminate
```

However, we can obtain something useful from this module by controlling rewriting with the strategy language. The command for executing a strategy expression

$\alpha$ on a term $t$ is `srewrite t using` $\alpha$ and its output enumerates all terms that are obtained by this controlled rewriting. Multiple solutions are possible, since strategies are not required to completely remove nondeterminism. The elementary building block of the strategy language is the application of a rule, as cannot be otherwise, whose most basic form is the strategy `all` that applies any rule in the module.

```
Maude> srewrite i t using all .
```

```
Solution 1                          Solution 3
rewrites: 1                         rewrites: 3
result Word: t i                    result Letter: i

Solution 2                          No more solutions.
rewrites: 2                         rewrites: 3
result Letter: t
```

The previous fragment evaluates `all` on the term `i t` yielding three different solutions, one for `swap` and two for `remove`. This is equivalent to the command `search t =>1 W:Word` that looks for all terms reachable by a single rewrite from $t$, but the strategy language allows for more flexibility. For instance, if we want to apply only rules with a given label, say `swap`, we can simply write `swap`.

```
Maude> srewrite i t using swap .
```

```
Solution 1
rewrites: 1
result Word: t i

No more solutions.
rewrites: 1
```

Rules are applied in any position of the term by default, as seen in the second and third solutions of the first `srewrite` command, or in application of `swap` to the word w o n with result {o w n, n o w, w n o}. If this is not desired, the `top` modifier can be used to limit their application to the whole term, like in `top(swap)` or `top(all)`, whose only result is n o w. For being more precise when applying rules anywhere, we can also specify an initial substitution to be applied to both sides of the rule and its condition before matching. For example, `swap[L <- w]` would instantiate the rule L W R => R W L to w W R => R W w and yield n o w and o w n as solutions. Similarly, `swap[L <- w, W <- nil]` would turn the rule into w R => R w and produce the single solution o w n.

Substitutions are essential when dealing with nonexecutable rules, like `append` in the `WORDS` module, whose unbound variables can then be instantiated. We can execute `top(append[L <- a])` ; `top(append[L <- t])` on the word g o to turn it into a g o a t. In addition to using `top` for ensuring that the letter is appended at the end of the word, the previous strategy introduced a new combinator ; that executes a strategy on the results of the previous one, like functional composition or concatenation. Its identity element is the strategy constant `idle` that returns the original term unchanged as only solution. Another pervasive combinator is the disjunction or nondeterministic choice of strategies $\alpha_1 \mid \cdots \mid \alpha_n$, whose results are the union of the results of its operands. For example, `remove[L <- w] | remove[L - n]` evaluates on `w o n` to {`o n`, `w o`}. The identity element of the disjunction is `fail`, which does not produce any solution at all. In a broader sense, we say that a strategy *fails* when it does not produce any solution.

Suppose we want to calculate all permutations of a given word. This can be achieved by accumulating the words obtained by successive swaps,

$$\texttt{swap} \mid (\texttt{swap} ; \texttt{swap}) \mid (\texttt{swap} ; \texttt{swap} ; \texttt{swap}) \mid \cdots,$$

until no new words are obtained. The iteration combinator $\alpha*$, which can be inductively described as `idle` $\mid \alpha ; \alpha *$, expresses this common pattern. Observe the correspondence between the last strategy combinators and the constructors of regular expressions:

| Regular expressions | $\varepsilon$ | $\emptyset$ | $\alpha \mid \beta$ | $\alpha\beta$ | $\alpha^*$ |
|---|---|---|---|---|---|
| Strategy language | `idle` | `fail` | $\alpha \mid \beta$ | $\alpha ; \beta$ | $\alpha *$ |

As formalized in [38], the full strategy language is able to describe any recursively enumerable subset of the executions of the original rewrite system, over both finite and infinite words, but regular languages are specially easily expressed with these constructs. Coming back to the example, the expression `swap *` gives all permutations of the original word, so 24 solutions for `g o a t` after a total of 81 rewrites. If we only need the solutions that start with `g` and finish with a letter other than `a`, we can execute the strategy `swap * ;` `match g W R s.t. R =/= a` where `match` $P$ `s.t.` $C$ is an operator that filters the terms that match a pattern $P$ and satisfy a condition $C$. Indeed, it works like an `idle` when the conditions hold and like a `fail` when they do not. Other test variants, `xmatch` and `amatch`, exist for matching with extension for structural axioms (i.e. matching fragments of the flattened associative and/or commutative operators) or inside subterms, respectively.

In Spanish, the letter `h` is not pronounced except when preceded by `c`, so texters and tweeters sometimes obviate it against the criteria of the Royal Spanish Academy. If we do likewise, we would reduce `h o l a` to its homophone `o l a` with `remove[L <- h]`. However, we do not want to transform `b r o c h a` into `b r o c a`, because they are pronounced differently. We need a new tool to restrict the application to a specific context, and this is the subterm rewriting `matchrew` operators. Their syntax is similar to that of tests

$$\texttt{matchrew } P \texttt{ s.t. } C \texttt{ by } x_1 \texttt{ using } \alpha_1, \ldots, x_n \texttt{ using } \alpha_n$$

but the subterms matched by the variables $x_1, \ldots, x_n$ in its pattern $P$ are rewritten with strategies $\alpha_1, \ldots, \alpha_n$. The solutions of this operator are the combinations of all solutions obtained for every subterm, which are rewritten independently. For example, `matchrew L W by L using remove[L <- h]` will safely remove the first letter of the word if it is an `h`. For removing `h` in the middle of the word, we write xmatchrew L R by s.t. L =/= c by R using remove[L <- h] to ensure that the previous letter is not a c. Notice that we have used xmatchrew instead of matchrew, because we do not want to match the whole term but a fragment of the associative list of letters. These two strategies can be combined with a nondeterminsitic choice $\alpha \mid \beta$ to remove any silent h letters in a word. For example, applying this strategy to h e c h o yields e c h o but not h e c o. Unfortunately, the word h i p o c l o r h i d r i a is not rewritten to i p o c l o r i d r i a, because only one h is removed at a time. In order to normalize a term with respect to a strategy, i.e. to apply a strategy until it cannot be executed further, the language includes the $\alpha$ ! combinator. Putting the previous strategy under this normalization operator we obtain an expression that removes all silent h from a word.

Writing strategies as standalone expressions becomes unmanageable as they grow in size. Strategy modules are available to give them name and define them modularly. For example, the following strategy module WORDS-STRAT extends the system module WORDS with two new strategies, rmh and rmh-one, declared with the strats statement.

```
smod WORDS-STRAT is
  protecting WORDS .

  strats rmh rhm-one @ Word .

  vars L R : Letter .
  var  W   : Word .

  sd rmh := rmh-one ! .
  sd rmh-one := matchrew L W by L using remove[L <- h] .
  sd rmh-one := xmatchrew L R s.t. L =/= c
                 by R using remove[L <- h] .
endsm
```

The sort after the @ sign indicates which terms are intended to be rewritten by the strategy, although it does not have any practical effect. Each named strategy is assigned zero or more strategy expressions with definitions that start by the sd keyword or by csd if they are conditional. In the module above, rmh is the strategy that removes every silent h in a word, while the two definitions of rmh-one remove a single h at initial and inner position, respectively. When the strategy rmh-one is called in rmh, the two definitions for rmh-one are executed nondeterministically, as if their expressions where joined by the disjunction | operator.

One of the greatest advantages of strategy modules is the possibility of defining recursive strategies. For example, the following strategy module WORDS-REPEAT declares a single recursive strategy remove($l$, $n$) with two parameters that removes exactly $n$ occurrences of the letter $l$ in the subject term.

```
smod WORDS-REPEAT is
  protecting WORDS .

  strat remove : Letter Nat @ Phrase .
```

```
var N : Letter . var N : Nat .
```

Its semantics is given by two definitions with disjoint matching patterns. For removing zero letters, we simply do nothing with `idle`.

```
sd remove(L, 0) := idle .
```

Otherwise, one occurrence of the letter L is deleted with the `remove` rule and the strategy itself is called recursively with a decremented counter.

```
sd remove(L, s N) := remove[L <- L] ; remove(L, N) .
endsm
```

For example, rewriting `b a z a a r` with `remove(a, 2)` gives `b z a r` and `b a z r`. However, `rewrite(a, 4)` would fail because of the attempt to call `remove` for the fourth time. We can make `remove(l, n)` erase as many occurrences of $l$ as possible but no more than $n$ with the following change on the second definition:

```
sd remove(L, s N) := remove[L <- L] ? remove(L, N)
                                     : idle .
```

We have used the conditional operator $\alpha ? \beta : \gamma$ that evaluates $\beta$ on the results of $\alpha$ or $\gamma$ on the original term if $\alpha$ yields no solution. This way, we only invoke the recursive strategy if the `remove` rule succeeds, and the execution is finished when it fails. Conditional operators are quite general since its condition is an arbitrary strategy and recurrent conditional patterns are given dedicated syntax. For example, $\alpha$ `or-else` $\beta$ executes $\beta$ only if $\alpha$ fails, and it is equivalent to $\alpha ?$ `idle` $: \beta$.

## 3   Some Examples

In this section, we further illustrate the language with three examples. At the same time, we cite other published works where applications of the language have been presented.

### 3.1   Deduction Procedures

In deductive reasoning, inference rules should be carefully applied to reach the desired conclusions in an efficient way. A free or inadequate application of the rules may loop or lead to a poor performance in many examples of inference systems. For instance, the Davis-Putnam-Logemann-Loveland (DPLL) system for deciding the satisfiability of a Boolean formula has a natural brute-force *split* rule that generates two subproblems, where the variable $x$ is respectively assumed true and false.

$$\text{(split)} \quad \frac{\Delta \vdash \Gamma, x \vee C}{\Delta, x \vdash \Gamma \qquad \Delta, \neg x \vdash \Gamma, C} \quad \text{if } x, \neg x \notin \Delta$$

Of course, repeatedly applying this rule will solve the satisfiability problem, but at an exponential cost in the best case. The inference system includes other rules that are better applied first. For example, *subsume* removes pending clauses with a satisfied atom.

$$\text{(subsume)} \quad \frac{\Delta \vdash \Gamma, x \vee C}{\Delta \vdash \Gamma} \quad \text{if } x \in \Delta$$

Hopefully, this may remove some variables in $C$ that do not appear elsewhere, avoiding some superfluous case distinctions. A first rudimentary strategy for SAT solving with these rules would be (subsume | ···) or-else split where the dots are occupied by the other simplification rules. A second one can be more selective and apply split to the variable that cancels the most possible clauses. More serious strategies for the DPLL rules are programmed in the Maude strategy language in [20].

Implementations of deduction procedures do not usually individualize the rules in their code, but rule-based systems like Maude can easily separate the basic logic and its control using strategies. In the literature, this has been encouraged by the Kowalski's motto *Algorithm = Logic + Control* [22] or Lescanne's *Rule + Control* approach [24]. This latter work implements in Caml four equational completion procedures on top of the inference rules by Bachmair and Dershowitz [5], decoupling at some extent the rules from their control. These same completion procedures have also been specified using the initial prototype of the Maude strategy language in [42] and an improved redesign of this specification is available in [32]. In this latter version, we have clearly separated the inference rules in a system module COMPLETION and the four deduction procedures in four strategy modules being protected extensions of COMPLETION, as depicted in Fig. 1. Each procedure is a recursive strategy that maintains the inference state in its call arguments without modifying the term or adding more rules.

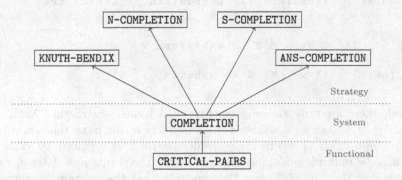

**Fig. 1.** Equational procedure specification with clear separation of concerns.

Following similar ideas, other examples of deduction procedures are programmed using the Maude strategy language like congruence closure [16], the Martelli-Montanari unification procedure [1], and a Sudoku solver [40].

## 3.2   Semantics of Programming Languages

Strategies are also meaningful when dealing with semantics of programming languages. Structural operational semantics define the small-step behavior of programs through inference rules whose premises are steps for the constituents of the program. In this sense, they are not much different to the deduction procedures seen in the previous section, and the Maude strategy language can be useful to describe them [9]. Strategies are particularly useful to generalize semantic rules with negative premises or rule precedence, which are not easily captured otherwise. For example, negation in Prolog is described by the following rule

$$\text{(split)} \quad \frac{\Gamma \vdash g \not\rightarrow^* \text{nil}}{\Gamma \vdash (\backslash{+}g), gs \rightarrow gs}$$

that removes the negated goal $n+g$ from the goal list if $g$ cannot be solved. This premise can be expressed with the strategy combinator $\texttt{not}(\alpha) \equiv \alpha$ ? $\texttt{fail}$ : $\texttt{idle}$. Indeed, we have specified an executable Prolog interpreter in [13] where negation and cuts are described with strategies.

Let us illustrate the relation between strategies and programming with two simple strategies for the untyped $\lambda$-calculus. We specify the basics of this formalism in the following module LAMBDA.

```
mod LAMBDA is
  sorts Var LambdaTerm .
  subsort Var < LambdaTerm .

  op \_._ : Var LambdaTerm -> LambdaTerm [ctor...] .
  op __ : LambdaTerm LambdaTerm -> LambdaTerm [ctor...] .

  op subst : LambdaTerm Var LambdaTerm -> LambdaTerm .
  *** the equational definition of subst is ommitted

  var x : Var . vars M N : LambdaTerm .

  rl [beta] : (\ x . M) N => subst(M, x, N) .
endm
```

As usual, there are only two constructors of $\lambda$-terms, abstraction $\lambda x.M$ and application $M\,N$, and we consider a single reduction rule beta that transforms $(\lambda x.M)N$ into $M[x/N]$ where every occurrence of $x$ is replaced by $N$ in $M$. There may be multiple positions where to apply the $\beta$ rule in a $\lambda$-term, called $\beta$-redexes, but the Church-Rosser theorem tells that the calculus is confluent, i.e. if we can reduce $t \rightarrow^* t_1$ and $t \rightarrow^* t_2$, there exists a $t'$ such that $t_1 \rightarrow^* t'$ and $t_2 \rightarrow^* t'$. Nevertheless, how rules are applied still matters, since some reductions may lead to a normal form while others may diverge for the same term (see Fig. 2). Another classical result of the $\lambda$-calculus tells that repeatedly reducing the outer leftmost redex always leads to a normal form in case it exists, and this can be expressed as a strategy.

$$\mathbf{(KI)}\,\Omega \begin{array}{c} \nearrow \\ \searrow \end{array} \begin{array}{l} \mathbf{(KI)}\,\Omega\;\circlearrowleft \\ (\lambda y.\mathbf{I})\,\Omega \longrightarrow \mathbf{I} \end{array} \qquad\qquad \begin{array}{l} \mathbf{K} = \lambda x.\,(\lambda y.x) \\ \mathbf{I}\; = \lambda x.\,x \\ \Omega = (\lambda x.\,xx)(\lambda x.\,xx) \end{array}$$

**Fig. 2.** Two reduction paths from the $\lambda$-term $\mathbf{(KI)}\,\Omega$.

In the strategy module LAMBDA-STRAT, we extend LAMBDA with two strategies normal and applicative for reducing $\lambda$-terms, but more variants can be defined like call-by-value and call-by-name. These are covered in an extended specification of this example [32,34].

```
smod LAMBDA-STRATS is
   protecting LAMBDA .
   strats normal applicative ··· @ LambdaTerm .
   vars x y z t : Var .          vars M N : LambdaTerm .
```

The definition of normal describes a single reduction step of the normalization strategy mentioned in the previous paragraph, i.e. applying beta on the outer leftmost redex.

```
sd normal := matchrew \ x . M by M using normal
   | top(beta) or-else matchrew M N by M using normal
               or-else matchrew M N by N using normal .
```

For completely reducing a term, we can simply write normal ! with the normalization operator. Alternatively, $\lambda$-terms can be reduced in the usual applicative order, by selecting the inner rightmost redex.

```
sd applicative := matchrew \ x . M by M using applicative
   | matchrew M N by N using applicative
     or-else matchrew M N by M using applicative
     or-else top(beta) .
```

However, this new strategy does not ensure that a normal form is reached if it exists. We can see it by running the K I Omega term of Fig. 2 under both strategies.

```
Maude> srew K I Omega using normal ! .

Solution 1
rewrites: 17
result LambdaTerm: \ x . x

No more solutions.
rewrites: 17
```

The normal form $\mathbf{I} \equiv \lambda x.x$ is reached with normal, but it is not with applicative. Notice that the srewrite command finishes even though the strategy does not terminate.

```
Maude> srew K I Omega using applicative ! .
```

```
No solution.
rewrites: 11
```

This is because the `srewrite` infrastructure is able to detect execution cycles and interrupt the evaluation of the strategy, but the absence of solutions is the evidence that applicative reduction does not terminate for this term.

The semantics of other programming languages have been addressed with the Maude strategy language like Eden [21], the REC language of Glynn Winskel's textbook [34], the ambient calculus [31], CCS [27], and the Maude strategy language itself [38].

### 3.3  Games

Strategies are pervasive in games, most usually for specifying how players can solve or win them. Besides the Sudoku solver [40], already mentioned, the strategy language has been used to work out the 15-puzzle [32], the Hanoi tower's puzzle [15], to compare different player strategies for Tic-Tac-Toe by model checking [37], and to solve other smaller games [1].

In addition to expressing procedures for solving a game, strategies can also specify intrinsic restrictions that are rather difficult to express with rules. For example, in the river-crossing problem formalization in [36], we use strategies to enforce a precedence that is part of the rules of the game. Here, we briefly describe this example without going into details about the data representation, which are available in the referenced article and in the repository of examples [1]. In the classical river-crossing puzzle, a *shepherd* needs to cross a river with a `wolf`, a `goat`, and a `cabbage` using a boat with room for two passengers, the shepherd included. The problem is that the wolf would eat the goat and the goat would eat the cabbage as soon as they are left alone without the shepherd in any side of the river. Our representation of the river is `left` $L$ | `right` $R$ where $L$ and $R$ are sets of characters and `left shepherd wolf goat cabbage` | `right` is the initial state. Four rewrite rules, `alone`, `wolf`, `goat`, and `cabbage`, let the shepherd cross alone or with the corresponding passenger to the other side. Two more rules, `wolf-eats` and `goat-eats`, carry out the threat of the mentioned animal over its "prey". Moreover, a key restriction is that the wolf and the goat will never miss the opportunity to eat, so eating must happen eagerly before moving. For instance, the `wolf` rule rewrites the initial state to `left goat cabbage` | `right shepherd wolf`, where the goat and the cabbage are left alone. In this situation, the `goat-eats` rule must be applied to yield `left goat`| `right shepherd wolf`, but moving `alone` is also allowed in the uncontrolled rewrite system. Indeed, we can try to use the `search` command to solve the problem, by looking for the final state.

```
Maude> search initial
           =>* left | right shepherd wolf goat cabbage .

Solution 1 (state 31)
states: 32   rewrites: 60
```

```
empty substitution

No more solutions.
states: 36   rewrites: 89
```

This answer would make us think that the problem is solvable. It is indeed, but this command is not an evidence, since recovering the path to this solution gives an invalid sequence of moves.

```
Maude> show path 31 .
state 0, River: right | shepherd wolf goat cabbage left
===[ wolf ]===>
state 2, River: goat cabbage left | shepherd wolf right
===[ alone ]===>
state 7, River: shepherd goat cabbage left | wolf right
===[ goat ]===>
state 15, River: cabbage left | shepherd wolf goat right
===[ alone ]===>
state 23, River: shepherd cabbage left | wolf goat right
===[ cabbage ]===>
state 31, River: left | shepherd wolf goat cabbage right
```

In the second state the **goat-eats** rule should be applied, but **alone** is applied instead.

In order to enforce the precedence of eating over moving we can use the Maude strategy language. The following recursive strategy **eagerEating** applies rules under this restriction until the final state is reached.

```
sd eagerEating :=
   (match left | right shepherd wolf goat cabbage) ? idle
   : ((eating or-else oneCrossing) ; eagerEating) .
sd eating       := wolf-eats | goat-eats .
sd oneCrossing := shepherd | wolf | goat | cabbage .
```

Notice that nonterminating executions are also admitted by the strategy, but they are not a problem for the strategy execution engine because of its cycle detector. We can use the experimental **search** command controlled by a strategy[1] to find a valid solution for the problem.

```
Maude> search initial =>* left | right shepherd wolf
          goat cabbage using eagerEating .
Solution 1 (state 30)
states: 36   rewrites: 72
empty substitution

No more solutions.
states: 36   rewrites: 75
Maude> show path 30 .
state 0, River: shepherd wolf goat cabbage left | right
```

_____
[1] The **search-using** command is not currently available in the official version of Maude, but in an extended version with the strategy-aware model checker [38].

```
===[ goat ]===>
state 23, River: wolf cabbage left | shepherd goat right
===[ alone ]===>
state 24, River: shepherd wolf cabbage left | goat right
===[ wolf ]===>
state 25, River: cabbage left | shepherd wolf goat right
===[ goat ]===>
state 27, River: shepherd goat cabbage left | wolf right
===[ cabbage ]===>
state 28, River: goat left | shepherd wolf cabbage right
===[ alone ]===>
state 29, River: shepherd goat left | wolf cabbage right
===[ goat ]===>
state 30, River: left | shepherd wolf goat cabbage right
```

### 3.4   Other Examples

Beyond the examples already cited in the previous sections, other applications of the Maude strategy language have been published like specifications of the Routing Information Protocol [38], membrane systems with several extensions and model checking [39], the simplex algorithm and a parameterized backtracking scheme with instances for finding solutions to the labyrinth, 8-queens, graph $m$-coloring, and Hamiltonian cycle problems [34], semaphores and processor scheduling policies [38], a branch and bound scheme [1], Bitcoin smart contracts [4], neural networks [41], and more [27].

## 4   Related Tools and Extensions

In this section, we briefly describe three extensions and related tools for the strategy language: an extended model checker for strategy-controlled systems, the support for reflective manipulation of strategies with some applications, and a probabilistic extension of the language.

### 4.1   Model Checking

Model checking [11] is an automated verification technique based on the exhaustive exploration of the execution space of the model. The properties to be checked are usually expressed in temporal logics like Linear-Time Temporal Logic (LTL) or Computation Tree Logic (CTL). Its integrated model checker for LTL is one of the most widely used features of Maude [17]. However, it cannot be applied to strategy-controlled specifications, since it does not know anything about strategies. In order to solve this, we implemented a strategy-aware extension [33,38] of this LTL model checker, which has been extended for branching-time temporal logics in subsequent works [36].

Intuitively, a strategy describes a subset of the executions of the original model or a subtree of the original computation tree, so the satisfaction of a linear-time or branching-time temporal property in a strategy-controlled system should be evaluated on these representations of its restricted behavior. For example, in the river-crossing puzzle of Sect. 3.3, the LTL formula $\square\,(risky \rightarrow \square\,\neg\,goal)$ (once a risky state –where an animal is able to eat– is visited, the goal is no longer reachable) does not hold in the uncontrolled system

```
Maude> red modelCheck(initial, [] (risky -> [] ~ goal)) .
rewrites: 43
result ModelCheckResult: counterexample(..., ...)
```

but it does hold when the system is controlled by the `eagerEating` strategy.

```
Maude> red modelCheck(initial, [] (risky -> [] ~ goal),
                      'eagerEating) .
rewrites: 178
result Bool: true
```

However, the property $\diamond\,goal$ (the goal is eventually reached) does not hold in any case, since the shepherd may keep moving in cycles, for example.

```
Maude> red modelCheck(initial, <> goal, 'eagerEating) .
rewrites: 24
result ModelCheckResult: counterexample(..., ...)
```

Counterexamples returned by the strategy-aware model checker are executions allowed by the strategy, which are often shorter or easier to understand. The usage of the strategy-aware model checker is documented in [38] and it can be downloaded from maude.ucm.es/strategies, along with examples and documentation. Branching-time properties in CTL, CTL*, and the $\mu$-calculus can also be checked with the `umaudemc` tool [36], also available at this website. For example, we can check the CTL* property $\mathbf{A}(\square\,\neg\,risky \rightarrow \mathbf{E}\,\diamond\,goal)$, saying that we can eventually reach the goal by avoiding risky states, which holds both with and without strategy.

```
$ umaudemc check river.maude initial
    'A ([] ~ risky -> E <> goal)'
The property is satisfied in the initial state
(36 system states, 197 rewrites, holds in 18/36 states)
$ umaudemc check river.maude initial
    'A ([] ~ risky -> E <> goal)' eagerEating
The property is satisfied in the initial state
(35 system states, 176 rewrites, holds in 17/35 states)
```

## 4.2   Reflective Manipulation of Strategies

Even though the strategy language was introduced to avoid the complications of the metalevel when controlling rewriting, reflection is still useful in the context of strategies. Like any other Maude feature, the strategy language, strategy modules, and the associated operations are reflected at the metalevel [13, §17.3]. First, every combinator of the language is declared as a term of sort `Strategy` or its subsorts in the `META-STRATEGY` module of the Maude prelude.

```
ops fail idle : -> Strategy [ctor] .
op _[_]{_} : Qid Substitution StrategyList
                -> RuleApplication .
op match_s.t._ : Term EqCondition -> Strategy .
op _?_:_ : Strategy Strategy Strategy -> Strategy [...] .
op _[[_]] : Qid TermList -> CallStrategy [ctor prec 21] .
*** and more
```

Then, strategy modules and their statements are defined as data in `META-MODULE`.

```
op sd_:=_[_]. : CallStrategy Strategy AttrSet
                   -> StratDefinition .
op csd_:=_if_[_]. : CallStrategy Strategy EqCondition
                      AttrSet -> StratDefinition .
op smod_is_sorts_._____endsm : ··· -> StratModule .
```

Finally, the `srewrite` and `dsrewrite`[2] commands are metarepresented in the `META-LEVEL` module.

```
sort SrewriteOption .
ops breadthFirst depthFirst : -> SrewriteOption [ctor] .
op metaSrewrite : Module Term Strategy SrewriteOption
                    Nat ~> ResultPair? [...] .
```

Strategies can be reflectively generated and transformed using these tools with interesting applications. In [37], we explain metaprogramming of strategies with several examples, from a theory-dependent normalization strategy for context-sensitive rewriting [25] to extensions of the strategy language itself. For instance, the similar strategy languages ELAN [8] and Stratego [10] include some constructs that are not available in Maude, like congruence operators $f(\alpha_1, \ldots, \alpha_n)$ for applying strategies to every argument of a symbol $f$. However, these absences are not substantial, since most can be easily expressed using the combinators of the Maude strategy language, for which an automated translation can be programmed at the metalevel.

Multistrategies is another more complex extension that allows distributing the control of the system in multiple strategies $\alpha_1, \ldots, \alpha_n$ orchestrated by another one $\gamma$. Typically, each strategy $\alpha_k$ describes the behavior of a component, agent, or player of the system, while $\gamma$ specifies how their executions are interleaved. Namely, $\gamma$ can make them execute concurrently at almost rule-application granularity, by turns, or in other arbitrary ways. Systems controlled by multistrategies can be executed and model checked with an implementation that relies on the metarepresentation of the strategy language.

Yet another example is an extensible small-step operational semantics of the Maude strategy language, already mentioned in Sect. 3.2. It is specified with

---

[2] `dsrewrite` is the depth-first search variant of `srewrite`, which does a fair breadth-first-like search.

rules and strategies that manipulate terms and strategies at the metalevel [38]. Of course, running strategies or model checking under this semantics is not useful in practice, since the builtin implementation of the language is much more efficient. However, experimentation is easier with this specification. For instance, a synchronized rewrite or intersection operator $\alpha \wedge \beta$ denotes the rewriting paths allowed by both $\alpha$ and $\beta$, which cannot be expressed in terms of the original combinators. Nevertheless, $\alpha \wedge \beta$ can be implemented with a pair of two execution states of the semantics that are advanced in parallel as long as they represent the same term.

## 4.3 A Probabilistic Extension

In addition to qualitative properties, quantitative aspects like time, cost, and probabilities are relevant when analyzing the behavior of systems. Statistical methods are often used to estimate them by simulation, that is, by evaluating the measures on many executions generated at random. However, for this analysis to be sound, all sources of nondeterminism must be quantified. We argued before that strategies are a useful resource to restrict nondeterminism, but they are also suitable for quantifying it. Indeed, probabilistic choice operators have been proposed for ELAN [7] and are available in Porgy [18]. In the context of Maude, PSMaude [7] proposes a restricted strategy language for quantifying the choice of positions, rules, and substitutions. These latter specifications can be simulated and model checked against PCTL properties.

For the specification of probabilities in the Maude strategy language, new combinators have been added. The first one is equivalent to those of ELAN and Porgy.

- A quantified version of non-deterministic choice $\alpha_1 \mid \cdots \mid \alpha_n$ where each alternative is associated a weight

$$\texttt{choice}(w_1 : \alpha_1, \ldots, w_n : \alpha_n)$$

Weights $w_k$ are terms of sort Nat or Float that are evaluated in the context where the strategy is executed. The probability of choosing the alternative $\alpha_k$ is $\sigma(w_k)/\sum_{i=1}^{n} \sigma(w_i)$ where $\sigma$ is the current variable context.
- A sampling operator from a probabilistic distribution $\pi$ to a variable X that can be used in a nested strategy $\alpha$

$$\texttt{sample } X := \pi(t_1, \ldots, t_n) \texttt{ in } \alpha$$

The repertory of available distributions includes bernoulli($p$), uniform($a$, $b$), norm($\mu$, $\sigma$), exp($\lambda$) (for the exponential distribution), and gamma($\alpha$, $\lambda$). Their parameters are also evaluated in the current variable context.

These operators are not currently available in the official version of Maude, but in the extended version including the strategy-aware model checker in Sect. 4.1. They can be used in the usual srewrite and dsrewrite commands, and in

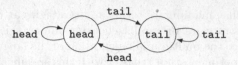

**Fig. 3.** Coin toss module.

the `metaSrewrite` function. When a `sample` operator is evaluated, a variable is sampled at random and the nested $\alpha$ is executed with this random value. When a `choice` is executed, one of the strategies is chosen at random according to their probabilities.

We can apply both statistical and probabilistic model-checking methods on these specifications enhanced with probabilities. For instance, suppose we model tossing a coin like in Fig. 3, with two constants `head` and `tail`, two homonym rules, and two homonym atomic propositions. A fair coin can then be modeled with the strategy `choice(1 : head, 1 : tail)`. The expected number of steps until the first tail is obtained can be estimated with the `scheck` subcommand of `umaudemc`.

```
$ umaudemc scheck coin head firstTail.quatex
  --assign strategy 'choice(1 : head, 1 : tail) !'
Number of simulations = 46530
μ = 6.00143993122 σ = 5.50060692624 r = 0.0499808311155
```

The simulation is driven by an expression in the QUATEX language of PMaude [2] specified in the `firstTail.quatex` file, where # means *in the next step*.

```
FirstTail() = if (s.rval("C == tail") == 1) then
                s.rval("steps") else # FirstTail() fi ;
eval E[FirstTail()] ;
```

However, statistical model checking is more useful when continuous-time aspects are involved, i.e., when using the `sample` operator. For discrete models like this one, we can also use probabilistic model-checking techniques. This is available through the `pcheck` subcommand of `umaudemc` and relies on either the PRISM [23] or Storm [19] model checkers. The following command is equivalent to the previous one but using probabilistic methods.

```
$ umaudemc pcheck coin head '<> tail' --assign strategy
              'choice(1 : head, 1 : tail) !' --steps
Result: 6.0
```

As well as obtaining expected values, `pcheck` allows calculating the probabilities that a temporal formula in LTL, CTL, PCTL, and other logics holds.

```
$ umaudemc pcheck coin head '<> <= 5 tail'
  --assign strategy 'choice(1 : head, 1 : tail) !'
Result: 0.96875
```

For complementing this haphazard appetizer, more information on the probabilistic extensions can be found in [32] and the strategy language website.

## 5 Conclusions

In this tutorial, we have provided an overview of the Maude strategy language, illustrated with several examples, and explained some extensions and associated tools. We refer the interested reader to the works cited in the paper and to the maude.ucm.es/strategies website to complete the information about the strategy language and those tools.

As future work, we plan extending the probabilistic strategy language in Sect. 4.3 with an operator to quantify the choice of matches

matchrew $P$ s.t. $C$ with weight $w$ by $x_1$ using $\alpha_1, \ldots, x_n$ using $\alpha_n$

and new verification features. Another natural and interesting extension of the strategy language is its application to narrowing [3].

**Acknowledgments.** I would like to thank Kyungmin Bae for the invitation to give this tutorial at WRLA, as well as the coauthors of previous works on the strategy language, Steven Eker, Narciso Martí-Oliet, José Meseguer, Alberto Verdejo, Isabel Pita, and the other members of the Maude Team. Special thanks are due to Narciso Martí-Oliet for their helpful comments when revising this manuscript. This work was partially supported by the Spanish Ministry of Science and Innovation through projects TRACES (TIN2015-67522-C3-3-R) and ProCode (PID2019-108528RB-C22), and by the Spanish Ministry of Universities through grant FPU17/02319.

## References

1. Examples of the Maude strategy language (2022). https://fadoss.github.io/strat-examples
2. Agha, G.A., Meseguer, J., Sen, K.: PMaude: rewrite-based specification language for probabilistic object systems. In: Cerone, A., Wiklicky, H. (eds.) Proceedings of the Third Workshop on Quantitative Aspects of Programming Languages, QAPL 2005, Edinburgh, UK, 2–3 April 2005. Electronic Notes in Theoretical Computer Science, vol. 153(2), pp. 213–239. Elsevier (2006). https://doi.org/10.1016/j.entcs.2005.10.040
3. Aguirre, L., Martí-Oliet, N., Palomino, M., Pita, I.: Strategies in conditional narrowing modulo SMT plus axioms. Technical report 2/21, Departamento de Sistemas Informáticos y Computación. Universidad Complutense de Madrid (2021). https://eprints.ucm.es/68621/
4. Atzei, N., Bartoletti, M., Lande, S., Yoshida, N., Zunino, R.: Developing secure bitcoin contracts with BitML. In: Dumas, M., Pfahl, D., Apel, S., Russo, A. (eds.) Proceedings of the ACM Joint Meeting on European Software Engineering Conference and Symposium on the Foundations of Software Engineering, ESEC/SIGSOFT FSE 2019, Tallinn, Estonia, 26–30 August 2019, pp. 1124–1128. ACM (2019). https://doi.org/10.1145/3338906.3341173

5. Bachmair, L., Dershowitz, N.: Equational inference, canonical proofs, and proof orderings. J. ACM **41**(2), 236–276 (1994). https://doi.org/10.1145/174652.174655
6. Balland, E., Brauner, P., Kopetz, R., Moreau, P.-E., Reilles, A.: Tom: piggybacking rewriting on Java. In: Baader, F. (ed.) RTA 2007. LNCS, vol. 4533, pp. 36–47. Springer, Heidelberg (2007). https://doi.org/10.1007/978-3-540-73449-9_5
7. Bentea, L., Ölveczky, P.C.: A probabilistic strategy language for probabilistic rewrite theories and its application to cloud computing. In: Martí-Oliet, N., Palomino, M. (eds.) WADT 2012. LNCS, vol. 7841, pp. 77–94. Springer, Heidelberg (2013). https://doi.org/10.1007/978-3-642-37635-1_5
8. Borovanský, P., Kirchner, C., Kirchner, H., Ringeissen, C.: Rewriting with strategies in ELAN: a functional semantics. Int. J. Found. Comput. Sci. **12**(1), 69–95 (2001). https://doi.org/10.1142/S0129054101000412
9. Braga, C., Verdejo, A.: Modular structural operational semantics with strategies. In: van Glabbeek, R., Mosses, P.D. (eds.) Proceedings of the Third Workshop on Structural Operational Semantics, SOS 2006, Bonn, Germany, 26 August 2006. Electronic Notes in Theoretical Computer Science, vol. 175, no. 1, pp. 3–17. Elsevier (2007). https://doi.org/10.1016/j.entcs.2006.10.024
10. Bravenboer, M., Kalleberg, K.T., Vermaas, R., Visser, E.: Stratego/XT 0.17. A language and toolset for program transformation. Sci. Comput. Program. **72**(1–2), 52–70 (2008). https://doi.org/10.1016/j.scico.2007.11.003
11. Clarke, E.M., Henzinger, T.A., Veith, H., Bloem, R. (eds.): Handbook of Model Checking. Springer, Cham (2018). https://doi.org/10.1007/978-3-319-10575-8
12. Clavel, M.: Strategies and user interfaces in Maude at work. In: Gramlich, B., Lucas, S. (eds.) Proceedings of the 3rd International Workshop on Reduction Strategies in Rewriting and Programming, WRS 2003, Valencia, Spain, 8 June 2003. Electronic Notes in Theoretical Computer Science, vol. 86, no. 4, pp. 570–592. Elsevier (2003). https://doi.org/10.1016/S1571-0661(05)82612-X
13. Clavel, M., et al.: Maude manual v3.2.1 (2022-02). http://maude.cs.illinois.edu
14. Clavel, M., Meseguer, J.: Reflection and strategies in rewriting logic. In: Meseguer, J. (ed.) Proceedings of the First International Workshop on Rewriting Logic and its Applications, WRLA 1996, Asilomar, California, 3–6 September 1996. Electronic Notes in Theoretical Computer Science, vol. 4, pp. 126–148. Elsevier (1996). https://doi.org/10.1016/S1571-0661(04)00037-4
15. Durán, F., et al.: Programming and symbolic computation in Maude. J. Log. Algebraic Methods Program. **110** (2020). https://doi.org/10.1016/j.jlamp.2019.100497
16. Eker, S., Martí-Oliet, N., Meseguer, J., Verdejo, A.: Deduction, strategies, and rewriting. In: Archer, M., de la Tour, T.B., Muñoz, C. (eds.) Proceedings of the 6th International Workshop on Strategies in Automated Deduction, STRATEGIES 2006, Seattle, WA, USA, 16 August 2006. Electronic Notes in Theoretical Computer Science, vol. 174, no. 11, pp. 3–25. Elsevier (2007). https://doi.org/10.1016/j.entcs.2006.03.017
17. Eker, S., Meseguer, J., Sridharanarayanan, A.: The Maude LTL model checker. In: Gadducci, F., Montanari, U. (eds.) Proceedings of the Fourth International Workshop on Rewriting Logic and its Applications, WRLA 2002, Pisa, Italy, 19–21 September 2002. Electronic Notes in Theoretical Computer Science, vol. 71, pp. 162–187. Elsevier (2004). https://doi.org/10.1016/S1571-0661(05)82534-4
18. Fernández, M., Kirchner, H., Pinaud, B.: Strategic port graph rewriting: an interactive modelling framework. Math. Struct. Comput. Sci. **29**(5), 615–662 (2019). https://doi.org/10.1017/S0960129518000270

19. Hensel, C., Junges, S., Katoen, J.P., Quatmann, T., Volk, M.: The probabilistic model checker STORM. Int. J. Softw. Tools Technol. Transf. **23**(4), 1–22 (2021). https://doi.org/10.1007/s10009-021-00633-z

20. Hernández Cerezo, A.: Strategies for implementing SAT algorithms in rewriting logic. Bachelor's thesis, Universidad Complutense de Madrid (2020). https://eprints.ucm.es/63693

21. Hidalgo-Herrero, M., Verdejo, A., Ortega-Mallén, Y.: Using Maude and its strategies for defining a framework for analyzing Eden semantics. In: Antoy, S. (ed.) Proceedings of the Sixth International Workshop on Reduction Strategies in Rewriting and Programming, WRS 2006, Seattle, WA, USA, 11 August 2006. Electronic Notes in Theoretical Computer Science, vol. 174, no. 10, pp. 119–137. Elsevier (2007). https://doi.org/10.1016/j.entcs.2007.02.051

22. Kowalski, R.A.: Algorithm = logic + control. Commun. ACM **22**(7), 424–436 (1979). https://doi.org/10.1145/359131.359136

23. Kwiatkowska, M., Norman, G., Parker, D.: PRISM 4.0: verification of probabilistic real-time systems. In: Gopalakrishnan, G., Qadeer, S. (eds.) CAV 2011. LNCS, vol. 6806, pp. 585–591. Springer, Heidelberg (2011). https://doi.org/10.1007/978-3-642-22110-1_47

24. Lescanne, P.: Implementation of completion by transition rules + control: ORME. In: Kirchner, H., Wechler, W. (eds.) ALP 1990. LNCS, vol. 463, pp. 262–269. Springer, Heidelberg (1990). https://doi.org/10.1007/3-540-53162-9_44

25. Lucas, S.: Context-sensitive rewriting. ACM Comput. Surv. **53**(4), 78:1–78:36 (2020). https://doi.org/10.1145/3397677

26. Marin, M., Kutsia, T.: Foundations of the rule-based system $\rho$Log. J. Appl. Non Class. Log. **16**(1–2), 151–168 (2006). https://doi.org/10.3166/jancl.16.151-168

27. Martí-Oliet, N., Meseguer, J., Verdejo, A.: Towards a strategy language for Maude. In: Martí-Oliet, N. (ed.) Proceedings of the Fifth International Workshop on Rewriting Logic and its Applications, WRLA 2004, Barcelona, Spain, 27 March–4 April 2004. Electronic Notes in Theoretical Computer Science, vol. 117, pp. 417–441. Elsevier (2004). https://doi.org/10.1016/j.entcs.2004.06.020

28. Martí-Oliet, N., Meseguer, J., Verdejo, A.: A rewriting semantics for Maude strategies. In: Roşu, G. (ed.) Proceedings of the Seventh International Workshop on Rewriting Logic and its Applications, WRLA 2008, Budapest, Hungary, 29–30 March 2008. Electronic Notes in Theoretical Computer Science, vol. 238, no. 3, pp. 227–247. Elsevier (2009). https://doi.org/10.1016/j.entcs.2009.05.022

29. Meseguer, J.: Conditional rewriting logic as a unified model of concurrency. Theor. Comput. Sci. **96**(1), 73–155 (1992). https://doi.org/10.1016/0304-3975(92)90182-F

30. Meseguer, J.: Twenty years of rewriting logic. J. Log. Algebr. Program. **81**(7–8), 721–781 (2012). https://doi.org/10.1016/j.jlap.2012.06.003

31. Rosa-Velardo, F., Segura, C., Verdejo, A.: Typed mobile ambients in Maude. In: Cirstea, H., Martí-Oliet, N. (eds.) Proceedings of the 6th International Workshop on Rule-Based Programming, RULE 2005, Nara, Japan, 23 April 2005. Electronic Notes in Theoretical Computer Science, vol. 147(1), pp. 135–161. Elsevier (2006). https://doi.org/10.1016/j.entcs.2005.06.041

32. Rubio, R.: Model checking of strategy-controlled systems in rewriting logic. Ph.D. thesis, Universidad Complutense de Madrid (2022). https://eprints.ucm.es/71531

33. Rubio, R., Martí-Oliet, N., Pita, I., Verdejo, A.: Model checking strategy-controlled rewriting systems. In: FSCD 2019 (2019). https://doi.org/10.4230/LIPIcs.FSCD.2019.31

34. Rubio, R., Martí-Oliet, N., Pita, I., Verdejo, A.: Parameterized strategies specification in Maude. In: Fiadeiro, J.L., Tutu, I. (eds.) WADT 2018. LNCS, vol. 11563, pp. 27–44. Springer, Cham (2019). https://doi.org/10.1007/978-3-030-23220-7_2
35. Rubio, R., Martí-Oliet, N., Pita, I., Verdejo, A.: The semantics of the Maude strategy language. Technical report 01/21, Departamento de Sistemas Informáticos y Computación, Universidad Complutense de Madrid (2021). https://eprints.ucm.es/67449
36. Rubio, R., Martí-Oliet, N., Pita, I., Verdejo, A.: Strategies, model checking and branching-time properties in Maude. J. Log. Algebr. Methods Program. **123**, 100700 (2021). https://doi.org/10.1016/j.jlamp.2021.100700
37. Rubio, R., Martí-Oliet, N., Pita, I., Verdejo, A.: Metalevel transformation of strategies. J. Log. Algebr. Methods Program. **124**, 100728 (2022). https://doi.org/10.1016/j.jlamp.2021.100728
38. Rubio, R., Martí-Oliet, N., Pita, I., Verdejo, A.: Model checking strategy-controlled systems in rewriting logic. Autom. Softw. Eng. **29**(1), 1–62 (2021). https://doi.org/10.1007/s10515-021-00307-9
39. Rubio, R., Martí-Oliet, N., Pita, I., Verdejo, A.: Simulating and model checking membrane systems using strategies in Maude. J. Log. Algebr. Methods Program. **124**, 100727 (2022). https://doi.org/10.1016/j.jlamp.2021.100727
40. Santos-García, G., Palomino, M.: Solving Sudoku puzzles with rewriting rules. In: Denker, G., Talcott, C. (eds.) Proceedings of the 6th International Workshop on Rewriting Logic and its Applications, WRLA 2006, Vienna, Austria, 1–2 April 2006. Electronic Notes in Theoretical Computer Science, vol. 176, no. 4, pp. 79–93. Elsevier (2007). https://doi.org/10.1016/j.entcs.2007.06.009
41. Santos-García, G., Palomino, M., Verdejo, A.: Rewriting logic using strategies for neural networks: An implementation in Maude. In: Corchado, J.M., Rodríguez, S., Llinas, J., Molina, J.M. (eds.) International Symposium on Distributed Computing and Artificial Intelligence (DCAI 2008). Advances in Soft Computing, vol. 50, pp. 424–433. Springer, Heidelberg (2009). https://doi.org/10.1007/978-3-540-85863-8_50
42. Verdejo, A., Martí-Oliet, N.: Basic completion strategies as another application of the Maude strategy language. In: Escobar, S. (ed.) WRS 2011 (2012). https://doi.org/10.4204/EPTCS.82.2

# Teaching Formal Methods
# to Undergraduate Students Using Maude

Peter Csaba Ölveczky[(✉)]

University of Oslo, Oslo, Norway
peterol@ifi.uio.no

**Abstract.** I have been teaching an introductory formal methods course based on Maude—first to third- and fourth-year students, and lately to second-year students—at the University of Oslo for a number of years. The first part of the course introduces functional modules in Maude and covers basic topics in term rewriting, whereas the second part of the course uses Maude to formally model and analyze a number of classic distributed systems, including: transport protocols such as the alternating bit and the sliding windows protocols, the two-phase commit protocol for distributed atomic commitment, distributed algorithms for mutual exclusion and leader election, and authentication protocols.

In this invited "experience report" I briefly motivate the use of Maude for an introductory formal methods course, outline the course content, and summarize student feedback and my own impressions about the course.

## 1   Introduction

Too many years ago I had to design an introductory formal methods course for third-year students at the University of Oslo. The main question was, and remains: *How* to teach an elective introductory formal methods course in an environment where students have never heard about formal methods, and where our colleagues are not overly receptive to the usefulness and beauty of a giving formal treatment to computer systems?

In this "invited experience report" I briefly describe the setting and some challenges when it comes to teaching introductory formal methods courses, and how these challenges might be overcome (Sect. 2). In Sect. 3 I discuss how some papers argue that formal methods should be taught. In Sect. 4 I argue that—based on the criteria for teaching formal methods—rewriting logic [14] and its accompanying Maude tool [10] should provide a suitable framework for introducing formal methods to undergraduate students.

I have taken my own medicine and have been teaching formal methods based on Maude for twenty years; first to third- and fourth-year students, and since 2019 to second-year students. When the course had reached a certain stability and maturity, I wrote a textbook, called "Designing Reliable Distributed Systems: A Formal Methods Approach Based on Executable Formal Modeling in Maude," which was published in 2018 in Springer's *Undergraduate Topics in Computer Science* series [21]. In Sect. 5 I give an overview of the content of the

© Springer Nature Switzerland AG 2022
K. Bae (Ed.): WRLA 2022, LNCS 13252, pp. 85–110, 2022.
https://doi.org/10.1007/978-3-031-12441-9_5

| 4. semester | IN2000 – Software Engineering med prosjektarbeid | | IN2140 – Introduksjon til operativsystemer og datakommunikasjon /IN2080 – Beregninger og kompleksitet/IN2100 – Logikk for systemanalyse |
| 3. semester | IN2010 – Algoritmer og datastrukturer | IN2120 – Informasjonssikkerhet - | IN2090 – Databaser og datamodellering |
| 2. semester | IN1010 | IN1030 – Systemer, krav og konsekvenser | IN1150 – Logiske metoder |
| 1. semester | IN1000 – Introduksjon i objektorientert programmering og HMS-emner | IN1020 – Introduksjon til datateknologi | EXPHIL03 – Examen philosophicum |

**Fig. 1.** The structure of the "Programming and Networks" bachelor degree at my university. The third year is devoted to freely selected courses and is not shown.

course and of this book. Finally, Sect. 6 summarizes my experiences and the results of the anonymous student evaluation throughout the years.

The longer paper [19] on the same topic gives more details, and presents a broader case for using Maude for teaching, since—in contrast to this paper—it is aimed at the formal methods community without expertise in Maude.

## 2  Setting and Challenges

In this section I discuss some challenges involved in trying to teach formal methods to undergraduate students at a place like the University of Oslo.

When Turing Award winner and department founder Ole-Johan Dahl was at the department, formal methods/verification was a mandatory course in the Bachelor program on "Programming," and hence around 80 students took the formal methods course every year. However, since then my esteemed colleagues have relegated formal methods to an elective course in the periphery of that Bachelor program, shown in Fig. 1, which shows the courses that the students should take in the first two years. (The program is in Norwegian, so there is no English version.) The formal methods course ("IN2100–Logikk for Systemanalyse") has to compete with a course introducing operating systems and computer networks and one on computability and complexity for the final 10 credits in this Bachelor program. In such a setting, would an 18–20-year-old student, who has no idea what formal methods are, choose to take the formal methods course instead of a (supposedly good, from what I hear) course on operating systems and computer networks? I am pretty sure that as a 19-year-old student I would have taken the OS course instead, and would never have been exposed to formal methods during my studies. However, I was lucky enough to study while the above-mentioned Ole-Johan Dahl was still teaching, so we *had to* take the verification course, which led me to my current path.

This problem is compounded by the fact that students at the University of Oslo study "Informatics" to quickly get a good job, and therefore prefer to take more "practical," seemingly more work-relevant, courses. In a recent *Communications of the ACM* blog post [25], Daniel G. Schwartz at the Florida State University writes that this does not just apply to Norway: "Another issue is that most CS students are primarily only interested in acquiring the skills that will enable them to find jobs as software developers. Few have any interest in pursuing graduate studies and research. For this reason, they see no purpose in studying theoretical topics." If this were not enough, our students tend to have very limited background in mathematics, and tend not to study too much.

How can we overcome such "structural" challenges? Unless the Bachelor curriculum changes, or the course again becomes a third-year course, it would seem hard to attract students. Therefore, the main hope is to create *such a good course* that students recommend the course on an unknown subject to their peers. Indeed, most students taking the course this year do it because they heard it was a good course. The problem with this "word-of-mouth" strategy is that students mostly socialize with students at the same stage in their studies. Because of this, and because students "try out" many courses at the beginning of the semester, there is a need to *quickly* demonstrate the power and usefulness on relevant problems and applications. Furthermore, the lack of mathematical background[1] also means that the course should not be very hard or "theoretic."

Related to the above challenges, and maybe the reason why the formal methods course has been relegated to the purgatory of elective courses, is the following misconception, quoted from [17]:

> In industry, formal methods have a reputation for requiring a huge amount of training and effort to verify a tiny piece of relatively straightforward code, so the return on investment is justified only in safety-critical domains (such as medical systems and avionics).

Fortunately, formal methods and their tools have matured quite a lot, and we also have a better understanding of what formal methods can and cannot do well. We need to advertise the success stories of formal methods; for example, in my course I discuss in some depth: the paper "How Amazon Web Services Uses Formal Methods" [17] by engineers developing the key cloud computing systems at Amazon Web Services; the work of Ralf Sasse and others to find previously unknown flaws in the Internet Explorer web browser using Maude [9]; and the work by David Basin and Ralf Sasse and others who use "Maude-related" methods to find serious flaws in the 5G standard [23] and, in particular, in the

---

[1] I once got complaints from the head of studies for supposedly having shown a quantifier in a lecture!

VISA and MasterCard payment systems [5,6]. The main Norwegian newspaper
has even made a short video about the latter, which I show to my students.

Another misconception, that we ourselves quite often perpetuate, is that
formal methods are aimed at *safety-critical* systems. It *is* true that society is
increasingly dependent on such systems (from self-driving cars to airplanes and
power distribution systems). However, in a country like Norway, I do not think
that many students will end up developing safety-critical systems. Selling formal
methods for safety-critical systems could therefore be self-defeating. Fortunately,
in contrast to 20–30 years ago, when everybody developed their own systems for
local use, these days cloud computing has led to world-wide services, where "the
winner takes it all" in each kind of "service," with the profit for being that winner potentially enormous. Together with increasing system complexity, this need
to develop the highest-quality system implies that an additional up-front investment in system quality really pays off in "mainstream" software development;
this is also the main message of the above AWS paper [17].

Another challenge is the worse and worse mathematical background, and
skepticism toward mathematics, among students. In [25] Schwartz writes that
"most of CS undergrads don't like mathematics and so-called 'theory' courses,
and would prefer to not take them," and quotes Leslie Lamport, who argues that
"while good programming really requires mathematical precision, [Lamport] also
acknowledges that 'basically, programmers and many (if not most) computer
scientists are terrified by math.' " I guess that the solution to this problem is to
use *accessible/intuitive formal methods* that do not require much mathematical
background, and/or to make formal methods look more like "programming,"
which they like and master.

Another issue that sometimes pops up is that formal methods are not integrated with other courses. Therefore, showing the strength of, or at least exemplify the use of, formal methods to model and analyze systems encountered
in other courses would show students—and maybe also our colleagues defining
study plans—the usefulness of formal methods. This could include examples
from security, networking/communication, databases and distributed transactions, operating systems, etc.

The paper [12] discusses the problem of addressing appropriate systems. The
authors write that formal methods courses use examples and case studies that
are either "constructed and thus do relate to practice" or are "based on projects
of industry partners and are thus, too involved for students." Again, we need to
address problems which look relevant, in fields such as social media, online shop-
ping and other cloud applications (i.e., distributed transactions), and/or in authentication. To be able address relevant problems in different courses/domains
we need an *expressive formalism*.

## 3   How to Teach Formal Methods?

Section 2 listed some challenges involved in making students take formal methods
courses when they are not mandatory, and listed some possible "solutions" to

these challenges. In this section I first briefly discuss a few key papers on teaching formal methods (see, e.g., [19] and [8] for longer discussions on papers on the topic), and then try to distill some requirements for courses in formal methods.

### 3.1   A Few Papers on Teaching Formal Methods.

As its title suggests, in their paper "Teaching Formal Methods for Software Engineering: Ten Principles," Cerone, Roggenbach, Schlingloff, Schneider, and Shaikh list and elaborate on ten principles for teaching formal methods, which in my view boil down to the following "principles:"

- Formal methods are too large to gain encyclopedic knowledge; we should just use a few formal methods, since "there is loads to gain by intensively studying [a] few methods."
- Formal methods need tools, which "teach the method," and lab classes, which should imply that we need a high-quality and fairly stable tool.
- Formal methods are best taught by examples.

In their paper "Teaching Concurrency: Theory in Practice" [1], Aceto, Ingolfsdottir, Larsen, and Srba also share the view that "less is more," and that we should repeatedly convey key concepts, instead of providing a broad overview. They also advocate using *automatic* verification tools and very expressive and flexible, yet mathematically simple, executable modeling formalisms, as well as using modal and temporal logics to specify system requirements.

In the paper with the promising title "Teaching Formal Methods in the Context of Software Engineering," Liu, Takahashi, Hayashi, and Nakayama take a somewhat contrarian view [13]. They propose using VDM, refinement, and Hoare logic, but admit that "none of these techniques is easy to use by ordinary practitioners to deal with real software projects." In another divergence from teaching-formal-methods orthodoxy, they claim that "most effective for students [...] is to write formal specifications by hand, as they learn English as a foreign language." Like others, they also argue that "each course should not be too ambitious; instead it should be focused." Finally they admit that "there is little hope to apply refinement calculus in practice."

There is also a "white paper" on teaching formal methods, "Rooting Formal Methods Within Higher Education Curricula for Computer Science and Software Engineering: A White Paper" [8] by a number of participants, including me, at the *First International Workshop on "Formal Methods – Fun for Everybody"* in 2019. This paper advocates that a formal methods course must be mandatory for all Bachelor students in computer science. In addition to also being a proponent for using small "games" to teach formal methods, this paper emphasizes tool use, but not industrial tools, which "can cause frustration."

### 3.2   What to Teach?

We can try to summarize the various requirements for an introductory course in formal methods, where such a course is not mandatory, as follows:

1. It should repeatedly convey key formal methods concepts. But what are these key concepts? Certainly mathematical *modeling/formalization* of both *systems/designs* and of *requirements*, and of course *reasoning* about the models in terms of *model checking* and *verification*, and preferably also model-based *performance estimation*. Another key, slightly orthogonal use of formal methods, is the mathematical analysis of *code*. More generally, one might also want to introduce students to *logical reasoning* in general, dealing with logics, deduction rules, models, satisfaction, and may include key folklore results.
2. It should be fun for the students.
3. It should use relevant applications/examples, also related to other courses the student take, and should seem relevant to today's systems. To be able to this, the modeling formalism must be fairly general and expressive.
4. It should use *few*, but mature, tools, which should seem industry-relevant.
5. We must motivate with industrial success stories.
6. It should be simple and intuitive, and not require much mathematical background.
7. It should support automatic model checking methods.
8. The formalism must be executable, expressive, and general.

## 4   Why Teaching Formal Methods Using Maude?

In my view, rewriting logic and its Maude language and tool should be very well suited to introduce formal methods to undergraduates, as I think that Maude satisfies the "requirements" in the previous sections as follows:

1. (Repeatedly convey main formal methods concepts.) Maude primarily deals with modeling systems/designs in rewriting logic. It also supports formalizing systems requirements in the most elegant and intuitive temporal logic [27], linear temporal logic (LTL), and provides an LTL model checker. While Maude's primary focus is on modeling and model checking of said models, rewriting logic has been used to define the semantics of many programming languages [15,16], and is the foundation of the $\mathbb{K}$ programming language semantics and analysis framework [24]. $\mathbb{K}$ is a leading tool for formalizing the semantics of programming languages, and then analyzing programs, including Ethereum contracts [22]. For model-based performance analysis, rewriting logic has extensions to timed [18,20] and probabilistic [2] systems, such that the performance of the resulting probabilistic (and possibly timed) models can be analyzed by statistical model checking using, e.g., the PVeStA tool [3]. Finally, teaching Maude also provides an excuse to introduce simple logics (equational, rewriting, and temporal logics) and their deduction systems, satisfaction, models, and some folklore (un)decidability proofs.
2. (Fun for students.) What does "fun for students" mean? The students probably study computer science because they like programming. When I was a student, I loved functional programming. Maude modeling is essentially (first-order) functional programming in an object-oriented style.

3. (Relevant examples.) While simple, the Maude specification formalism is expressive and general. Therefore, relevant examples from different fields of computer science can easily be specified by undergraduates; as explained later, I use examples from security/cryptographic protocols, database courses, key communication protocols and distributed algorithms in my course.

4. (Few and mature tools.) My course only uses Maude, which is a mature and high-quality tool.

5. (Motivate with industrial success stories.) There are probably other tools with more industrial success stories. Maude has been used to find a number of previously unknown errors in Internet Explorer, as well as to model (aspects of) industrial systems such as Google's Megastore, Apache Cassandra, Apache ZooKeeper, and so on. However, the Tamarin tool [4] has been used to break the EMV card payment standard [6] and the 5G standard [23], and is based on multiset rewriting, and even includes some parts of the Maude implementation. Furthermore, the rewriting-logic-based $\mathbb{K}$ framework has been applied commercially to analyze electronic contracts on the blockchain.

6. (Simple and intuitive.) Maude is based on equational and rewriting logic. Equations—like $(x + y)^2 = x^2 + 2xy + y^2$—and their use to simplify an expression by replacing equals for equals, is something that all students are familiar with from school. Rewriting is fairly similar, so this simplicity and intuitive logic is one main strengths of rewriting logic, and should make it an ideal formalism for introducing formal methods to undergraduates.

7. (Model checking.) Maude provides a range of automatic analysis methods, including rewriting for quick simulation/prototyping and automatic model checking methods such as reachability analysis and LTL model checking.

8. (Executable expressive and intuitive formalism.) As elaborated above, the Maude formalism is both expressive, executable, and simple and intuitive.

Reasons for skepticism include a lack of good documentation for beginners; the manual is very nice and comprehensive, but is not well suited to learn the formalism for a formal methods novice. Furthermore, I use Full Maude for object-based modeling, but the lack of error messages in Full Maude is a significant problem. I understand that others, including Francisco Durán, teach object-based modeling by "encoding" classes and object rules directly in Maude.

## 5    Course and Textbook Content

In this section I give an overview of the content of the second-year introductory formal methods course I teach at the University of Oslo, and of its aforementioned textbook, "Designing Reliable Distributed Systems: A Formal Methods Approach Based on Executable Modeling in Maude" (Fig. 2). I also mention interesting exam problems I have given, and some exercises in the book, which might be useful for professors looking for exam problems.

The course consists of 14–15 90-minute lectures, and of the same number of 90-minute problem-solving sessions. I cannot assume much mathematical knowledge, even though many (but far from all) students have taken a basic course in logic before taking my course.

The course is divided into two parts: *Part I* deals with equational specification in Maude, and covers basic theory of algebraic specifications and term rewrite systems, in addition to defining equational specifications using Maude. *Part II* deals with specifying and analyzing various distributed systems in Maude using rewriting logic. It is implicitly also meant to introduce some fundamental algorithms in distributed systems.

### 5.1   Part I: Equational Specification in Maude and Term Rewrite Theory

**Equational Specification in Maude (3 lectures).** These lectures introduce basic equational specifications in Maude, starting with many-sorted ones, followed by order-sorted and then membership specifications. We exemplify such specifications with Peano natural numbers, with a wide

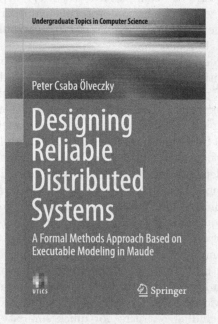

**Fig. 2.** Course textbook.

range of functions on such numbers, and then turn to Boolean values, lists, multisets, binary trees, and so on.

I then cover the built-in Maude modules BOOL, NAT, INT, STRING, CONVERSION, and RANDOM, and specification modulo structural axioms such as associativity, commutativity, and identity. This book includes a section on parametrized modules/programming in Maude, although I do not teach this in class.

Examples include sorting algorithms such as quicksort, merge-sort, insertion-sort, and bubble-sort.

*Example 1.* Lists of natural numbers can be defined as follows in Maude:

```
fmod LIST-NAT is protecting NAT .
  sorts List NeList .
  subsort Nat < NeList < List .

  op nil : -> List [ctor] .
  op _::_ : List List -> List [ctor assoc id: nil] .
  op _::_ : NeList NeList -> NeList [ctor assoc id: nil] .

  op length : List -> Nat .
```

```
  ops first last : NeList -> Nat .
  op reverse : List -> List .

  vars M N K : Nat .     var L : List .

  eq length(nil) = 0 .      eq length(N :: L) = 1 + length(L) .
  eq first(N :: L) = N .    eq last(L :: N) = N .
  eq reverse(nil) = nil .   eq reverse(N :: L) = reverse(L) :: N .
endfm
```

The module LIST-NAT defines a sort List, for lists of natural numbers, and a sort NeList, for non-empty such lists. Lists are constructed by the constructors nil and an infix associative "list concatenation" function _::_, so that a list $\langle 6, 2, 8, 4, 6 \rangle$ is represented as the term 6 :: 2 :: 8 :: 4 :: 6. Since the concatenation function is declared to be associative, parentheses are not needed in this term. Furthermore, since the concatenation constructor is declared to have identity element nil, any list $l$ is considered identical to the lists $l$ :: nil and nil :: $l$, explaining why the equations above do not explicitly consider the case of singleton lists.

The well-known *merge-sort* algorithm can then be specified as follows, where the merge function is declared to be commutative:

```
fmod MERGE-SORT is protecting LIST-NAT .
  op mergeSort : List -> List .
  op merge : List List -> List [comm] .

  vars L1 L2 : List .     vars NEL1 NEL2 : NeList .     vars M N : Nat .

  eq mergeSort(nil) = nil .
  eq mergeSort(N) = N .

  ceq mergeSort(NEL1 :: NEL2) = merge(mergeSort(NEL1), mergeSort(NEL2))
      if length(NEL1) == length(NEL2)
         or length(NEL1) == length(NEL2) + 1 .

  eq merge(nil, L1) = L1 .
  ceq merge(M :: L1, N :: L2) = M :: merge(L1, N :: L2) if M <= N .
endfm
```

I also introduce some classic NP-complete problems (Knapsack, Subset Sum, Traveling Salesman, Hamiltonian Circuit, Clique, etc.) and show in the book how Subset Sum and Hamiltonian Circuit can be solved in Maude.

*Example 2.* In the Subset Sum problem the question is: Given a multiset $MS$ of positive natural numbers and a number $K > 0$, is there a subset of $MS$ whose elements have the sum $K$? This problem can be solved by the following function subsetSum, where the module also declares a data type Mset of multisets of nonzero natural numbers:[2]

---

[2] The function sd gives the difference between two natural numbers, since subtraction is not defined on natural numbers.

```
fmod SUBSET-SUM is protecting NAT .
  sort Mset .     --- multisets of non-zero natural numbers
  subsort NzNat < Mset .
  op none : -> Mset [ctor] .     --- empty multiset
  op _;_ : Mset Mset -> Mset [ctor assoc comm id: none] . --- mset union

  op subsetSum : Mset NzNat -> Bool .

  vars N K : NzNat .    var MS : Mset .

  eq subsetSum(none, K) = false .
  eq subsetSum(N ; MS, K) =
      if N == K then true
      else (if N > K then subsetSum(MS, K)
            else subsetSum(MS, sd(K,N)) or subsetSum(MS, K) fi)
      fi .
endfm
```

**Termination (1+ lecture).** This is one of my favorite topics. This part presents the basics of classic theory on termination of rewriting à la Dershowitz, in the simple, unsorted, and unconditional case without function attributes. The book shows one of the well-known proofs for the undecidability of termination based on reducing the uniform halting for Turing machines to a term rewrite system termination problem. It then discusses methods for proving termination using "weight functions" on well-founded domains, before presenting the elegant theory of simplification orders. Finally, it introduces two such simplification orders: the lexicographic and the multiset path order.

One exercise—used in two exams, to the chagrin of the students—is defining the lexicographic path order in Maude. This gives a taste of meta-programming: how an equational specification can be represented as a Maude term.

I loved to teach the theory of simplification orders, but since I started teaching the course to second-year students, I no longer deal with this nice theory, or with representing Turing machines as term rewrite systems. The grateful second-year undergraduate students are taught temporal logic instead.

**Confluence (1- lecture).** I continue the term rewriting basics by devoting a (short) lecture and book chapter to introducing students to confluence, again, in the most basic setting. I cover the expected bases: Newman's Lemma, unification, and checking (local) confluence using the Critical Pair's Lemma. I am not sure I convey this topic in a particularly interesting way, and I do not believe that confluence is the favorite topic of most students.

## 5.2   Equational Logic (1 lecture)

In one, probably quite heavy, lecture I cover equational logic: the deduction system (again in the unsorted and unconditional case), undecidability, and the usual equivalences between deduction in equational logic and equational simplification.

Then I discuss validity in *all* structures satisfying the equations $E$ versus validity in the *"intended"* structure. Our Maude specifications (also) define the domains of our data types, so we are mostly interested in properties holding in models with those elements we have so painstakingly defined. This leads us to inductive theorems. I start by excusing myself that I cannot give a (finitary) sound and complete proof systems for inductive validity, since such a proof system for cannot exist due to the negative solution to Hilbert's Tenth Problem (thankfully for the students, the argument why that solution leads to the non-existence of the desired proof system for inductive theorems has been relegated to a long footnote). I present the general induction principle for data types—which can be seen as special case of the induction principle for natural numbers—and show the usual examples (binary trees, lists, natural numbers, etc.). I also show how Maude sometimes can be used to automatically prove induction theorems (or at least discharge the proof obligations):

*Example 3.* We can let Maude prove by induction that our addition function (defined in the module NAT-ADD) is associative:

```
fmod NAT-ADD is
  sort Nat .
  op 0 : -> Nat [ctor] .
  op s : Nat -> Nat [ctor] .
  op _+_ : Nat Nat -> Nat .

  vars M N : Nat .
  eq 0 + M = M .
  eq s(M) + N = s(M + N) .
endfm

fmod NAT-ASSOC-IND-PROOF is including NAT-ADD .
  ops t t2 t3 : -> Nat .
  eq (t + t2) + t3 = t + (t2 + t3) .        --- induction hypothesis
endfm

red (0 + t2) + t3 == 0 + (t2 + t3) .        --- base case
red (s(t) + t2) + t3 == s(t) + (t2 + t3) .  --- inductive case
```

Both **red** commands returns **true**, proving associativity of our addition function.

**Models of Equational Specifications.** We are doing "mathematical modeling" because we use an equational specification to precisely specify a mathematical model/structure. For equational specifications the models are algebras. Although I do not teach this to second-year students, I have devoted a chapter of the textbook to the classics of algebras in the context of algebraic specifications. This chapter covers many-sorted $\Sigma$-algebras, then $(\Sigma, E)$-algebras, leading to the algebras $\mathcal{T}_{\Sigma,E}$ and *normal form algebras*. I present the proof of Birkhoff's Completeness Theorem. Finally, I discuss initial algebras and why they are the ones we really wanted to specify, and that $\mathcal{T}_{\Sigma,E}$ and the normal form algebra both are this desired mathematical model specified by an equational specification.

## 5.3    Part II: Modeling and Analysis of Dynamic/Distributed Systems Using Rewriting Logic

In the second part of the course/book, we leave the static world of equations, and the classic theory of algebraic specification and (term) rewrite systems, and move to modeling and analyzing distributed, and dynamic systems in general. I am not aware of any other textbook (in English, at least) that gives an introduction to the modeling and analysis of distributed systems using Maude.

As mentioned elsewhere, additional goals include:

– giving a brief introduction to fundamental distributed algorithms and other folklore systems (such as the dining philosophers problem) that the students should know for "computer science literacy"; and
– looking at systems that are relevant to other courses that students take.

**Rewriting Logic and Analysis in Maude (1 lecture).** I start by explaining why equations are not suitable for modeling dynamic systems, and then introduce rewriting logic and its proof system, including the definition of concurrent steps. This can be introduced by small games (since some papers on teaching formal methods advocate that).

*Example 4.* In the following simple model of a soccer game, the term "Malmo FF" - "Barcelona" 3 : 2 models a state in an (ongoing) game, whereas a state "Malmo FF" - "Barcelona" finalScore: 4 : 2 represents a finished game.

```
mod GAME is protecting NAT . protecting STRING .
  sort Game .
  op _-__:_ : String String Nat Nat -> Game [ctor] .
  op _-_finalScore:_:_ : String String Nat Nat -> Game [ctor] .

  vars HOME AWAY : String .    vars M N : Nat .

  rl [homeTeamScores] : HOME - AWAY M : N => HOME - AWAY M + 1 : N .
  rl [awayTeamScores] : HOME - AWAY M : N => HOME - AWAY M : N + 1 .
  rl [finalWhistle] : HOME - AWAY M : N => HOME - AWAY finalScore: M : N .
endm
```

*Example 5.* In the *whiteboard game*, some natural numbers are written on a whiteboard. In each step of this exciting game, any two numbers $n$ and $m$ can be replaced by their arithmetic mean $n + m$ quo 2. Importing the data type Mset for multisets of numbers from Example 2, this game can be specified as follows:

```
mod WHITEBOARD is protecting SUBSET-SUM .
  vars M N : NzNat .
  rl [replace] : M ; N => (M + N) quo 2 .
endm
```

There is a treasure trove of small examples on this topic in the textbook and among the exam problems. As a running example, I use modeling the life of a person—her age and civil status. This model is then extended to a population; i.e., a multiset of persons, who can communicate synchronously to get engaged, and use message passing to separate. Other examples include classics such the towers of Hanoi, tic-tac-toe, the coffee bean game, modeling all traveling salesman trips and using search to find short trips, packing "suitable" knapsacks (instead of just knowing that *there is* a suitable knapsack), and simulating the behaviors of Turing machines.

This lecture also covers the rewrite and search commands of Maude.

*Example 6.* We can use the rewrite command to simulate one behavior of the whiteboard game from a given initial state, and search to find all reachable *final* states where the resulting number is less than 13. We then exhibit the path to one such desired final state:[3]

```
Maude> rew 6 ; 33 ; 99 ; 1 ; 7 .
result NzNat: 59

Maude> search 6 ; 33 ; 99 ; 1 ; 7 =>! M such that M < 13 .

Solution 1 (state 151)
M --> 12

Solution 2 (state 153)
M --> 11

No more solutions.

Maude> show path 153 .
state 0, Mset: 1 ; 6 ; 7 ; 33 ; 99
===[ rl N ; M => (N + M) quo 2 [label replace] . ]===>
state 10, Mset: 1 ; 6 ; 7 ; 66
===[ rl N ; M => (N + M) quo 2 [label replace] . ]===>
state 51, Mset: 1 ; 7 ; 36
===[ rl N ; M => (N + M) quo 2 [label replace] . ]===>
state 125, Mset: 1 ; 21
===[ rl N ; M => (N + M) quo 2 [label replace] . ]===>
state 153, NzNat: 11
```

**Object-Based Modeling of Distributed Systems (1 lecture).** This lecture first shows that "objects" can be modeled as standard Maude terms, and explains that the state of a distributed system naturally can be seen as a multiset of objects and messages.

After showing that all this can be modeled in Maude, I make the possibly problematic decision to use Full Maude's support for very convenient object-oriented syntax, including for subclasses. I prefer a clean theory and un-cluttered models—at a high cost of lot of frustration when Full Maude.

---

[3] The command echo and some other Maude output are not shown.

The running example, populations of persons, is ideal for illustrating many notions of distributed object-based models: dynamic object creation (birth of children) and deletion (death of a person), rules only involving a single object (like birthdays), synchronous communication (getting engaged) and message-based communication (getting separated), and e subclasses to model that some people are Christians and others are Muslims.

This chapter/lecture also models the dining philosophers problem in an object-oriented style, and uses Maude's pseudo-random number generator to model different variations of *blackjack*, where the next card is drawn pseudo-randomly from the remaining cards. We can then perform randomized simulations to simulate how much money I have left after a day in the casino.

**Modeling Communication and Transport Protocols (1 lecture).** The goal of Part II is to model sophisticated distributed systems. To achieve this we need to model different forms and variations of communication: unicast, multicast, broadcast and "wireless" broadcast, message loss, ordered communication through links, and so on.

The first "larger" applications are a range of well-known transport protocols used to achieve ordered and reliable message communication on top of an unreliable and unordered communication infrastructure. we begin with a TCP-like sequence-number-based protocol. When the underlying infrastructure provides ordered (but lossy) communication between pairs of nodes, the sequence numbers in the TCP-like protocol can be reduced to 0 and 1, giving us the alternating bit protocol (ABP). Generalizing the TCP-like protocol and ABP so that a node can send any one of $k$ different messages at any time, instead of only the same message, gives us the sliding windows protocol, supposedly the most used protocol in distributed systems. I like sliding windows for a homework exercise/project, since the search commands take some time to finish, which I think is useful for students who are used to programs always giving immediate feedback.

**Distributed Algorithms (1 lecture).** As mentioned, one of the goals of the course is to give a flavor of distributed systems, which we do using a number of fundamental distributed algorithms that are still used in state-of-the-art cloud-based transaction systems.

These algorithms are also easy to motivate using modern distributed transactions. For example, in today's cloud-based world the same *eBay* item could have been sold (at the dying moments of an auction) to two different persons; one through a server in Munich, and one through a server in Vanuatu. Or we can imagine an online travel agency with the following distributed transaction:

> *reserve*($X$, OSL-CDG, KLM, Dec 6 to 15);
> *reserve*($X$, Ritz, Imperial Suite, Dec 6 to 15);
> *reserve*($X$, Chez M, dinner, Dec 9);
> *pay*($X$, *6000*, MasterCard, *1234567891234567*, *11/20*, ...);

These examples can motivate the two-phase commit (2PC) protocol: In the *eBay* example, Vanuatu can veto Munich's commit request if it has also sold the item. In the Paris vacation example, if each single operation (at its own site) goes through, the entire transaction should be committed. If, however, one of the operations cannot be performed (there is no money on the credit card, or the Imperial Suite at the Ritz is not available those days), the entire distributed transaction must be aborted. 2PC ensures this.

Multiple distributed operations on the same data could lead to "lost updates." This can motivate the use of distributed mutual exclusion algorithms.

*Example 7.* In the *token-ring-based* distributed mutual exclusion algorithm, the nodes are organized in a ring structure. There is one *token*, and a node must hold the token to enter the critical section; when it exits the critical section, or when a node receives the token without wanting to enter the critical section, it sends the token to the next node in the ring.

This algorithm, where each node alternates forever between executing outside the critical section, and (if possible) executing inside the critical can be modeled using objects a class **Node**, whose attribute **status** can have the values **outsideCS** (the node is executing outside the critical section), **waitForCS** (the node is waiting to enter the critical section), and **insideCS** (the node is executing inside the critical section). The attribute **next** denotes the "next" node in the ring. The token is being sent around as a message:

```
load model-checker
load full-maude31

(omod TOKEN-RING-MUTEX is
  sorts Status MsgContent .
  ops outsideCS waitForCS insideCS : -> Status [ctor] .
  op msg_from_to_ : MsgContent Oid Oid -> Msg [ctor] .

  class Node | next : Oid, status : Status .

  op token : -> MsgContent [ctor] .

  vars O O1 O2 : Oid .

  rl [wantToEnterCS] :
    < O : Node | status : outsideCS >
  =>
    < O : Node | status : waitForCS > .

  rl [rcvToken1] :
    (msg token from O1 to O)
    < O : Node | status : waitForCS >
  =>
    < O : Node | status : insideCS > .

  rl [rcvToken2] :
```

```
      (msg token from O1 to O)
      < O : Node | status : outsideCS, next : O2 >
      =>
      < O : Node | >
      (msg token from O to O2) .

   rl [exitCS] :
      < O : Node | status : insideCS, next : O2 >
      =>
      < O : Node | status : outsideCS >
      (msg token from O to O2) .
endom)

(omod INITIAL is including TOKEN-RING-MUTEX .
   ops a b c d : -> Oid [ctor] .  --- object names
   op init : -> Configuration .   --- an initial state
   eq init
    = (msg token from d to a)
      < a : Node | status : outsideCS, next : b >
      < b : Node | status : outsideCS, next : c >
      < c : Node | status : outsideCS, next : d >
      < d : Node | status : outsideCS, next : a > .
endom)
```

We can then check whether it is possible to reach a state in which two nodes are executing in the critical section at the same time:

```
Maude> (search [1] init =>*  REST:Configuration
                         < O1:Oid : Node | status : insideCS >
                         < O2:Oid : Node | status : insideCS > .)

No solution.
```

Distributed mutual exclusion algorithms are ideal exam problems. The book presents the central server algorithm, the token ring algorithm, and Maekawa's voting algorithm, and I have used Lamport's bakery algorithm and the interesting Suzuki-Kasami algorithm as exam problems.

2PC solves the "same item sold twice problem" by aborting the whole transaction, since one site will veto another site's attempt to commit a (conflicting) transaction. A better idea is to sell the item to one of the buyers, which leads us to distributed leader election and distributed consensus. Leader election is a key part of distributed consensus algorithms, such as Paxos, which again are key components in many of today's cloud-based systems, like Google's Megastore. The textbook describes the Chang and Roberts ring-based distributed leader election algorithm and a spanning-tree based useful for wireless systems. I also introduce distributed consensus, but leave modeling Paxos as an exercise.

Staying on the cloud computing track, a very nice exam problem that illustrates how a cloud-based replicated data store can compromise between desired levels of consistency, performance, and fault tolerance is inspired by Apache Cassandra: Your data are stored at $n$ replicas; a read or a write request is sent to all replicas. A client gets the answer ("ok" for writes and the most recent value of the data item for reads) when $k$ replicas have responded. A lower $k$ gives improved performance (shorter waiting time for the client) and fault tolerance (since $n - k$ replicas can crash) improves, while consistency suffers (not even "read-your-writes" holds when $k$ is low). The system should satisfy "eventual consistency," but other transaction guarantees depend on the value of $k$.

Finally, this chapter presents useful techniques for analyzing fault tolerance by modeling failures and repairs.

**Cryptographic Protocols: Breaking NSPK (1 lecture + 1 guest lecture).** One of my favorite chapters/lectures introduces public-key and shared-key cryptography. We then model the well-known Needham-Schroeder Public-Key (NSPK) authentication protocol. This is a great example to motivate formal analysis. The protocol is super small, only three lines, yet its flaws went undetected for 17 years before they were found by formal methods. This demonstrates that even very small distributed protocols are hard to understand, and that formal methods are useful to find subtle bugs in distributed systems.

It is very easy and natural to model NSPK with four intuitive rewrite rules. Another 13 or so simple rules model Dolev-Yao intruders. A plain Maude search for an unwanted "trusted" connection then breaks NSPK in around 100 minutes on my laptop; the standard search for compromised keys takes a few seconds. The students understand these models, and can modify them (e.g., to analyze Lowe's fix of NSPK) without problems.

NSPK is still a simpler older protocol, and Maude is not a cryptanalysis tool (although Maude-NPA [11] is a leading one). However, a tool like the Tamarin prover [4] is based on multiset rewriting, and has been in the news in Norway and elsewhere for breaking our card payment systems [5,6]. Ralf Sasse from ETH Zürich has generously given a guest the last two years where he talks about using Maude to find news flaws in the Internet Explorer web browser as a summer intern at Microsoft [9], and, especially, how they have used Tamarin to break the EMV protocol [6] and the 5G standard [23]. This guest lecture has been mentioned by some students as a highlight of the course.

**System Requirements (1 lecture).** I devote one lecture to introduce "system requirements" informally. What are invariants, eventually, until, and response properties? I explain how to analyze invariants by searching for bad states, and how to inductively prove (by hand) that something is an invariant for *all* initial states. I also discuss state-based versus action-based requirements, and various kinds of fairness assumptions needed to prove "eventually" properties.

**Formalizing and Checking System Requirements Using Temporal Logic (1 lecture).** I introduce linear temporal logic (LTL) and the use of Maude's

LTL model checker to formalize and then model check system requirements.
We then have a wealth of examples to model check.

*Example 8.* Consider the token-ring mutual exclusion algorithm in Example 7.
The key liveness property we want to prove is that each node executes in its
critical section infinitely often. This cannot be proved using search, but can easily
be done using LTL model checking. We define a parametric atomic proposition
inCS(*o*) to hold if node *o* is currently executing inside its critical section:

```
(omod MODEL-CHECK-MUTEX is protecting INITIAL . including MODEL-CHECKER .
  subsort Configuration < State .
  op inCS : Oid -> Prop [ctor] .
  var REST : Configuration .  var S : Status .  var O : Oid .
  eq REST < O : Node | status : S > |= inCS(O) = (S == insideCS) .
endom)
```

We check if each node in **init** executes infinitely often in its critical section:[4]

```
Maude> (red modelCheck(init,  ([] <> inCS(a))  /\  ([] <> inCS(b))  /\
                              ([] <> inCS(c))  /\  ([] <> inCS(d))) .)
```

```
result ModelCheckResult : counterexample(...)
```

The property does not hold: the model checker returns a counterexample where
node **d** never wants to enter its critical section. We therefore add the following
*justice fairness* assumption for the first rule: *for each node o*, if, from some point
on, the first rule is continuously enabled for *o* (that is, *o*'s **status** is **outsideCS**),
then the first rule must also be taken infinitely often for *o* (i.e., *o*'s **status** must
be **waitForCS**). We add the following declarations to the above module to define
the formula **justAll** that encodes this justice assumption:

```
ops waiting outside : Oid -> Prop [ctor] .
eq REST < O : Node | status : S > |= waiting(O) = (S == waitForCS) .
eq REST < O : Node | status : S > |= outside(O) = (S == outsideCS) .
op just : Oid -> Formula .
op justAll : -> Formula .
eq just(O) = (<> [] outside(O)) -> ([] <> waiting(O)) .
eq justAll = just(a) /\ just(b) /\ just(c) /\ just(d) .
```

We can check whether the justice fairness assumption **justAll** implies the
desired property:

```
Maude> (red modelCheck(init, justAll ->
                       (([] <> inCS(a))  /\  ([] <> inCS(b))  /\
                        ([] <> inCS(c))  /\  ([] <> inCS(d)))) .)
```

```
result Bool :  true                                                □
```

---

[4] '[]' and '<>' denote the temporal operators □ and ◊, respectively, and '/\' and '->'
denote logical conjunction and implication.

I have also proved by LTL model checking that all philosophers are guaranteed to eat infinitely often in one of the solutions to the dining philosophers problem. This required formalizing a number of fairness assumptions.

I briefly mention other logics like CTL and CTL*, LTL with past temporal operators, and Meseguer's *temporal logic of rewriting*, which allows us to reason about both state-based and action-based properties.

I included temporal logic for second-year students with trepidity. This is a completely new kind of logic for the students, which should require time and maturity to understand. I am pleasantly surprised that the students seem to master temporal logic with only one lecture: their exam solutions show that they understand temporal logic formulas and can judge whether such a formula holds in a model.

**Real-Time and Probabilistic Systems (not taught).** Up to this point, the models have been *untimed*. However, the *performance* of a system is also an important metric, whose analysis requires modeling *time*. Furthermore, fault-tolerant systems must detect message losses and node crashes, which is impossible in untimed asynchronous distributed systems. The course textbook introduces how real-time systems can be modeled and analyzed in Maude, and also discusses timed extensions of temporal logics.

Randomized simulations, such that those performed simulating playing *black-jack* with each card drawn pseudo-randomly, do not provide performance estimates with mathematical guarantees. I need more solid guarantees to quit my day job and move to Las Vegas. My textbook therefore indicates how probabilistic systems can be modeled in rewriting logic as *probabilistic rewrite theories* [2]. Such probabilistic models can then be subjected to *statistical model checking* (SMC) using Maude-connected tools such as PVESTA [3] and MultiVesta [26], which estimate the expected value of a path expression up to certain confidence intervals. Although, in contrast to precise *probabilistic model checking*, SMC does not give absolute guarantees, it is considered to be a *scalable* formal method, which, since it is based on simulating single paths until the desired confidence level has been reached, can be easily parallelized.

In contrast to the other chapters in the book, the book only gives a flavor of these subjects, and does not give details about how to run Real-Time Maude or PVESTA. I have sometimes taught this part to fourth-year students, but do not currently teach it to second-year students.

**Using Maude on Cloud Systems and the Use of Formal Methods at Amazon (1 lecture).** To give students the impression that Maude can be applied to analyze industrial designs, in the last lecture I give an overview of the use of Maude (and PVESTA) to model and analyze both the correctness and performance of cloud transaction systems such as Google's Megastore (which runs, e.g., Gmail and Google AppEngine), Apache Cassandra (developed at Facebook and used by, e.g., Amadeus, CERN, Netflix, Twitter), and the academic P-Store design, as well as our own extensions of these designs (see [7] for an overview).

The last lecture should summarize the course: What have you learnt? What is it useful for? Instead of singing the praises of formal methods myself, I end the course by quoting the experiences of engineers at Amazon Web Services, who used formal methods while developing their *Simple Storage System* and *DynamoDB* data store, which are key components of Amazon's profitable cloud computing business. The engineers at Amazon used Lamport's TLA+ formalism with its model checker TLC. They report that formal methods have been a big success at Amazon, and describe their experiences in the previously mentioned paper "How Amazon Web Services Uses Formal Methods" [17] as follows:

- Formal methods found serious "corner case" bugs in the systems that were not found with any other method used in industry.
- A formal specification is a valuable precise *description* of an algorithm, which, furthermore, can be directly tested.
- Formal methods can be learnt by engineers in short time and give good return on investment.
- Formal methods makes it easy to quickly explore design alternatives and optimizations.

My textbook does not contain a chapter on the topics covered in this lecture.

## 6    Evaluation

That I have worked hard on designing what I think *should* be a good and accessible introduction to formal methods by using Maude does not help much if the students disagree. The all-important question is therefore: What do the students think? Unfortunately, I have not solicited their feedback. Instead, the students have the possibility to provide feedback anonymously on courses signed up for. Most students do not bother to do this. Therefore, although I am trying to summarize the students' experiences the best I can, this evaluation is unscientific, anecdotal, and may suffer from selection bias.

### 6.1    Summary of Student Feedback

I have gathered anonymous student feedback, administered by the department, from 2007. In general, only 10%–15% of the students submit responses, and those include students who quit the course during the semester.

The following tables show the cumulated response to the all-important questions "How do you rate this course in general?" and "How do you rate the level (difficulty) of the course?" Since 2019 was the first time the course was given at the second-year level, I also show the results from 2019 in separate columns. Furthermore, since 2020 and 2021 were destroyed by/taught online due to Covid-19, I also separate out the results from those years. In particular, I believe, again without evidence, that the lack of (physical) lectures that make the curriculum understandable is a larger problem for harder-to-access theoretical courses than the more "practical" courses that students usually take. Or is Covid-19 just a convenient scapegoat for the 2020–2021 feedback?

| How do you rate this course in general? | | | |
|---|---|---|---|
| | 2007–2019 | 2019 | 2020–2021 |
| Exceptionally good | 15 | 4 | 8 |
| Very good | 23 | 3 | 4 |
| Good | 8 | 0 | 6 |
| OK (neither good nor bad) | 6 | 1 | 2 |
| Not that good | 1 | 0 | 3 |
| Not good | 0 | 0 | 0 |

| Difficulty/level of the course | | | |
|---|---|---|---|
| | 2007–2019 | 2019 | 2020–2021 |
| Too difficult | 1 | 0 | 2 |
| Somewhat difficult | 38 | 4 | 18 |
| OK/Average | 38 | 4 | 3 |
| Easy | 0 | 0 | 0 |
| Too easy | 0 | 0 | 0 |

An overwhelming majority (75–80%) of the student report that the workload is "OK" (or average) for the number of credits (10) given.

## 6.2 Selected Student Comments

The evaluation form allows students to comment on the course in free-text. Below I quote some student opinions about the course content from 2015 to 2021. What students liked about the course:

- "Very interesting course where we learnt a lot. A unique course at the bachelor level in informatics in Norway."
- "Different and powerful method for system analysis. Creative textbook."
- "Learn a different kind of programming language. Learn about algorithms, and how to model them to check security vulnerabilities. After finishing the course you have relevant knowledge that some of the world's leading companies are looking for."
- "Programming was fun."
- "Introduction to a different programming paradigm."
- "Interesting, but not too extensive, curriculum."
- "Fun curriculum."
- "Course content."
- "IN2100 is the best course I have taken at the University of Oslo."
- "Showed the importance of the topic."
- "Interesting topic."
- "It allows to develop complex systems, and test safety and security of critical systems as well."
- "Strong foundations, applicable to real systems, useful for developing robust systems."

- "All in all I think this was a very fun course, clearly one of those I remember the most from my bachelor. Maude essentially worked well, and even though I don't think that I will ever use it after the course, I have learnt a lot by using it."
- "I did not choose this course [...] but I loved every week and content."
- "The assignments are really well balanced between theory and the entertaining Maude programming parts."
- "One of the best of the ten courses I have taken at the department."

What the students liked less:

- "Language that is not used much or at all."
- "Course might be difficult for many of us."
- "Need more real world critical systems for analysis. [...] Lack of applicability in industry."
- "Maude is very frustrating because of bad or (in Full Maude) missing error messages."
- "The theory part was more difficult than the rewrite rules part."
- "Lectures crashed with the lectures in a more "important" course."
- "Difficult. Unnecessary. Unnecessarily complicated language. Irrelevant."

Main complaints concern Full Maude and its "peculiarities" (lack of robustness and good error messages) and that there are too few resources about Maude. Finally, as expected, a number of students do not understand why they need to learn a programming language that is not widely used. When teaching, we have to emphasize again and again that we use a convenient language to teach and illustrate *general* formal methods principles, so that you could easily work with more "industrial" tools, like TLA+, after taking this course.

## 7   Concluding Remarks

In this paper I surveyed a few papers on, and distilled some requirements for, teaching formal methods. I claimed that rewriting logic and Maude provide an ideal framework that seems to satisfy these criteria. I have given an overview of the topics I cover in my second-year course and in its accompanying textbook. Finally, I summarized the feedback that students provide anonymously to the university. I end this paper by trying to address some obvious questions, and by making a suggestion to the organizers of WRLA 2024.

*Is Maude really a suitable framework for introducing formal methods to second-year undergraduate students?* This is really a two-pronged question: is Maude a good tool for teaching formal methods, and is the second year too early to introduce formal methods?

Concerning the first question, I still think that Maude is a great choice, as I argue in Sect. 4. It provides an intuitive functional programming style, which I think students enjoy. Furthermore, despite taking in a lot of well-established term rewriting theory, we still manage to model and analyze fundamental distributed algorithms and protocols, such as sliding windows, all those distributed mutual exclusion, algorithms for distributed transaction systems, and also cryptographic protocols like NSPK, which is very easy to model and analyze, even for students. Maude also encourages us to teach temporal logic, which is quintessential for formalizing requirements of distributed systems. The by far main problem is that I teach object-oriented modeling using Full Maude. The lack of (useful) error messages in Full Maude understandably frustrates students, and takes away the pleasures of modeling in Maude. It might well be a mistake to use Full Maude for modeling object-based distributed systems; even Francisco Durán teaches object-based modeling using (core) Maude. (Core) Maude will supposedly provide support for object-based specification in the near future; that would make my course *much better* for the students, and cannot happen soon enough.

Regarding the second question, I have no answer or good methodology to answer it. Results from the first exam for second-year students were encouraging, and student feedback has been as positive as in previous years. However, it is hard to conclude anything from exam results and other feedback in 2020 and 2021, because of Covid-19. As usual, many students quit the course during the semester. However, I am not sure that one can gain much insight from this. It is common in Oslo, since signing up for classes is free, so students "try out" many courses. Furthermore, competing against a (supposedly good) introductory course on operating systems and networks for the only optional slot in the Bachelor program is challenging.

*Is the course a success?* This is the million-dollar question, and, again, the jury is still out on this one. I believe that it is fair to say that student evaluations generally are positive. Is this due to the topics covered and the textbook, or does the quality of the lectures and exercise seminars also play a role? Furthermore, just a small fraction of the students reply to these surveys, with a possible selection bias. Eventually, the proof is in the pudding, as they say: do the students take the course? In a related paper [19] from 2020, I wrote that the course—due to its precarious place in the Bachelor program—crucially relies on word-of-mouth recommendation by other students. Then the trend looked good, going from the usual 15–20 students to 42 students who took the exam in 2020. But with two years of Covid19-induced closure of the university, the interaction between students has essentially been non-existent, removing the potential for word-of-mouth recommendation. So we are back where we were before: between 15 and 20 student will take the exam this year.

*Is my way the right way to teach Maude to undergraduates?* If we want to teach Maude, is the way I do it the right way? Based on general evaluation and what I hear from students, it is tempting to significantly reduce the material on classic term rewriting and equational logic theory, which takes almost half the lectures. One could then add more Part II stuff: more fundamental distributed systems, and/or real-time and probabilistic systems. Maybe meta-programming? Strategies? Programming "web applications" with Maude's support for external objects, e.g., via sockets and file systems? Or develop Maude semantics for simple multi-threaded imperative programming languages, which I think would be fun for the students.

Should I remove this theory? I like the theory on termination, but cannot convey as much enthusiasm for confluence; and students are not always enamored of equational logic and inductive theorems either. Can I drop the theory and make the course even more "practical"? If I drop confluence then also the equational logic part will suffer; furthermore, Maude requires your specifications to be confluent, so students should know about this. What should I do? If you, dear reader, teach Maude or related methods, I would love to hear your opinion and experiences. I would also love to know of interesting distributed systems that could be included in the course, or given as exam problems.

*A suggestion to the organizers of WRLA 2024.* With Maude now a mature tool with an impressive range of applications, it should be ripe for teaching formal methods. I think that multiple groups around the world are using Maude in teaching (mostly at the graduate level?). It would be enormously important for our community to know about each other's experiences, curricula, and ways of teaching Maude-based courses. I would therefore like to wrap up this WRLA 2022 "invited experience report" by proposing that the organizer of WRLA 2024 organize a special session on using Maude for teaching, where we can share our experiences on this important topic.

**Acknowledgments.** I would like to thank Kyungmin Bae for inviting me to give an invited talk at WRLA 2022, and for patiently waiting for this paper to be finished.

# References

1. Aceto, L., Ingolfsdottir, A., Larsen, K.G., Srba, J.: Teaching concurrency: theory in practice. In: Gibbons, J., Oliveira, J.N. (eds.) TFM 2009. LNCS, vol. 5846, pp. 158–175. Springer, Heidelberg (2009). https://doi.org/10.1007/978-3-642-04912-5_11
2. Agha, G.A., Meseguer, J., Sen, K.: PMaude: rewrite-based specification language for probabilistic object systems. Electr. Notes Theor. Comput. Sci. **153**(2), 213–239 (2006)
3. AlTurki, M., Meseguer, J.: PVeStA: a parallel statistical model checking and quantitative analysis tool. In: Corradini, A., Klin, B., Cîrstea, C. (eds.) CALCO 2011. LNCS, vol. 6859, pp. 386–392. Springer, Heidelberg (2011). https://doi.org/10.1007/978-3-642-22944-2_28
4. Basin, D.A., Cremers, C., Dreier, J., Sasse, R.: Tamarin: verification of large-scale, real-world, cryptographic protocols. IEEE Secur. Priv. **20**(3), 24–32 (2022)
5. Basin, D.A., Sasse, R., Toro-Pozo, J.: Card brand mixup attack: bypassing the PIN in non-Visa cards by using them for Visa transactions. In: 30th USENIX Security Symposium, USENIX Security 2021, pp. 179–194. USENIX Association (2021)
6. Basin, D.A., Sasse, R., Toro-Pozo, J.: The EMV standard: break, fix, verify. In: 42nd IEEE Symposium on Security and Privacy, SP 2021. IEEE (2021)

7. Bobba, R., et al.: Survivability: design, formal modeling, and validation of cloud storage systems using Maude. In: Assured Cloud Computing, chap. 2, pp. 10–48. Wiley-IEEE Computer Society Press (2018)
8. Cerone, A., et al.: Rooting formal methods within higher education curricula for computer science and software engineering: a white paper. In: Cerone, A., Roggenbach, M. (eds.) FMFun 2019. CCIS, vol. 1301, pp. 1–26. Springer, Cham (2021). https://doi.org/10.1007/978-3-030-71374-4_1
9. Chen, S., Meseguer, J., Sasse, R., Wang, H.J., Wang, Y.M.: A systematic approach to uncover security flaws in GUI logic. In: IEEE Symposium on Security and Privacy, pp. 71–85. IEEE Computer Society (2007)
10. Clavel, M., et al.: All About Maude. LNCS, vol. 4350. Springer, Heidelberg (2007). https://doi.org/10.1007/978-3-540-71999-1
11. Escobar, S., Meadows, C.A., Meseguer, J.: Maude-NPA: cryptographic protocol analysis modulo equational properties. In: Aldini, A., Barthe, G., Gorrieri, R. (eds.) FOSAD 2007/2008/2009. LNCS, vol. 5705, pp. 1–50. Springer, Heidelberg (2009). https://doi.org/10.1007/978-3-642-03829-7_1
12. Krings, S., Körner, P.: Prototyping games using formal methods. In: Cerone, A., Roggenbach, M. (eds.) FMFun 2019. CCIS, vol. 1301, pp. 124–142. Springer, Cham (2021). https://doi.org/10.1007/978-3-030-71374-4_6
13. Liu, S., Takahashi, K., Hayashi, T., Nakayama, T.: Teaching formal methods in the context of software engineering. ACM SIGCSE Bull. 41(2), 17–23 (2009)
14. Meseguer, J.: Conditional rewriting logic as a unified model of concurrency. Theor. Comput. Sci. 96, 73–155 (1992)
15. Meseguer, J., Rosu, G.: The rewriting logic semantics project. Theor. Comput. Sci. 373(3), 213–237 (2007)
16. Meseguer, J., Roşu, G.: The rewriting logic semantics project: a progress report. Inf. Comput. 231, 38–69 (2013)
17. Newcombe, C., Rath, T., Zhang, F., Munteanu, B., Brooker, M., Deardeuff, M.: How Amazon Web Services uses formal methods. Commun. ACM 58(4), 66–73 (2015)
18. Ölveczky, P.C.: Real-Time Maude and its applications. In: Escobar, S. (ed.) WRLA 2014. LNCS, vol. 8663, pp. 42–79. Springer, Cham (2014). https://doi.org/10.1007/978-3-319-12904-4_3
19. Ölveczky, P.C.: Teaching formal methods for fun using Maude. In: Cerone, A., Roggenbach, M. (eds.) FMFun 2019. CCIS, vol. 1301, pp. 58–91. Springer, Cham (2021). https://doi.org/10.1007/978-3-030-71374-4_3
20. Ölveczky, P.C., Meseguer, J.: The Real-Time Maude tool. In: Ramakrishnan, C.R., Rehof, J. (eds.) TACAS 2008. LNCS, vol. 4963, pp. 332–336. Springer, Heidelberg (2008). https://doi.org/10.1007/978-3-540-78800-3_23
21. Ölveczky, P.C.: Designing Reliable Distributed Systems: A Formal Methods Approach Based on Executable Modeling in Maude. UTCS, Springer, London (2017). https://doi.org/10.1007/978-1-4471-6687-0
22. Park, D., Zhang, Y., Saxena, M., Daian, P., Roşu, G.: A formal verification tool for Ethereum VM bytecode. In: Proceedings of the ESEC/FSE 2018, pp. 912–915. ACM (2018)
23. Peltonen, A., Sasse, R., Basin, D.A.: A comprehensive formal analysis of 5G handover. In: 14th ACM Conference on Security and Privacy in Wireless and Mobile Networks, WiSec 2021, pp. 1–12. ACM (2021)
24. Roşu, G., Şerbănuţă, T.F.: An overview of the K semantic framework. J. Logic Algebraic Program. 79(6), 397–434 (2010)

25. Schwartz, D.G.: Rethinking the CS curriculum. Blog at the Communications of the ACM, May 2022. https://cacm.acm.org/blogs/blog-cacm/261380-rethinking-the-cs-curriculum/fulltext
26. Sebastio, S., Vandin, A.: MultiVeStA: statistical model checking for discrete event simulators. In: ValueTools, pp. 310–315. ICST/ACM (2013)
27. Vardi, M.Y.: Branching vs. linear time: final showdown. In: Margaria, T., Yi, W. (eds.) TACAS 2001. LNCS, vol. 2031, pp. 1–22. Springer, Heidelberg (2001). https://doi.org/10.1007/3-540-45319-9_1

# Regular Papers

# Business Processes Analysis
# with Resource-Aware Machine Learning
# Scheduling in Rewriting Logic

Francisco Durán[1], Daniela Martínez[2], and Camilo Rocha[2(✉)]

[1] ITIS Software, Universidad de Málaga, Málaga, Spain
[2] Department of Electronics and Computer Science,
Pontificia Universidad Javeriana, Cali, Colombia
camilo.rocha@javerianacali.edu.co

**Abstract.** A significant task in business process optimization is concerned with streamlining the allocation and sharing of resources. This paper presents an approach for analyzing business process provisioning under a resource prediction strategy based on machine learning. A timed and probabilistic rewrite theory specification formalizes the semantics of business processes. It is integrated with an external oracle in the form of a long short-term memory neural network that can be queried to predict how traces of the process may advance within a time frame. Comparison of execution time and resource occupancy under different parameters is included for a case study, as well as details on the building of the machine learning model and its integration with Maude.

## 1 Introduction

Business process optimization is the practice of increasing organizational efficiency by improving processes. The main motto of this area in business process management is that optimized processes lead to optimized business goals. Since efficiency is one of major quantitative tools in industrial decision making, the most common goals in process optimization are maximizing throughput and minimizing costs. For instance, companies that use a business process for a long time can greatly benefit from increasing the usage of resources within reasonable limits of redundancy and costs, and streamlining workflows. However, business process optimization tends to be a multi-objective optimization problem in which many variables can be involved. One main challenge is to integrate predictive tools at design stages for business process optimization.

Deep learning models are becoming increasingly important in business applications because they serve as a basis for monitoring and predicting process behavior (see, e.g., [10,12]). In particular, applications concerned with resource allocation, time and cost optimization, fault monitoring, and process discovery are using event logs to train these models and predict variables of interest. Logs usually contain information about processes in an organization as a list of events. Each event is associated to a process instance, which is identified by a case number. A case is seen as a collection of activities or tasks with attributes. Typically,

© Springer Nature Switzerland AG 2022
K. Bae (Ed.): WRLA 2022, LNCS 13252, pp. 113–129, 2022.
https://doi.org/10.1007/978-3-031-12441-9_6

they include a name or case number, a timestamp, and the resources or costs associated to it, among other attributes. Recurrent neural networks [14] have been widely used for performing sequence prediction. They are trained to learn from information in event logs and predict the next events that are more likely to occur based on the gained knowledge.

This paper proposes a two-layered approach to formal process optimization. The first layer uses the prediction power of deep learning models to help anticipate the demand of resources (i.e., number of replicas) in a business process from a partial execution. The second layer, on top of the first one, formally specifies the concurrent behavior of many instances of the process that compete for the same collection of resources and replicas. The result is then the integration of these two layers, resulting in a sophisticated heuristic-oriented technique for the formal analysis and optimization of business processes. Their quantitative analysis can then be carried out by, e.g., computing the best combination of parameter values reducing the costs or processing time.

Long short-term memory neural networks (LSTM) [14], a type of recurrent neural network, are used for sequence prediction as proposed in [11]. More precisely, given a business process $B$ and a partial trace $t$ of tasks in $B$, an LSTM can predict an extension of $t$ that conforms to $B$ based on a (previous) training with event logs obtained for $B$. A rewriting logic semantics of the Business Process Modeling Notation (BPMN) simulates the concurrent behavior of many instances of $B$, under a given set of constraints over the resources, by querying the LSTM as a scheduler and adjusting the number of replicas accordingly. One novelty of the proposed approach is that resource allocation can happen at execution time based, not only on the current state of execution of a given instance of $B$, but also on some history of previous executions of the process.

For training an LSTM, the approach takes as input a BPMN process $B$ and a set of traces $T$ of $B$. The traces $T$ represent executions of $B$ in, e.g., a production environment. The process $B$ has information about the type of resources needed to complete the tasks. Therefore, the structure of $B$ and the traces are used to predict how resources are to be allocated/released in order to optimize their use: e.g., minimize the time a resource is not being used or, similarly, maximize the usage of resources meeting some budget and timing constraints associated to the process' execution.

The LSTMs used in the proposed approach have been implemented in Python with Keras and TensorFlow. The integration with Maude is designed via socket communication, where the rewriting logic semantics is the client of the prediction server written in Python. From the partial concurrent execution of a given number of instances of a BPMN process in the Maude semantics, the neural network is queried with a time window. It then returns a sequence of events that extends the traces of the given instances, i.e., make a prediction on their continuation. This prediction is then used to adjust the number of replicas per resource at runtime. The analysis is based on the simulation of the execution guided by such trace. The percentage usage of resources and the number of replicas during the time span of all replicas, among other, are monitored and summarized for each experiment.

**Outline.** Sections 2 and 3 present overviews, respectively, of BPMN and LSTM neural networks. Section 4 presents the rewriting semantics of BPMN and its interaction with the LSTMs. A case study is presented in Sect. 5, while Sect. 6 concludes the paper.

## 2   The Business Process Modeling Notation (BPMN)

BPMN is a graphical notation for modeling business processes as collections of related tasks that produce specific services or products. In BPMN, processes are modeled using graphical representations for tasks and gateways, which are connected through flows and events. In this work, the focus is on its control flow constructs, including the most common types of tasks, events, and gateways.

To introduce and illustrate the use of the supported BPMN constructs, and the analysis techniques presented in this work, the process depicted in Fig. 1 is used. It describes a parcel ordering and delivery. The process consists of three lanes: one for clients, one for the order management, and one for the delivery management. In this process, the client first signs in and then repeatedly looks for products. Eventually, the client can decide to give up or to make an order by submitting it to the order management lane. The client then waits for a response (i.e., acceptance or refusal of this order). However, the client waits for a response for a maximum amount of time, as is represented by a timer-event branch. If the order can be completed, then the parcel is received and the client pays for it. Otherwise (i.e., timeout or order refused), the client fills in a feedback form. As far as the management lane is concerned, the first task aims at verifying whether the goods ordered by the client are available. If they are not available, then the order is canceled; otherwise, the order is confirmed. The order management takes care of the payment of the order whereas the delivery lane is triggered to prepare the parcel to be delivered. The delivery may be carried out by car or by drone.

The initiation and finalization of processes are represented by initial and final events. Events are also used to represent the sending of messages and the firing of timers. A task represents an atomic activity that has exactly one incoming and one outgoing flow. A sequence flow describes two nodes executed one after the other, i.e., imposing an execution order between these nodes. Tasks may send messages, which in such a case activate the corresponding message flows.

Gateways are used to control the divergence and convergence of the execution flows. In this work, *exclusive*, *inclusive*, *parallel*, and *event-based* gateways are supported. Gateways with one incoming branch and multiple outgoing branches are called *splits* (e.g., split inclusive gateway). Gateways with one outgoing branch and multiple incoming branches are called *merges* (e.g., merge parallel gateway). An exclusive gateway chooses one out of a set of mutually exclusive alternative incoming or outgoing branches. For an inclusive gateway, any number of branches among all its incoming or outgoing branches may be taken. A parallel gateway creates concurrent flows for all its outgoing branches or synchronizes concurrent flows for all its incoming branches. For an event-based gateway, it takes one of its outgoing branches or accepts one of its incoming branches based on events.

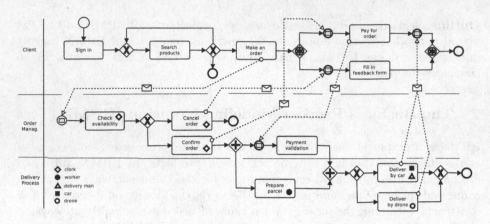

**Fig. 1.** Running example: parcel delivery.

In addition to the description of specific tasks and their sequencing, collaboration diagrams also involve *pools* and *lanes*, which are structuring elements that split processes into pieces. In BPMN, each lane in a collaboration diagram corresponds to a specific role or resource. However, other resources may also be involved, and tasks could require multiple resources or instances of the same resource. Therefore, instead of implicitly associating resources to lanes, in our approach, resources are explicitly defined at the task level. Hence, a task that requires resources for its execution can include, as part of its specification, the required resources. To do it graphically, symbols are associated to each resource type, and these symbols are depicted inside the corresponding tasks. For example, the process in Fig. 1 relies on clerks for the handling of customers' orders, workers for parcel packing, and couriers for car delivery. In addition, cars and drones are used to deliver the parcels. For instance, the diamonds at the right-top corners of the Check availability, Cancel order, and Confirm order tasks indicate that one instance of the clerk resource is required for the execution of the tasks. Task Deliver by car requires instances of the car and courier resources. To avoid dealing with multiple units of measurement, resources are counted as instances or replicas, and if more than one instance of a certain resource type is required, they are depicted as a number of icons in the task.

The process evolves by successively executing its tasks. However, the execution of a task requires the specified amounts of resources, which may lead to a competition for such resources: multiple instances of the process may also run concurrently, and multiple tasks in the same run may require the same resources. In our running example, e.g., clerks are used in several tasks, and multiple customers may be trying to simultaneously purchase products.

## 3   Using Long Short-Term Memory Neural Networks

Recurrent neural networks (RNNs) are a type of neural networks used to process and predict sequential data [14]. They consist of a set of neurons connected with each other, where each neuron has an input $x_t$ and output $h_t$ at a specified timestamp $t$, as well as a feedback loop that provides information of the previous timestamp. LSTM is a type of recurrent neural network used for sequence prediction. In particular, they are useful for solving problems involving long sequences that require previous information for larger periods of time. The main characteristic of this type of networks is their architecture composed by feedback loops to maintain information over time and a set of gates that control the flow of information into a memory cell. Memory cells enables the learning of longer patterns using a group of gates including the forget, input, output gates.

The model used in this work consists of a 2-layer LSTM with one shared layer. The main layers are in charge of activities and timestamp predictions. The data from event logs is transformed into a vector representation using one-hot encoding (i.e., a sequence of bits among which the legal combinations of values are only those with a single 1): each vector has a 1 in the location that matches the corresponding task; its maximum length is given by the number of unique tasks found in the event logs. The vectors are then grouped in a sequence based on the time of their occurrence and this sequence is divided in a prefix and suffix. The prefix represents the activities known to have executed, and the suffix represent the activities to be predicted.

Once the data has been adapted to train the LSTM model, 80% of the available traces are used as training set and the remaining 20% as the testing set to evaluate the accuracy of the model. In each prediction, the LSTM model assigns a probability to all the tasks to decide which one will happen next. Furthermore, the prediction is adapted to take as input a number of tasks to predict.

## 4   Rewriting Logic Semantics with LSTM Integration

Two alternative ways of evolving business process models are presented. Namely, one guided by event logs, obtained from executions of the system, and another one, equipped with time information and probabilities modeling the actual behavior of the system, for its simulation. The second one is responsible for the simulation of the system and the managing and analysis of resources. It is also in charge of communicating with the Python process performing the predictions. The Maude process submits the event log of the activities carried out and, periodically, requests predictions. At this time the Maude BPMN process creates a replica of itself, which will be guided by the event sequence submitted by the predictor. Once the guidance consumes all the predicted events, the status of the resources is analyzed to decide on the allocation/releasing of replicas, and the simulation is restored. In summary, sharing a common representation of the core elements of processes, one specification defines the evolution of the process in accordance to the events in the log trace, and the other non-deterministically advances using the information with which models are annotated.

The information required by each of the specifications is not the same. To simulate the execution of processes, the process specifications are enriched with quantitative information. This additional information is added as annotations to

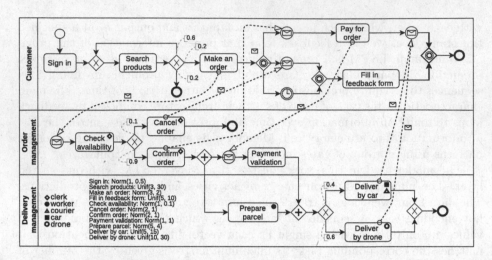

**Fig. 2.** Running example: parcel delivery with durations and probabilities.

the process model. Specifically, durations and delays associated to tasks and flows are expressed as stochastic expressions. Similarly, alternative constructs (split exclusive and inclusive gateways) are extended with probabilities associated to outgoing flows. Figure 2 shows the process given in Fig. 1 enriched with such information.

Data-based conditions for split gateways are modeled using probabilities associated to outgoing flows of exclusive and inclusive split gateways. For instance, notice the exclusive split after the Search products task in the customer lane of the running example, which has outgoing branches with probabilities 0.6, 0.2, and 0.2, specifying the likelihood of following each corresponding path. The probabilities of the outgoing flows in an exclusive split must sum up to 1, while each outgoing flow in an inclusive split can be equipped with a probability between 0 and 1 without a restriction on their total sum.

The timing information associated to tasks and flows (durations or delays) is described either as a literal value (a non-negative real number) or sampled from a probability distribution function according to some meaningful parameters. The probability distribution functions currently available include exponential, normal/Gauss, and uniform (see, e.g., [13]). To simplify the reading of the process in Fig. 2, the specification of task durations has been placed apart from the process description, at the bottom-left corner. In the modelling tool, these parameters would be specified as properties of the corresponding elements. For instance, the duration of the Sign in task is specified as Norm(1, 0.5), which means that it follows a normal distribution with mean 1 and variance 0.5, and the Search products task follows a uniform distribution in the interval [3, 30], that is specified as Unif(3, 30). Also to simplify the specification of the process, the delays in all flows are set to Norm(1.0, 0.2) to express that it takes some time to move from one task to the following one(s).

The specification builds on a specification that has evolved along different extensions through time [2–9].

## 4.1 The Specification of BPMN Processes

In the Maude specification of BPMN, a process is represented as an object with sets of nodes and flows as attributes. The representation of each node type includes the necessary information to describe its structure and to contribute to the overall process analysis. For instance, a task node involves an identifier, a description, two flow identifiers (input and output), a stochastic function modeling its duration, a set of resources required for its execution, and a set of messages to be delivered after its completion. A split node includes a node identifier, a gateway type (exclusive, parallel, inclusive, or event-based), an input flow identifier, and a set of output flow identifiers. A merge node includes a node identifier, a gateway type, a set of input flow identifiers, and an output flow identifier. The representation of a flow includes a probability distribution function specifying its delay, and an optional message or timer. The message blocks the flow until it is received, whereas the timer represents a delay after which the execution is triggered.

Given unique identifiers for nodes, flows, resources, and events the process of the running example can be specified as shown in the excerpt in Fig. 3. It shows how a Process object has attributes with the definition of its nodes and flows connecting them. For example, the exclusive split id("n005") (lines 5–6) has id("f004") as incoming flow, and id("f005"), id("f006"), and id("f007"), with associated probabilities 0.6, 0.2, and 0.2, respectively, as outgoing flows. Furthermore, the event-based split gate id("n007") (line 7) has id("f008") as incoming flow, and id("f009"), id("f010"), and id("f011") as outgoing flows. Note the definition of these flows in lines 17–19; after the corresponding delay, they become active upon the reception of the corresponding messages or by the id("timeout") timer firing. Finally, note that the specifications of tasks and flows also include their duration or delays as stochastic functions. For example, the duration of the Prepare parcel task follows a normal distribution with average 5 and variance 4, which is specified by the term Norm(5.0, 4.0). All flows are specified with Norm(1.0, 0.2) as second argument, stating that they all have a delay that follows a normal distribution with given parameters.

## 4.2 Autonomous Processes

A set of rewrite rules specifies how *tokens* evolve through a process. Each move of a token inside a BPMN process is modeled as a rewrite rule. E.g., one of the actions that may occur, and that is modeled by a corresponding rewrite rule, is that when there is a token in the incoming flow of an exclusive split, the token is moved to one of the outgoing flows of the gate, with its timer set to the value resulting from evaluating the stochastic expression of the flow, which represents the delay of the flow. Objects of classes Simulation, Workload, and Supervisor manage different aspects of the simulations.

```
1 < pid : Process |
2     nodes :
3       (start(id("n001"), id("f001")),
4        merge(id("n003"), exclusive, (id("f002"), id("f005")), id("f003")),
5        split(id("n005"), exclusive, id("f004"),
6          ((id("f005"), 0.6) (id("f006"), 0.2) (id("f007"), 0.2))),
7        split(id("n007"), eventbased, id("f008"), (id("f009"), id("f010"), id("f011"))),
8        task(id("n020"), "Prepare parcel", id("f023"), id("f025"), Norm(5.0, 4.0),
9          id("worker"), empty),
10       task(id("n023"), "Deliver by car", id("f027"), id("f029"), Unif(5.0, 15.0),
11         (id("car"), id("courier")), id("parcel delivered")),
12       task(id("n024"), "Deliver by drone", id("f028"), id("f030"), Unif(10.0, 30.0),
13         id("drone"), id("parcel delivered")),
14       ...),
15     flows :
16       (flow(id("f001"), Norm(1.0, 0.2)), ...,
17        flow(id("f009"), Norm(1.0, 0.2), message(id("order confirmed"), "order confirmed")),
18        flow(id("f010"), Norm(1.0, 0.2), message(id("order canceled"), "order canceled")),
19        flow(id("f011"), Norm(1.0, 0.2), timer(id("timeout"), 60)),
20       ...) >
```

**Fig. 3.** Running example: Maude representation of the parcel delivery process.

While process objects represent static processes, and they do not change along simulations, all the information on process execution is kept in simulation objects. Specifically, a Simulation object stores a collection of tokens (in a tokens attribute), a global time (gtime), a set of events (events, including messages and timers), and a set of resources (resources). It also keeps track of the metrics being computed. For analysis purposes, during the execution of a process some information is collected in the corresponding attributes: time stamps, task durations, and waiting time at parallel and inclusive merge gateways. This information is necessary for guiding the execution of the process, periodically evaluating the amount of resource instances, and for presenting the results to the user for possible optimizations.

Tokens are used to represent the evolution of the workflow under execution. Since there may be several simultaneous executions of a process, each execution is identified with a unique identifier, which is used to associate tokens to executions. Thus, a token is represented as a term token(TId, Id, T), where TId is the execution instance the token belongs to, Id is the identifier of the flow or node it is attached to, and T represents a timer, of sort Time, modeling a delay of the token, which represents the duration of a task or the delay associated to a flow. Once its timer becomes 0, a token can be consumed.

Tokens are stored in the tokens attribute of the Simulation object—implemented as a priority queue, so that tokens are processed according to their due time. However, even if a token is at the front of the queue with timer 0, it may be required to delay its execution. For example, consider a task that requires some resource that is not available, or a parallel merge for which some incoming flow is not yet active. To avoid deadlocks, the scheduler implements a *shifting* mechanism that identifies the first active token to the front of the queue in case the current head needs to be delayed.

For each resource type, a number of instances or replicas are provisioned. At each moment during a simulation, some of these instances can be in use

```
1 rl [initTask] :
2   < PId : Process |
3       nodes : (task(NId, TaskName, FId1, FId2, SE, RIds, SEI), Nodes), Atts >
4   < SId : Simulation |
5       tokens : (token(TId, FId1, 0) Tks),
6       task-tstamps : TTSs, gtime : T, resources : Rs, Atts1 >
7   < CId : Counter | counter : N >
8 => if allResourcesAvailable(RIds, Rs)
9 then < PId : Process |
10          nodes : (task(NId, TaskName, FId1, FId2, SE, RIds, SEI), Nodes), Atts >
11        < SId : Simulation |
12          tokens : insert(Tks, token(TId, NId, time(eval(SE, N)))),
13          task-tstamps : if TTSs[TId][NId] == undefined
14                         then insert(TId, insert(NId, T, TTSs[TId]), TTSs)
15                         else TTSs
16                         fi,                ---- for loops, stamps get overwritten
17          gtime : T,
18          resources : grabResources(RIds, Rs, time(eval(SE, N)), T), Atts1 >
19        < CId : Counter | counter : int(eval(SE, N)) >
20 else ...                                 ---- if necessary, the scheduler is updated
21 fi .
```

**Fig. 4.** Task initiation rule.

and others can be available for tasks to use them. Given a number of provisioned resources, if a running task requires resources and they are available, it blocks them and initiates execution immediately. Indeed, whenever a task requires several resource types, it *atomically* picks them, or waits for all of them to be available. If the required resources are not all available, resource requests are submitted, and the task remains blocked until its requests are satisfied. To support this, each resource type keeps a queue of requests.

Each resource type is represented by a resource operator that gathers all required information: an identifier, the minimum and maximum number of allocatable replicas (0, if unlimited), its allocation time, the total number of allocated replicas, the number of available replicas, the total amount of time the replicas of this resource type have been in use, and some historical information on resource usage, request queues, etc., which are handy for analysis purposes.

The execution of a task is modeled with two rules. The first rule, the initTask rule shown in Fig. 4, represents the task initiation, which is applied when a token with zero time is available at the incoming flow (line 5). If all the resources required by this task are available, which is checked with the allResourcesAvailable function (line 8), then a new token is generated with the task identifier and the task duration (line 12). Otherwise, the shifting mechanism is invoked (line 20)—note the ellipsis. If available, all required resources are removed from the resource set (grabResources function, line 18). Note also that rules update the information on execution times, task durations, etc. (see, e.g., the update of the task-tstamps attribute, lines 13–16).

A second rule, which models task completion, is triggered when there is a token for that task with zero time. In that case, the token is consumed and a new one is generated for the outgoing flow. All resources are released, and all the message events associated to that task, if any, are added to the set of events.

The Simulation object is in charge of collecting the data on the chosen metric for the specified window of time (history length). A supervisor then analyzes the collected information and, if necessary, decides to update (increase or decrease) the number of resource instances. Simulation-based analysis techniques are typically parameterized by the workload, which defines the rate at which new instances of a given process are executed. In a closed workload, a fixed number of tokens will be injected in the process, corresponding to the number of times the process is to be executed.

Finally, a class CtrlSocket is in charge of the interaction with the predictor component. Every time a process starts or terminates, or the execution of a task begins or terminates, an event is sent to the predictor. As we will see in Sect. 4.4, when a prediction is due, the execution of the system stops, and a special event is sent to the predictor component to notify that a prediction is due.

## 4.3   Event-Guided Processes

Even though the representation of processes presented in Sect. 4.1 includes elements like probabilities, durations and delays that are not needed when the system is guided by a provided event log, the fact that both stages of the simulations—the autonomous execution described in Sect. 4.2 and the one described in this section—use the same representation greatly simplifies its specification, since although the control will be different, the process will just make a copy of itself to evolve. In fact, the main difference is that whilst in an autonomous process the execution is guided by tokens, inserted in the process by a workload manager object, in an event-guided process, it is the events who guide the execution.

In the context of business processes, event logs are collections of time-stamped events produced by the execution of business processes. Each event indicates the execution of a task of the process. For example, an event may specify that a given task started or completed at a given time. Event logs are used for different purposes, including process mining, conformance checking, etc. They may be represented using different formats, like CSV, XES, MXML, XLSX or Parquet. Independently of the format chosen, they typically include fields for date and time, event identifier, source, and possibly others, in one way or another. Today, XES (eXtensible Event Stream) [1] is the most-widely-used standard for storing event logs.

Since an event may belong to any of the on-going execution sessions, and we do not require dates or any other information, instead of using any of the existing formats, our events are represented by sequences of events that include three values, separated by commas: a session identifier (a number), an event description, and a time-stamp. Event identifiers may be either initial, for the beginning of a process session, final, for the final node of a process session, or a task identifier followed by either -init or -end, representing the beginning and the end, respectively, of the execution of a task. For example, the following is a fragment of a sequence of events of our example:

```
crl [initTask] :
    < PId : Process | nodes: (task(NId, TaskName, FId1, FId2, SE, RIds, SEI), Nodes), Atts >
    < SId : Simulation | task-tstamps: ..., gtime: T, resources: Rs, Atts1 >
    < LId : Ctrl | events: (event(TId, NId, event("init"), T') EL), Atts2 >
=> < PId : Process | nodes: (task(NId, TaskName, FId1, FId2, SE, RIds, SEI), Nodes), Atts >
    < SId : Simulation |
        task-tstamps: ...
        gtime: T',
        resources: grabResources(TId, NId, RIds, Rs, T'),
        Atts1 >
    < LId : Ctrl | events: EL, Atts2 > .
```

**Fig. 5.** Rule initTask for processes guided by events.

```
298, n004-init, 3703784394059892335/9007199254740992
297, n014-init, 3705754467844272657/9007199254740992
356, n004-end, 1854135580310736077/4503599627370496
282, initial, 463663424722316089/1125899906842624
299, n004-end, 9286532598107735583/2251799813685248
281, initial, 464495724782831233/1125899906842624
297, n014-end, 3716244327443022323/9007199254740992
```

These event sequences are received through a socket, are parsed, and then represented using appropriate declarations. For example, the event

```
298, n004-init, 3703784394059892335/9007199254740992
```

is represented as

```
event(id("298"), id("n004"), event("init"), 3703784394059892335/9007199254740992)
```

An object of class Ctrl keeps the list of events, as is in charge of the interaction with the predictor, reading the sequence of events throw a socket, and then guiding the execution in accordance with such events.

```
class Ctrl | events: List{LogEvent}, socket: Maybe{Oid}, buffer: String .
```

Once in its events attribute, events will guide the execution by activating rules specifying the different actions that may occur in the system. For example, rule initTask in Fig. 5 specifies the initiation of a task when an init task event is at the front of the event sequence. Although not shown in the rule to simplify the presentation, all the information on the execution (time stamps, resources, etc.) is gathered as in the autonomous simulations presented in Sect. 4.2. This information will be used, when the execution consumes all the events in the prediction, to update the number of instances of the resources. Note that the initRule mirrors quite closely that for the autonomous execution.

## 4.4   Resource Adaptation Based on LSTM Predictions

The scheme proposed in [8] is followed for the definition of adaptation strategies. Among others, in [8], a strategy based on predictions was proposed. However, in that case, the prediction was carried out by using the execution of autonomous process itself, looking ahead before making a decision. Here, a process advances on its execution, but instead log traces are used for the prediction.

The general scheme assumes that resource instances are taken from a pool when required. However, instead of assuming a fixed number of instances, new instances may be allocated or released to adjust the offer of available resources to their demand, and in this way minimizing costs. The general scheme consists of periodically evaluating the amount of resources, by looking at different metrics on the recent history or the current state. To specify such a mechanism, a class Supervisor provides attributes time-between-checks, time-to-next-check, to keep a timer, and check-interval to specify the length of the history to look at.

```
class Supervisor | time-between-checks: Time, time-to-next-check: Time, check-interval: Time .
```

This general procedure assumes that decisions are taken in accordance to some given thresholds, which are also provided as parameters. The algorithm periodically checks if the value of the considered property is greater than the upper-bound threshold, in which case a new instance of the resource is allocated to the set of available resources; if it is smaller than the lower bound, then an instance is removed so that it is no longer available for use.

The Supervisor class is extended in a subclass SupervisorPrediction to handle the new strategy. In addition to the thresholds attribute, with ranges for each resource type, it adds attributes look-ahead-time, to be able to consider different prediction sizes, and forked-state, to create an event-guided process to be executed on the prediction to be received from the Python predictor component.

```
class SupervisorPrediction |
    thresholds: Map{Id, Tuple{Float, Float}}, ---- usage thresholds
    look-ahead-time: Time
    forked-state: Maybe{System} .
subclass SupervisorPrediction < Supervisor .
```

The rules specifying the behavior of the supervisor object are shown in Fig. 6. The supervisor-initiate-prediction rule is fired when the value of the time-to-next-check attribute is zero. It creates a copy of the part of the state needed for the event-guided execution (Sect. 4.3): the Simulation object collects information on the execution, including time-stamps and measure of resource usage, and the Process object. A new object of class CtrlSocket is created to read from the socket and collect the events to guide the execution. On the right-hand side of the rule there is a send message: a "PREDICT" event notifies the predictor that it is time to use the trace submitted until that time to feed the neural network, generate the prediction, and submit it through the socket.

To mark the end of the prediction, the predictor component will send an END event. When the CtrlSocket object in the forked-state attribute finds the END event, the second rule, supervisor-prediction-completed, is fired. It terminates the event-guided system and updates the resources using the update function. This operation basically analyzes the resources along the execution of the prediction and decides whether changing the number of instances of each resource or not using the thresholds provided. Finally, notice that the tokens are restored in the Simulation object so that the simulation can be resumed. The time-to-next-check timer is reset with the value of the time-between-checks attribute.

```
crl [supervisor-initiate-prediction] :
    < SId : Simulation | tokens: Tks, Atts1 >
    < PId : Process | Atts2 >
    < Sup : SupervisorPrediction |
        time-to-next-check: 0,              ---- check is due
        forked-state: null,
        look-ahead-time: T,
        Atts5 >
    < SH : SocketHandler | socket: SOCKET, Atts6 >
 => < SId : Simulation | Atts1 >           ---- tokens are removed to stop the simulation
    < PId : Process | Atts2 >
    < Sup : SupervisorPrediction |
        time-to-next-check: 0,
        forked-state:                      ---- a copy of the state is used to predict
          { < SId : Simulation | Atts1 >   ---- no tokens, so regular rules cannot apply
            < PId : Process | Atts2 >
            < cs : CtrlSocket | socket: SOCKET, buffer: "", events: nil >
            Receive(SOCKET, cs) },
        look-ahead-time: T,
        tokens: Tks,                       ---- save tokens to restore the simulation
        Atts5 >
    < SH : SocketHandler | socket: SOCKET, Atts6 >
    send(SOCKET, SH, "PREDICT " + string(PREDICTION-TIME, 10) + "\n")
if Tks =/= nil .

rl [supervisor-prediction-completed] :
    < SId : Simulation | resources: Rs, gtime: T, Atts1 >
    < Sup : SupervisorPrediction |
        time-between-checks: TBC,
        time-to-next-check: 0,
        check-interval: CI,
        thresholds: Thds,
        forked-state:
          { < SId : Simulation | resources: Rs', gtime: T', Atts3 >
            < cs : CtrlSocket | buffer: "", events: END, Atts4 >
            Conf },
        tokens: Tks, ---- frozen tokens
        Atts2 >
 => < SId : Simulation |
        resources: update(Rs, Rs', Thds, TBC, CI, T, T'),
        gtime: T,
        tokens: Tks,                       ---- tokens are restored to resume the simulation
        Atts1 >
    < Sup : SupervisorPrediction |
        time-between-checks: TBC,
        time-to-next-check: TBC,
        check-interval: CI,
        thresholds: Thds,
        forked-state: null,
        Atts2 > .
```

**Fig. 6.** Supervisor's rules.

# 5 Case Study

Table 1 presents experimental results, including the average and variance of the execution times, total cost, and resource usage for different parameters of the running example. In all these executions, (1) the population is 500; (2) the ranges for the different resources is $[1, 2]$, except for drones, for which a range $[1, 4]$ was chosen; (3) the allocation times (AT) go from 2 to 5 for the different resources; (4) similarly, resource costs are in the range 20–60; and (5) thresholds are fixed to 50 and 85. Of course, many other combinations are possible, but considering all

**Table 1.** Outputs for some of the simulations carried out for the delivery example.

| | TBC | LAT | Resources | | | | | Usage (%) | Exec time (h) | | Cost (€) |
|---|---|---|---|---|---|---|---|---|---|---|---|
| | | | Name | Range | AT | Cost | Thrd | | Avg | Var | |
| 1 | 5 | 5 | car | [1, 2] | 5 | 60 | (50, 85) | 37.87 | 102.45 | 12.71 | 387 547.8 |
| | | | clerk | [1, 2] | 4 | 50 | | 43.44 | | | |
| | | | courier | [1, 2] | 3 | 40 | | 41.59 · | | | |
| | | | drone | [1, 4] | 2 | 20 | | 93.34 | | | |
| | | | worker | [1, 2] | 3 | 30 | | 52.77 | | | |
| 2 | 10 | 5 | car | [1, 2] | 5 | 60 | (50, 85) | 39.66 | 101.77 | 12.49 | 397 178.6 |
| | | | clerk | [1, 2] | 4 | 50 | | 43.28 | | | |
| | | | courier | [1, 2] | 3 | 40 | | 47.16 | | | |
| | | | drone | [1, 4] | 2 | 20 | | 93.13 | | | |
| | | | worker | [1, 2] | 3 | 30 | | 51.15 | | | |
| 3 | 15 | 5 | car | [1, 2] | 5 | 60 | (50, 85) | 36.33 | 115.96 | 17.38 | 386 596.2 |
| | | | clerk | [1, 2] | 4 | 50 | | 42.83 | | | |
| | | | courier | [1, 2] | 3 | 40 | | 40.73 | | | |
| | | | drone | [1, 4] | 2 | 20 | | 93.13 | | | |
| | | | worker | [1, 2] | 3 | 30 | | 48.38 | | | |
| 4 | 20 | 5 | car | [1, 2] | 5 | 60 | (50, 85) | 36.90 | 125.97 | 21.31 | 442 300.8 |
| | | | clerk | [1, 2] | 4 | 50 | | 41.75 | | | |
| | | | courier | [1, 2] | 3 | 40 | | 42.52 | | | |
| | | | drone | [1, 4] | 2 | 20 | | 94.84 | | | |
| | | | worker | [1, 2] | 3 | 30 | | 45.72 | | | |
| 5 | 25 | 15 | car | [1, 2] | 5 | 60 | (50, 85) | 41.46 | 90.08 | 9.07 | 465 235.8 |
| | | | clerk | [1, 2] | 4 | 50 | | 29.76 | | | |
| | | | courier | [1, 2] | 3 | 40 | | 41.46 | | | |
| | | | drone | [1, 4] | 2 | 20 | | 89.84 | | | |
| | | | worker | [1, 2] | 3 | 30 | | 40.35 | | | |
| 6 | 25 | 5 | car | [1, 2] | 5 | 60 | (50, 85) | 38.83 | 119.81 | 18.84 | 431 158.9 |
| | | | clerk | [1, 2] | 4 | 50 | | 40.29 | | | |
| | | | courier | [1, 2] | 3 | 40 | | 46.54 | | | |
| | | | drone | [1, 4] | 2 | 20 | | 92.42 | | | |
| | | | worker | [1, 2] | 3 | 30 | | 50.39 | | | |

of them involves many combinations, and should be handled as an optimization problem, and use some amenable technique for such a problem, such as genetic algorithms, or search-based algorithms like hill climbing or simulated annealing.

Although only a few combinations varying the time between checks (TBC) and the look-ahead time (LAT) are presented, some interesting observations can be made. TBC takes values 5, 10, 15, 20, and 25 in the different experiments. Given that the best average execution time is for case 5, with TBC 25, one

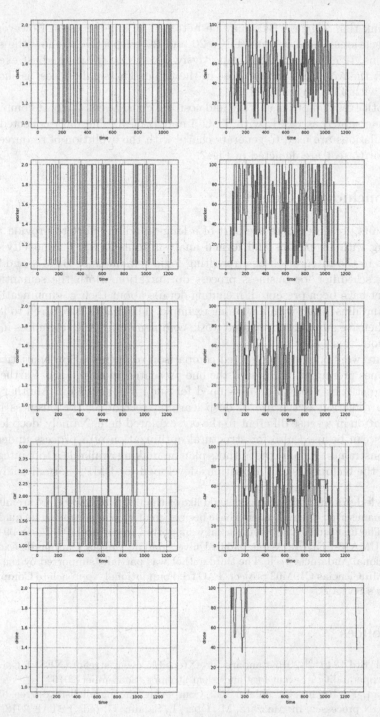

**Fig. 7.** Number of instances (left) and usage percentage (right) for each resource type for a simulation with the predictive-usage strategy, TBC = 5, LAT = 5, and Thds = (50, 85).

may think that bigger TBCs may be better than smaller ones. However, note
that executions 4 and 6, with TBCs 20 and 25, respectively, show the slowest
executions, 125.95 and 119.81, respectively. Execution 5 has the best execution
time but the worst cost, from those in the table. The smallest cost is shown by
Case 3, with TBS 15 and LAT 5.

If both execution time average and cost are considered, the best combination
is the one shown as case1, with TBC 5 and LAT 5. The data collected along
the simulations are used to generate charts with the evolution of resources. The
charts for Case 1 are depicted in Fig. 7.

## 6    Concluding Remarks

The results presented here are part of a long-standing effort to provide BPMN
modeling with extensions and formal analysis tools [2–9]. The novelty of this
paper is in the integration of a rewriting logic semantics of BPMN and a deep
learning scheduler for business process optimization. Both the semantics and
scheduler have been presented, including details about their communication and
usage, and illustrated with a running example. The reader is referred to [8] for a
comprehensive summary of related work, complementing the references included
throughout this paper.

Future work includes a detailed comparison of optimization heuristics, such
as the ones presented in [8], with the one presented in this paper. Furthermore,
new case studies need to be developed for such a comparison. Ahother future
research direction is the use of deep learning techniques for business process
optimization in a sense different to the one explored here. Namely, deep learning
methods can be used also for structural optimization of a process under some
given constraints. Finally, the authors plan on making available a tool integrating
most of the techniques and algorithms developed for BMPN formal analysis.

**Acknowledgments.** The authors would like to thank the reviewers for carefully reading the manuscript; their comments have been of great help in improving its quality and
clarity. The first author has been partially supported by projects PGC2018-094905-B-
100 and UMA18-FEDERJA-180, and by Universidad de Málaga, Campus de Excelencia
Internacional Andalucía Tech. The third author was partially supported by the ECOS-
NORD MinCiencias C19M03 project "FACTS: Foundational Approach to Computation
in Today's Society".

## References

1. IEEE Std 1849[TM]-2016 standard for eXtensible event stream (XES) for achieving
   interoperability in event logs and event streams, September 2016
2. Durán, F., Rocha, C., Salaün, G.: Computing the parallelism degree of timed
   BPMN processes. In: Mazzara, M., Ober, I., Salaün, G. (eds.) STAF 2018. LNCS,
   vol. 11176, pp. 320–335. Springer, Cham (2018). https://doi.org/10.1007/978-3-
   030-04771-9_24

3. Durán, F., Rocha, C., Salaün, G.: Stochastic analysis of BPMN with time in rewriting logic. Sci. Comput. Program. **168**, 1–17 (2018)
4. Durán, F., Rocha, C., Salaün, G.: Symbolic specification and verification of data-aware BPMN processes using rewriting modulo SMT. In: Rusu, V. (ed.) WRLA 2018. LNCS, vol. 11152, pp. 76–97. Springer, Cham (2018). https://doi.org/10.1007/978-3-319-99840-4_5
5. Durán, F., Rocha, C., Salaün, G.: Analysis of resource allocation of BPMN processes. In: Yangui, S., Bouassida Rodriguez, I., Drira, K., Tari, Z. (eds.) ICSOC 2019. LNCS, vol. 11895, pp. 452–457. Springer, Cham (2019). https://doi.org/10.1007/978-3-030-33702-5_35
6. Durán, F., Rocha, C., Salaün, G.: A rewriting logic approach to resource allocation analysis in business process models. Sci. Comput. Program. **183**, 102303 (2019)
7. Durán, F., Rocha, C., Salaün, G.: Analysis of the runtime resource provisioning of BPMN processes using maude. In: Escobar, S., Martí-Oliet, N. (eds.) WRLA 2020. LNCS, vol. 12328, pp. 38–56. Springer, Cham (2020). https://doi.org/10.1007/978-3-030-63595-4_3
8. Durán, F., Rocha, C., Salaün, G.: Resource provisioning strategies for BPMN processes: specification and analysis using maude. J. Log. Algebraic Methods Program. **123**, 100711 (2021)
9. Durán, F., Salaün, G.: Verifying timed BPMN processes using maude. In: Jacquet, J.-M., Massink, M. (eds.) COORDINATION 2017. LNCS, vol. 10319, pp. 219–236. Springer, Cham (2017). https://doi.org/10.1007/978-3-319-59746-1_12
10. Mehdiyev, N., Evermann, J., Fettke, P.: A novel business process prediction model using a deep learning method. Bus. Inf. Syst. Eng. **62**(2), 143–157 (2018). https://doi.org/10.1007/s12599-018-0551-3
11. Pasquadibisceglie, V., Appice, A., Castellano, G., Malerba, D.: Using convolutional neural networks for predictive process analytics. In: 2019 International Conference on Process Mining (ICPM), pp. 129–136, (2019)
12. Pfeiffer, P., Lahann, J., Fettke, P.: Multivariate business process representation learning utilizing gramian angular fields and convolutional neural networks. In: Polyvyanyy, A., Wynn, M.T., Van Looy, A., Reichert, M. (eds.) BPM 2021. LNCS, vol. 12875, pp. 327–344. Springer, Cham (2021). https://doi.org/10.1007/978-3-030-85469-0_21
13. Walck, C.: Hand-Book on Statistical Distributions for Experimentalists. Technical report SUF-PFY/96-01, Universitet Stockholms, Stockholm, September 2007
14. Wilmott, P.: Machine Learning: An applied Mathematics Introduction. Panda Ohana (2019)

# Modeling, Algorithm Synthesis, and Instrumentation for Co-simulation in Maude

Simon Thrane Hansen[1]([✉]) [iD] and Peter Csaba Ölveczky[2] [iD]

[1] DIGIT, Department of Electrical and Computer Engineering, Aarhus University, Aarhus, Denmark
sth@ece.au.dk
[2] Department of Informatics, University of Oslo, Oslo, Norway

**Abstract.** Simulation-based analysis of cyber-physical systems is vital in the era of Industry 4.0. Co-simulation enables composing specialized simulation tools via a co-simulation algorithm. In this paper, we provide a formal model in Maude of co-simulation for complex scenarios involving algebraic loops and step negotiation. We show not only how Maude can formally analyze co-simulations, but also how Maude can be used to synthesize co-simulation algorithms, port instrumentations, and parameter values so that the resulting co-simulation satisfies desired properties.

## 1 Introduction

Modern cyber-physical systems (CPSs), such as, e.g., nuclear power plants, cars, and airplanes, consist of multiple heterogeneous subsystems that are developed by different companies using different tools and formalisms [23]. Although these companies usually do not share their models for commercial reasons, there is nevertheless a need to determine how the different subsystems interact and to explore and analyze different design choices as early as possible. One way of addressing this need is to use, for each subsystem, an interface that provides an abstraction of that subsystem. *Simulation units* (SUs) provide such abstractions and are widely used in industry. A class of SUs are described by the Functional Mock-up Interface Standard [3] (FMI), which is used commercially and is supported by many tools [7]. An SU implements a well-defined interface and represents a subsystem by computing its behavioral trace using a dedicated solver.

*Co-simulation* [11,19] addresses the need to simulate a CPS given as the composition of such black-box SUs. Co-simulation transforms a continuous system to a discrete simulation with discrete interactions between the different SUs. Furthermore, a *digital twin* can be a co-simulation connected to a physical systems.

The objective of a co-simulation is to capture as accurately as possible the behavior of the modeled system. This is challenging due to discretization, cyclic dependencies between the SUs, and the fact that very few assumptions be made about the SUs: an SU may, e.g., be unable to predict its future state at the next desired point in time. A *co-simulation algorithm* is responsible for orchestrating the interaction of the SUs: it determines how and when the different SUs interact.

Since the co-simulation algorithm should make the virtual system correspond to its physical counterpart, the virtual system can be analyzed, and different

K. Bae (Ed.): WRLA 2022, LNCS 13252, pp. 130–150, 2022.
https://doi.org/10.1007/978-3-031-12441-9_7

design choices can be explored, to predict the behavior of the system to be built. However, the FMI standard is only informally described, and has been shown to be inconsistent [5]. For both of the above reasons, there is a need for formal methods to provide a formal semantics for co-simulation and to provide early model-based formal analysis of the co-simulations.

However, providing a formal semantics to co-simulation is challenging, due to, e.g., the complex behavior of the SUs, and the need to resolve cyclic dependencies between the SUs by fixed-point computations and to perform step negotiation to ensure that all SUs move in lockstep. Rewriting logic [21], with its modeling language and high-performance analysis tool Maude [6], should be a suitable formal method for co-simulation: Its expressiveness allows us to conveniently specify both complex dynamic behaviors and sophisticated functions (e.g., for detecting and resolving cyclic dependencies), and Maude provides automatic formal analysis capabilities for correctness analysis and design space exploration. Maude also supports connections to *external objects*, which means that Maude should be able to orchestrate the composition of real external components.

In this paper we present a formal framework for representing co-simulation in Maude. We give a formal model for co-simulation beyond the FMI 2.0 standard, also covering feed-through constraints, input instrumentations, and step rejection. We then use Maude to synthesize and symbolically execute suitable scenario-specific co-simulation algorithms, which enables the formal analysis of the resulting co-simulation. We also show how Maude can be used to synthesize instrumentations, parameter values, and co-simulation algorithms for such complex scenarios so that the resulting system satisfies desired properties. As discussed in Sect. 6, to the best of our knowledge this paper presents the first formal framework that covers design space exploration of complex co-simulation scenarios with algebraic loops and step rejection, and that also synthesizes correct-by-construction co-simulation algorithms and parameters for such scenarios.

Our framework currently does not connect to real-world SUs/FMUs; the interfaces of the SUs are abstractly represented in Maude. Nevertheless, as mentioned above, since Maude supports external objects, we believe that our framework can be naturally extended to perform co-simulation with real-world FMUs.

The rest of the paper is structured as follows. Section 2 provides necessary background to Maude and co-simulation. Section 3 presents a Maude model of co-simulation scenarios and SU behaviors. Section 4 shows how correct-by-construction co-simulation algorithms can be synthesized and executed in Maude. Section 5 describes how to synthesize instrumentation and parameter values such that the resulting co-simulation satisfies desired properties. Section 6 discusses related work and Sect. 7 gives some concluding remarks.

## 2   Preliminaries

### 2.1   Rewriting Logic and Maude

Maude [6] is a rewriting-logic-based executable formal specification language and high-performance analysis tool for object-based distributed systems.

A Maude module specifies a *rewrite theory* $(\Sigma, E \cup A, R)$, where:

- $\Sigma$ is an algebraic *signature*; i.e., a set of *sorts*, *subsorts*, and *function symbols*.
- $(\Sigma, E \cup A)$ is a *membership equational logic* theory, with $E$ a set of possibly conditional equations and membership axioms, and $A$ a set of equational axioms such as associativity, commutativity, and identity, so that equational deduction is performed *modulo* the axioms $A$. The theory $(\Sigma, E \cup A)$ specifies the system's states as members of an algebraic data type.
- $R$ is a collection of *labeled conditional rewrite rules* $[l] : t \longrightarrow t'$ **if** $cond$, specifying the system's local transitions.

A function $f$ is declared op $f : s_1 \ldots s_n$ -> $s$. Equations and rewrite rules are introduced with, respectively, keywords **eq**, or **ceq** for conditional equations, and **rl** and **crl**. A conditional rewrite rule has the form **crl** $[l] : t$ => $t'$ **if** $c_1 \wedge \ldots \wedge c_n$, where the conditions $c_1, \ldots, c_n$ are evaluated from left to right. A condition $c_i$ can be a Boolean term, an equation, a membership, or a *matching equation* $u(x_1, \ldots, x_n) := u'$ with variables $x_1, \ldots, x_n$ not appearing in $t$ and not instantiated in $c_1, \ldots, c_{i-1}$; these variables become instantiated by *matching* $u(x_1, \ldots, x_n)$ to the normal form of the (appropriate instance of) $u'$. $c_i$ can also be a *rewrite condition* $u_i$ => $u_i'$, which holds if $u_i'$ can be reached in zero or more rewrite steps from $u_i$. Mathematical variables are declared with the keywords **var** and **vars**, or can have the form $var : sort$ and be introduced on the fly.

A *class* declaration **class** $C \mid att_1 : s_1, \ldots, att_n : s_n$ declares a class $C$ of objects with attributes $att_1$ to $att_n$ of sorts $s_1$ to $s_n$. An *object instance* of class $C$ is represented as a term < $O : C \mid att_1 : val_1, \ldots, att_n : val_n$ >, where $O$, of sort $\mathtt{Oid}$, is the object's *identifier*, and where $val_1$ to $val_n$ are the current values of the attributes $att_1$ to $att_n$. A system state is modeled as a term of the sort $\mathtt{Configuration}$, and has the structure of a *multiset* made up of objects and messages (and *connections* in our case).

The dynamic behavior of a system is axiomatized by specifying each of its transition patterns by a rewrite rule. For example, the rule (with label l)

```
rl [l] :   < O : C | a1 : f(x, y), a2 : O', a3 : z >
    =>     < O : C | a1 : x + z,    a2 : O', a3 : z > .
```

defines a family of transitions in which the attribute **a1** of object **O** is updated to **x + z**. Attributes whose values do not change and do not affect the next state, such as **a2** and the right-hand side occurrence of **a3**, need not be mentioned.

*Formal Analysis in Maude.* Maude provides a number of analysis methods, including rewriting for simulation purposes, reachability analysis, and linear temporal logic (LTL) model checking. The rewrite command **frew** *init* simulates one behavior from the initial state/term *init* by applying rewrite rules. Given a state pattern *pattern* and an (optional) condition *cond*, Maude's **search** command searches the reachable state space from *init* for all (or optionally a given number of) states that match *pattern* such that *cond* holds:

```
search init =>! pattern [such that cond] .
```

The arrow =>! means that Maude only searches for *final* states (i.e., states that cannot be further rewritten) that match *pattern* and satisfies *cond*. If the arrow is =>* then Maude searches for all reachable states satisfying the search condition.

## 2.2  Co-simulation

Complex CPSs are composed of multiple communicating subsystems. For example, an autonomous car includes suspension, braking and collision avoidance subsystems. *Co-simulation* [12,19] is a technique enabling the discrete simulation of a continuous CPS, using multiple *simulation units* (SUs). Each such SU represents a subsystem and interacts with its environment through its ports.

**Co-simulation Scenarios.** A set of SUs can be composed into a *scenario* by *coupling* the input ports to output ports. A *coupling* connects an output port of an SU to an input port of another SU. The *coupling restriction* states that the value of an input and an output of a coupling must be the same at all times in the continuous system. However, in the discrete co-simulation, the coupling restrictions can only be satisfied at specific points in time called *communication points*. Therefore, each SU makes its own assumptions about the evolution of its input values *between* the communication points, which can introduce errors in the co-simulation [2]. An assumption about the evolution of an input can roughly be divided into two categories [14]:

- Interpolation (or *reactive*): the SU uses the current value at time $t$ and the *future* value at time $t + \Delta$ to predict the values in the interval $(t, t + \Delta)$.
- Extrapolation (or *delayed*): the SU uses the current value at time $t$ and the *previous* value at time $t - \Delta$ to predict the values in the interval $(t, t + \Delta)$.

*The orchestrator* computes the behavior of a scenario as a discrete trace, while it tries to satisfy the coupling restrictions, by exchanging values between the coupled ports. The orchestrator aims to find the communication points that minimize the co-simulation error while ensuring that the SUs move in lockstep by adapting to the behavior of the scenario. This is tricky, since the orchestrator needs to regard the SUs as nondeterministic blackboxes about which only few assumptions can be made. The optimal communication points furthermore depend on the approximation schemes used by the different SUs [13–15,22,24]. Unfortunately, most SUs will silently accept any given communication points, resulting in hard-to-debug erroneous results.

An example of the kind of nondeterministic behavior that the orchestrator needs to account for is *step rejection*, where an SU rejects a future state computation, since it implements error estimation and concludes that the desired step size may lead to an intolerable error. The FMI standard allows step rejections; however, they are generally unpredictable from the orchestrator's perspective. An SU implementing error estimation has a maximal step size $h$, defining the interval for which it can reliably compute its future state.

The orchestrator addresses step rejections using *step negotiation* [17]. A scenario can also contain *algebraic loops* (cyclic dependencies) between the SUs, which are resolved using fixed-point computations [17,19,22]. Scenarios with algebraic loops and step rejections are called *complex scenarios* and are notoriously hard to simulate, since the orchestrator must adapt to the behavior of the nondeterministic SUs to ensure an accurate simulation using "angelic nondeterminism."

The following definition of an SU is based on [4,16,17]:

**Definition 1 (Simulation Unit).** *A simulation unit (SU) is a tuple*

$$\mathcal{SU} \triangleq \langle S, U, Y, \mathcal{V}, \mathsf{set}, \mathsf{get}, \mathsf{step} \rangle,$$

*where:*

- $S$ *is a set, denoting the state space of the SU.*
- $U$ *and* $Y$ *are sets, of input and output ports, respectively. The union* $VAR = U \cup Y$ *of the inputs and outputs is called the ports of the SU.*
- $\mathcal{V}$ *is a set, denoting the values that a variable can hold.* $\mathcal{V}_T = \mathbb{R}_{\geq 0} \times \mathcal{V}$ *is the set of timestamped values exchanged between input and output ports.*
- *The functions* $\mathsf{set} : S \times U \times \mathcal{V}_T \to S$ *and* $\mathsf{get} : S \times Y \to \mathcal{V}_T$ *set an input and get an output, respectively.*
- $\mathsf{step} : S \times \mathbb{R}_{>0} \to S \times \mathbb{R}_{>0}$ *is a function;* $\mathsf{step}(s, H) = (s', h)$ *gives the state* $s'$ *after time* $h$, *where* $h$ *is either* $H$ *or the maximal time that the SU can progress from state* $s$.

**Definition 2 (Scenario).** *A scenario* $\mathcal{S}$ *is a tuple*

$$\mathcal{S} \triangleq \langle C, \{\mathcal{SU}_c\}_{c \in C}, L, M, R, F \rangle$$

- $C$ *is a finite set (of SU identifiers).*
- $\{\mathcal{SU}_c\}_{c \in C}$ *is a set of SUs, where each* $\mathcal{SU}_c = \langle S_c, U_c, Y_c, \mathcal{V}, \mathsf{set}_c, \mathsf{get}_c, \mathsf{step}_c \rangle$.
- $L$ *is a function* $L : U \to Y$, *where* $U = \bigcup_{c \in C} U_c$ *and* $Y = \bigcup_{c \in C} Y_c$, *and where* $L(u) = y$ *means that the output* $y$ *is connected to the input* $u$.
- $M \subseteq C$ *denotes the SUs that may reject a future state computation.*
- $R : U \to \mathbb{B}$ *is a predicate, which provides information about the SUs' input approximation functions.*
- $F$ *is a family of functions* $\{F_c : Y_c \to \mathcal{P}(U_c)\}_{c \in C}$. $u_c \in F_c(y_c)$ *means that the input* $u_c$ *feeds through to the output* $y_c$ *of the same SU.*

The function $R$ represents the *instrumentation* of the scenario. An input port $u$ is *reactive* if $R(u)$, and is *delayed* otherwise. Changing the instrumentation of a scenario changes the algorithm used to simulate the scenario. We assume that the instrumentation of a scenario is constant throughout the simulation, which is the case for most commercially used SUs [13]. Our definition extends the FMI 2.0 standard [3] with feed-through and port instrumentation. Figure 1 shows a way to graphically present co-simulation scenarios.

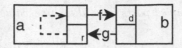

**Fig. 1.** A co-simulation scenario with two SUs $a$ and $b$. The dashed arrow denotes a feed-through connection, the ports are represented as small squares, the instrumentation of an input port is denoted by the letters $r$ (reactive) or $d$ (delayed). The solid arrows $f$ and $g$ represent couplings.

**Co-simulation Algorithms.** An *orchestrator* simulates a scenario by executing a *co-simulation algorithm*. A co-simulation algorithm consists of an initialization procedure and a co-simulation step [3]. This work focuses on the co-simulation step, which we refer to as "the algorithm" in the paper.

The state of a co-simulation scenario is defined as the combination of the states of its subcomponents:

**Definition 3 (Abstract SU State).** *The observable abstract state $s^R$ of an SU $\mathcal{SU}_c$ in a scenario $\mathcal{S}$ is an element of the set $S_c^R = \mathbb{R}_{\geq 0} \times S_{U_c}^R \times S_{Y_c}^R \times S_{V_c}^R$, where:*

- $S_{U_c}^R : U_c \to \mathbb{R}_{\geq 0}$ *is a function mapping each input port to a timestamp.*
- $S_{Y_c}^R : Y_c \to \mathbb{R}_{\geq 0}$ *is a function mapping each output port to a timestamp.*
- $S_{V_c}^R : VAR_c \to \mathcal{V}$ *is a function mapping each port to a value.*

*The first component of the abstract state denotes the time of the SU.*

We use the abstract state $s_c^R$ of an SU $c$ instead of the internal state $s_c$ since the orchestrator cannot observe the latter.

**Definition 4 (Abstract Co-simulation State).** *The abstract co-simulation state $s_{\mathcal{S}}^R$ of a scenario $\mathcal{S} = \langle C, \{\mathcal{SU}_c\}_{c \in C}, L, M, R, F \rangle$ is an element of the set $S_{\mathcal{S}}^R = time \times S_U^R \times S_Y^R \times S_V^R$ where:*

- $time : C \to \mathbb{R}_{\geq 0}$ *is a function, where $time(c)$ denotes the current simulation time of $\mathcal{SU}_c$. We denote by a time value $t \in \mathbb{R}_{\geq 0}$ the function $\lambda c.t$, which we use if all SUs are at the same time.*
- $S_U^R = \prod_{c \in C} S_{U_c}^R$ *maps all inputs of the scenario to a timestamp.*
- $S_Y^R = \prod_{c \in C} S_{Y_c}^R$ *maps all outputs of the scenario to a timestamp.*
- $S_V^R = \prod_{c \in C} S_{V_c}^R$ *maps all ports of the scenario to a value.*

A co-simulation step $P$ is a sequence of operations that takes a co-simulation from one consistent state to another consistent state. We write $s \xrightarrow{P} s'$ if executing the co-simulation step $P$ from the state $s$ results in the state $s'$.

**Definition 5 (Co-simulation Step).** *A co-simulation step $P$ is a sequence of SU actions that takes a consistent co-simulation state to another consistent co-simulation state. The state of the co-simulation is consistent if all input ports* *have a source, and all coupled ports have the same value. Formally:*

$$\left\langle t, s_U^R, s_Y^R, s_V^R \right\rangle \xrightarrow{P} \left\langle t', s_U^{R'}, s_Y^{R'}, s_V^{R'} \right\rangle$$

$$\implies (\texttt{consistent}(\left\langle t, s_U^R, s_Y^R, s_V^R \right\rangle) \implies (\texttt{consistent}(\left\langle t', s_U^{R'}, s_Y^{R'}, s_V^{R'} \right\rangle) \wedge t' > t))$$

*where consistent is defined as:*

$$\texttt{consistent}(\left\langle t, s_U^R, s_Y^R, s_V^R \right\rangle) \triangleq (\forall u_c \in U \exists y_d \in Y \cdot L(u_c) = y_d)$$

$$\wedge \, (\forall u_c, y_d \cdot L(u_c) = y_d \implies s_V^R(u_c) = s_V^R(y_d))$$

Informally, the co-simulation step advances the scenario from an initial state at time $t$ to a final state at time $t + H$, where $H > 0$, and ensures that the coupling restrictions are satisfied at both the initial and the final state.

Figure 2 shows three different co-simulation steps of the scenario in Fig. 1 that are allowed by the FMI standard 2.0 [3].

| Algorithm 1 | Algorithm 2 | Algorithm 3 |
|---|---|---|
| 1: $(s_A^{(H)}, H) \leftarrow \texttt{step}_A(s_A^{(0)}, H)$ | 1: $(s_B^{(H)}, H) \leftarrow \texttt{step}_B(s_B^{(0)}, H)$ | 1: $(s_B^{(H)}, H) \leftarrow \texttt{step}_B(s_B^{(0)}, H)$ |
| 2: $(s_B^{(H)}, H) \leftarrow \texttt{step}_B(s_B^{(0)}, H)$ | 2: $(s_A^{(H)}, H) \leftarrow \texttt{step}_A(s_A^{(0)}, H)$ | 2: $g_v \leftarrow \texttt{get}_B(s_B^{(H)}, y_g)$ |
| 3: $f_v \leftarrow \texttt{get}_A(s_A^{(H)}, y_f)$ | 3: $g_v \leftarrow \texttt{get}_B(s_B^{(H)}, y_g)$ | 3: $s_A^{(0)} \leftarrow \texttt{set}_A(s_A^{(0)}, u_g, g_v)$ |
| 4: $g_v \leftarrow \texttt{get}_B(s_B^{(H)}, y_g)$ | 4: $s_A^{(H)} \leftarrow \texttt{set}_A(s_A^{(H)}, u_g, g_v)$ | 4: $f_v \leftarrow \texttt{get}_A(s_A^{(0)}, y_f)$ |
| 5: $s_B^{(H)} \leftarrow \texttt{set}_B(s_B^{(s)}, u_f, f_v)$ | 5: $f_v \leftarrow \texttt{get}_A(s_A^{(H)}, y_f)$ | 5: $s_B^{(H)} \leftarrow \texttt{set}_B(s_B^{(H)}, u_f, f_v)$ |
| 6: $s_A^{(H)} \leftarrow \texttt{set}_A(s_A^{(H)}, u_g, g_v)$ | 6: $s_B^{(H)} \leftarrow \texttt{set}_B(s_B^{(H)}, u_f, f_v)$ | 6: $(s_A^{(H)}, H) \leftarrow \texttt{step}_A(s_A^{(0)}, H)$ |

**Fig. 2.** Three co-simulation algorithms of the scenario in Fig. 1 conforming to the FMI Standard (version 2.0).

Although the three algorithms satisfy Definition 5 and consist of the same actions, they are not equivalent, and simulating with one algorithm instead of one of the others could change the co-simulation result as shown in [16,18]. To differentiate between them, we need to consider the semantics of the different SU actions described in Definition 1.

The semantics described in Definitions 6 to 8 is based on [12,18] and operates on abstract states. It describes which assumptions the orchestrator can place on the behavior of SUs and restricts how actions can be composed to construct a co-simulation step.

**Definition 6 (Get Action).** *A value can be obtained from an output port $y$ of an SU at time $t$ using the action $\texttt{get}(s^{(t)}, y)$. The action changes the state of the SU according to:*

$$s^R \xrightarrow{\texttt{get}(s^{(t)}, y)} (v, s^{R'}) \implies \texttt{preGet}(y, s^R) \wedge \texttt{postGet}(y, s^R, s^{R'}, v)$$

*Where:*

$$\texttt{preGet}(y, \left\langle t, s_U^R, s_Y^R, s_V^R \right\rangle) \triangleq s_Y^R(y) < t \wedge \forall u \in F(y) \cdot s_U^R(u) = t$$

*The precondition above states that no value must have been obtained from the output y since the SU was stepped ($s_Y^R(y) < t$). Furthermore, it requires that all the inputs that feed through to y have been updated, so they are at time t. The following postcondition ensures that the output is advanced to time t:*

$$\texttt{postGet}(y, \langle t, s_U^R, s_Y^R, s_V^R \rangle, \langle t, s_U^R, s_Y^{R'}, s_V^R \rangle, v) \triangleq s_Y^{R'}(y) = t$$

$$\land \, \forall y_m \in (Y \setminus y) \cdot s_Y^{R'}(y_m) = s_Y^R(y_m)$$

**Definition 7 (Set Action).** *Setting a value $\langle t_v, x \rangle$ on the input port u of an SU using $\texttt{set}(s^{(t)}, u, \langle t_v, x \rangle)$ updates the time and value of the input port u such that they match $\langle t_v, x \rangle$:*

$$s^R \xrightarrow{\texttt{set}(s^{(t)}, u, \langle t_v, x \rangle)} s^{R'} \implies \texttt{preSet}(u, s^R) \land \texttt{postSet}(u, v, s^R, s^{R'})$$

*Where:*

$$\texttt{preSet}(u, \langle t_v, x \rangle, \langle t, s_U^R, s_Y^R, s_V^R \rangle) \triangleq s_U^R(u) < t_v$$

$$\land \, ((R(u_c) \land s_U^R(u) = t) \lor (\neg R(u_c) \land s_U^R(u) < t))$$

*The precondition says that the input must not have been assigned a new value since the SU was stepped ($s_U^R(u) < t_v$). Furthermore, it requires that the value $\langle t_v, x \rangle$ respects the instrumentation of the input. The following postcondition ensures that the value and time of the input u are updated so that they match the value assigned on the input:*

$$\texttt{postSet}(u, \langle t_v, x \rangle, \langle t, s_U^R, s_Y^R, s_V^R \rangle, \langle t, s_U^{R'} s_Y^R, s_V^{R'} \rangle) \triangleq t_v = s_U^{R'}(u)$$

$$\land \, (\forall u_m \in (U \setminus u) \cdot s_U^{R'}(u) = s_U^R(u)) \land s_V^{R'}(u) = x$$

**Definition 8 (Step Computation).** *Stepping an SU using $\texttt{step}(s^{(t)}, H)$ advances the state of the SU by at most H:*

$$s^R \xrightarrow{\texttt{step}(s^{(t)}, H)} s^{R'} \implies \texttt{preStep}(H, s^R) \land \texttt{postStep}(H, s^R, s^{R'})$$

*Where:*

$$\texttt{preStep}(H, \langle t, s_U^R, s_{Y, s_V^R}^R \rangle) \triangleq \forall u \in U \cdot ((R(u) \land t_{SU} + H = s_U^R(u))$$

$$\lor \, (\neg R(u) \land t_{SU} = s_U^R(u)))$$

*The above precondition states that all the SU's inputs have been updated according to their instrumentation. The following postcondition ensures that the time of the SU advances by at most H.*

$$\texttt{postStep}(H, \langle t, s_U^R, s_Y^R, s_V^R \rangle, \langle t', s_{U_c}^R, s_Y^R, s_{V_c}^{R'} \rangle) \triangleq t + h' = t' \land h' \leq H$$

An algorithm $P$ must satisfy Definition 5 while respecting the defined semantics. This means that Algorithm 3 is correct, while Algorithms 1 and 2 are incorrect since they do not respect the semantics. In particular, Algorithms 1 and 2 try to perform a $\texttt{step}_A$ action without respecting the reactive input $u_g$; the state of SU $A\ s_A^R = \langle 0, \{u_g \to 0\}, \{y_f \to 0\}, \_\rangle$ does not contain $\{u_g \to H\}$. Intuitively, we try to step SU $A$ without having provided it with a value on the reactive input $u_g$; this violates $\texttt{preStep}_A$.

**Problem Statement.** The two key problems in co-simulation that we address in this paper (in addition to the formalization of a co-simulation) are:

1. Given a scenario $\mathcal{S}$: Synthesize a co-simulation algorithm $P$ for $\mathcal{S}$. That is, find a sequence of SU actions $P$ which defines a valid co-simulation algorithm for $\mathcal{S}$. This involves solving possible algebraic loops and performing step negotiation to ensure that all SUs move in lockstep.
2. Given a *parametric* and *partially instrumented* scenario $\mathcal{S}$, where some SU parameters are unknown and where the instrumentation is incomplete, i.e., not all input ports are *reactive* or *delayed*: Find concrete values for the parameters, and concrete instrumentation of the input ports, such that the resulting *instrumented* scenario has desired properties.

## 3    Modeling Co-simulation Scenarios in Maude

This section describes how we model individual SUs and their composition in a co-simulation scenario in Maude. Due to space limitations, we only provide fragments of our Maude model. The entire model, including the synthesis and execution of co-simulation algorithms (Sect. 4) and the synthesis of instrumentations and parameters (Sect. 5) is available at https://github.com/SimplisticCode/Co-simulation_WRLA and consists of around 1400 LOC.

We formalize co-simulation scenarios in an object-oriented style. The state is a term *{SUs connections orchObjects}* of sort GlobalState, where *SUs* is set of objects modeling simulation units, *connections* denote the port couplings, and *orchObjects* are two additional objects used during synthesis and execution of co-simulation algorithms (see Sect. 4).

A simulation unit is modeled as an object instance of the following class:

```
class SU | time : Nat,              inputs : Configuration,
           outputs : Configuration, canReject : Bool,
           fmistate : fmiState,     parameters : LocalState,
           localState : LocalState .
```

The attribute time denotes the time of the SU; inputs and outputs denote the objects modeling the SU's input and output ports; canReject is true if the SU implements error estimation (i.e., is an element of the set $M$); fmistate denotes the *simulation mode* (see [3]) of the SU; localState denotes the SU's internal state; and parameters denotes the values of the SU's parameters.

Input and output ports are modeled as instances of the following classes:

```
class Port | value : FMIValue, time : Nat, status : PortStatus, type : FMIType .
class Input | contract : Contract .
class Output | dependsOn : OidSet .
subclasses Input Output < Port .
```

value and time denote, respectively, the value of the port and the time of its last set/get operation; status is true if the port was updated at the current time; contract denotes the input port's instrumentation (delayed or reactive); and dependsOn denotes the set of inputs that feed through to the output port.

*Example 1.* We illustrate our framework using a system where a *controller* controls the water level of a *water tank* with constant inflow of water, by opening and closing a valve in the tank. The system is modeled using one SU for the tank and one SU for the controller, and has the architecture in Fig. 1 without the feed-through. The water tank (in its initial state) is modeled as an object

```
< "tank" : SU | parameters : ("flow" |-> <5>), localState : ("waterlevel" |-> <0>),
                inputs : (< "valveState" : Input | value : <0>, time : 0, contract : delayed >),
                outputs : (< "waterlevel" : Output | value : <0>, time : 0,
                                                     status : Undef, dependsOn : empty >)
                time : 0,  canReject : false >
```

The tank has one delayed input port and one output port, and the local state indicates that the tank is empty. The parameter flow denotes the amount of water that flows into the tank per time unit.

To formalize the behaviors of an SU we formalize the operations set, get, and step in Definition 1. For example, the get operation that updates the time and status of a set of output ports is formalized as follows:[1]

```
op getAction : Object OidSet -> Object .
eq getAction(< SU1 : SU | >, empty) = < SU1 : SU | > .
eq getAction(< SU1 : SU | time : T,
                          outputs : (< O : Output | > OS) >, (O , P)) =
   getAction(< SU1 : SU | outputs :
                          (< O : Output | time : T, status : Def > OS) >, P) .
```

The application-specific behavior of an SU is given by defining its step function:

*Example 2.* The following definition of the step function in our running example defines how the water level of the tank changes as a function of the step duration STEP, the parameter flow, and the state (value) of the input valve:

```
eq step(< "tank" : SU | time : T, parameters : ("flow" |-> <FLOW>),
                        inputs : < "valve" : Input | value : <STATE> >,
                        outputs : < "waterlevel" : Output | time : T >,
                        localState : ("waterlevel" |-> <LEVEL>) >,
        STEP) =
 if STATE == 1 then    --- valve is open
   < "tank" : SU | time : (T+STEP), localState : ("waterlevel" |-> <0>),
         outputs : < "waterlevel" : Output | value : <0>, time : (T+STEP), status : Undef > >
```

---

[1] We do not show variable declarations, but follow the convention that variables are written with capital letters.

```
else                    --- valve is closed
  < "tank" : SU | time : (T+STEP),  localState : ("waterlevel" |-> <LEVEL+(STEP*FLOW)>),
                  outputs : < "waterlevel" : Output | value : < LEVEL + (STEP * FLOW) >,
                                                    time : (T + STEP), status : Undef > >
fi .
```

A connection/coupling connecting the output port $o$ of SU $su_1$ to the input port $i$ of SU $su_2$ is represented by the term $su_1 ! o ==> su_2 ! i$.

We define scenarios using constants `simulationUnits` and `external Connection` to denote, resp., the simulation unit objects and their connections.

*Example 3.* The SUs and their couplings in our example are defined as follows:

```
eq simulationUnits =
  < "tank" : SU | parameters : ("flow" |-> <100>), localState : ("waterlevel" |-> <0>),
                  time : 0, fmistate : Instantiated, canReject : false,
                  inputs : (< "valveState" : Input | value : <0>, type : integer, time : 0,
                                                   contract : delayed, status : Undef >),
                  outputs : (< "waterlevel" : Output | value : <0>, type : integer, time : 0,
                                                   status : Undef, dependsOn : empty >) >
  < "ctrl" : SU | parameters : (("high" |-> <5>) , ("low" |-> <0>)), canReject : false,
                  localState : ("valve" |-> <false>), fmistate : Instantiated, time : 0,
                  inputs : (< "waterlevel" : Input | value : <0>, type : integer, time : 0,
                                                   contract : reactive, status : Undef >),
                  outputs : (< "valveState" : Output | value : <0>, type : integer, time : 0,
                                                   status : Undef, dependsOn : empty >) > .

eq externalConnection = ("tank" ! "waterlevel" ==> "ctrl" ! "waterlevel")
                        ("ctrl" ! "valveState" ==> "tank" ! "valveState") .
```

The constant `setup` defines the initial state, and adds appropriate initialized orchestration objects to the scenario:

```
op setup : -> GlobalState .
ceq setup = {INIT}
  if SCENARIO := externalConnection simulationUnits
  /\ validScenario(SCENARIO)
  /\ LOOPS := tarjan(SCENARIO)
  /\ NeSUIDs := getSUIDsOfScenario(SCENARIO)
  /\ INIT := calculateSNSet(SCENARIO OData(1,LOOPS, NeSUIDs)) .
```

The function `validScenario` checks whether all inputs are coupled and that no input has two sources. The function `tarjan` returns (a possibly empty) set of algebraic loops in the scenario by searching for non-trivial strongly connected components in the graph constructed using the rules in [16]. The function `getSUIDsOfScenario` returns the set of all SU identifiers. Finally, `calculateSNSet` checks if step negotiation should be applied in the simulation of the scenario, and generates a global initial state with orchestration objects that store information about the discovered algebraic loops and whether step negotiation is needed.

## 4   Synthesizing and Executing Co-simulation Algorithms

This section describes how co-simulation algorithms for a given scenario can be synthesized and then executed in Maude.

## 4.1   Orchestration Data

The *orchestration* executes a given co-simulation algorithm on a scenario, and requires keeping track of the co-simulation algorithm and the execution state.

The following class `SimData` stores such data about the simulation:

```
class SimData | SNSet : OidSet,              defaultStepSize : NzNat,
         actualStepSize : NzNat,             unsolvedSCC : AlgebraicLoopSet,
         solvedSCC : AlgebraicLoopSet,       guessOn : PortSet,
         values : PortValueMap,              simulationTime : Nat,
         suids : NeOidSet .
```

The attribute `SNSet` denotes the set $M$ of SUs that may reject to step the desired step size (see Definition 2); `defaultStepSize` is the default step duration of the simulation, and the attribute `actualStepSize` is the negotiated step duration. The attributes `actualStepSize` and `defaultStepSize` are equal if $M = \emptyset$. The attributes `unsolvedSCC` and `solvedSCC` respectively denote the solved and unsolved algebraic loops. The attribute `guessOn` denotes the set of ports which are used to solve algebraic loops using the technique described in [17]; `values` is a map linking an input port to a value. The orchestration uses `values` to track which values it has obtained but not set on an input port. The attribute `simulationTime` describes the current time of the simulation, and `suids` denotes the identifiers of the SUs.

The following class `AlgoData` stores the co-simulation algorithm:

```
class AlgoData | CosimStep : ActionList,   Initialization : ActionList,
         Termination : ActionList, endTime : NzNat .
```

The attributes `Initialization` and `Termination` denote the initialization procedure and termination procedure, respectively. The attribute `CosimStep` denotes the co-simulation step procedure that the orchestration applies until it reaches the end time of the simulation (given by `endTime`). All elements of the algorithm are of the sort `ActionList`, which is a list of SU operations (where we do not show actions for handling complex scenarios):

```
ops Set Get Step Save : -> ActionType [ctor] .
op portEvent:_SU:_PId:_ : ActionType SUID OidSet -> Action [ctor] .
subsort Action < ActionList .
op emptyList : -> ActionList [ctor] .
op _;_ : ActionList ActionList -> ActionList [ctor assoc id: emptyList] .
```

## 4.2   Synthesis of Co-simulation Algorithms

We synthesize co-simulation algorithms for a scenario $\mathcal{S}$ by first performing and recording all possible SU actions, and then searching for consistent reachable final states. Any sequence of SU actions leading to such a state is a co-simulation algorithm.

A number of rewrite rules model the different SU actions. For example, the following rewrite rule describes a `get` operation:

```
crl [get-syn] :
   < SU1 : SU | fmistate : Simulation, inputs : IS,
                outputs : (< O : Output | time : T,  status : Undef,
                                         value : V, dependsOn : FT > OS) >
   (SU1 ! O ==> SU2 ! I)
   < OCH : SimData | values : PV >
   < ALG : AlgoData | CosimStep : ALGO >
   =>
   getAction(< SU1 : SU | >, SU1 ! O)
   (SU1 ! O ==> SU2 ! I)
   < OCH : SimData | values : insert((SU2 ! I), < T ; V >, PV) >
   < ALG : AlgoData | CosimStep : (ALGO ; EVENT) >
  if feedthroughSatisfied(FT, IS, T)
   /\ EVENT := portEvent: Get SU: SU1 PId: O .
```

A value V is obtained from the output O of SU1 if the state satisfies all feed-through constraints FT of the output O (checked by feedthroughSatisfied). The rule updates the output O using the operation getAction, inserts the output's value and time < T ; V > into values, and adds the performed action portEvent: Get SU: SU1 PId: O to its list CosimStep of performed actions.

All such "synthesis" rules in our model follow the same pattern: they rewrite the scenario while remembering how they did it.

We synthesize a co-simulation algorithm by starting with a consistent initial state and exploring how a consistent final state can be established. An algorithm for a given scenario is therefore synthesized using the following rewrite rule:

```
crl [getAlgorithm]: {INIT} => {getOrchestrator(FINALSTATE)}
if isInitialState(INIT)
   /\ LOOPS := tarjan(INIT)
   /\ SUIDsNE := getSUIDsOfScenario(INIT)
   /\ SIMDATA := initialOrchestrationData(1,LOOPS,SUIDsNE)
   /\ CONF := calculateSNSet(INIT) SIMDATA initialAlgorithmData(1)
   /\ {CONF} => {FINALSTATE}
   /\ allSUsinUnloaded(SUIDsNE, FINALSTATE) .
```

This rule checks whether the scenario INIT is a suitable initial state using the predicate isInitialState. Then we construct an initial simulation configuration CONF as in Sect. 3. The key condition that does most of the work is the rewrite condition {CONF} => {FINALSTATE}, which searches for states reachable from CONF until it finds a state FINALSTATE that satisfies the property allSUsinUnloaded, which ensures that all SUs have been properly simulated and unloaded. The function getOrchestrator extracts the synthesized co-simulation algorithm from this final state.

The following Maude command then synthesizes all valid co-simulation algorithms for a given *scenario*:

```
Maude> search scenario => FINALSTATE:GlobalState .
```

Many SU actions can happen independently at the same time, which means that multiple valid algorithms often can be synthesized for a scenario. For example, there are six different co-simulation algorithms for our water tank scenario.

## 4.3  Executing Co-Simulation Algorithms

This section describes how co-simulation algorithms can be executed. The state of such an execution is a term run: *algorithm* on: *scenario* with: *simData*:

```
op run:_on:_with:_ : Object Configuration Object -> SimState [ctor].
```

A co-simulation algorithm is executed by sequentially performing its actions, starting with performing all actions in the Initialization, then performing all actions of the CosimStep, and finally executing all actions in Termination.

The following rule shows the execution of the first action (Get) in CosimStep:

```
crl [get-exec] :
    run: < ALG : AlgoData | CosimStep : (action: Get SU: SU1 PId: O) ; ALGO >
    on: CONF
       < SU1 : SU | inputs : IS,
                         outputs : (< O : Output | time : T, value : V, dependsOn : FT > OS) >
       ( SU1 ! O ==> SU2 ! INPUT)
    with: < OCH : SimData | values : PV >
    =>
    run: < ALG : AlgoData | CosimStep : ALGO >
    on: CONF getAction(< SU1 : SU | >, SU1 ! O) ( SU1 ! O ==> SU2 ! INPUT)
    with: < OCH : SimData | values : insert((SU2 ! INPUT), < T ; V >, PV) >
 if feedthroughSatisfied(FT, IS, T) .
```

We can combine algorithm synthesis and execution into the following rewrite rule, so that rewriting the term runAnyAlgorithm *scenario* synthesizes *and* executes a co-simulation algorithm the for co-simulation scenario *scenario*:

```
crl [runAlg] : runAnyAlgorithm INIT => run: ORC on: INIT with: SIMDATA
  if LOOPS := tarjan(INIT)
  /\ SUIDsNE := getSUIDsOfScenario(INIT)
  /\ SIMDATA := initialOrchestrationData(1,LOOPS,SUIDsNE)
  /\ ALGO := initialAlgorithmData(1)
  /\ CONF := calculateSNSet(INIT ALGO) SIMDATA
  /\ {CONF} => {FINALSTATE}
  /\ ORC := getOrchestrator(FINALSTATE)
  /\ allSUsinUnloaded(SUIDsNE, FINALSTATE) .
```

This rule is similar to the rule getAlgorithm, and also extracts the resulting algorithm ORC and simulation data SIMDATA.

*Example 4.* The water tank scenario described in Example 3 (waterTankScenario below) can be simulated by rewriting:

```
Maude> frew (runAnyAlgorithm waterTankScenario) .
```

The command returns the final simulation state

```
run: < "Algorithm" : AlgorithmData | CosimStep : emptyList, Initialization : emptyList,
                                Termination : emptyList, endTime : 1 >
on: ("tank" ! "waterlevel" ==> "ctrl" ! "waterlevel")
    ("ctrl" ! "valveState" ==> "tank" ! "valveState")
    < "ctrl" : SU | canReject : false, fmistate : Unloaded, time : 1,
                    inputs : < "waterlevel" : Input | contract : reactive, status : Def,
                                                time : 1, type : integer, value : < 100 > >,
                    outputs : < "valveState" : Output | dependsOn : empty, status : Def,
                                                time : 1, type : integer, value : < 1 > >,
                    localState : "valve" |-> < true >,
                    parameters :"high" |-> < 5 >, "low" |-> < 0 >  >
    < "tank" : SU | canReject : false, fmistate : Unloaded, time : 1,
                    inputs : < "valveState" : Input | contract : delayed, status : Def,
                                                time : 1, type : integer, value : < 1 > >,
                    outputs : < "waterlevel" : Output | dependsOn : empty, status : Def,
                                                time : 1, type : integer, value : < 100 > >,
                    localState : "waterlevel" |-> < 100 >,
                    parameters : "flow" |-> < 100 > >
with: < "Orchestrator" : SimulationData | SNSet : empty,          actualStepSize : 1,
                                defaultStepSize : 1, guessOn : empty,
                                simulationTime : 1,  solvedSCC : empty,
                                unsolvedSCC : empty, values : empty,
                                suids :("ctrl", "tank") >
```

## 4.4   Checking Confluence of Synthesized Co-simulation Algorithms

Section 4.2 shows that multiple valid co-simulation algorithms can be synthesized for a given scenario. Executing all these valid co-simulation algorithms for a given scenario should give the same result, since all the SUs are deterministic in the sense that if we try to step an SU $A$ from some initial state $s$ with step size $h$, it will always produce the same final state $A'$. All this indicates that some actions are independent of each other. Therefore, their relative execution order is irrelevant, and an optimized algorithm can merge such independent actions and perform them in parallel.

The following Maude command checks whether all generated co-simulation algorithms for a scenario *scenario* result in the same final state:

```
Maude> search (runAnyAlgorithm scenario) =>! S:SimState .
```

This search command synthesizes and then executes all co-simulation algorithms for the scenario *scenario*. For our water tank scenario, the search produces a single result, which means that all synthesized algorithms give the same result.

## 5   Synthesizing Instrumentations and SU Parameters

Our framework makes possible different kinds of design space exploration to allow the practitioner to see how different design choices affect the behavior of the system. This section shows how our framework can be used to synthesize parameter values and instrumentations of the inputs that lead to desired simulations.

### 5.1   Instrumentation of a Scenario

Finding a good instrumentation of the input ports (i.e. deciding whether an input port should be reactive or delayed) is important not only to achieve accurate co-simulation results [14,18,22], but also because some instrumentations of a scenario may lead to algebraic loops while others do not.

We use reachability analysis to explore the consequences of different instrumentations of a scenario to find the instrumentation that yields the desired simulation results. To explore different instrumentations of a scenario, we create a *partially instrumented* scenario, where the some of the input ports have the contract `noContract` instead of `reactive` or `delayed`.

*Example 5.* In the following partially instrumented water tank scenario, the input port `"valveState"` of the SU `"tank"` and the `"waterLevel"` input port of the SU `"ctrl"` are not yet instrumented:

```
eq waterTankNotInstrumented =
< "tank" : SU | parameters : ("flow" |-> <100>), localState : ("waterlevel" |-> <0>),
        time : 0,  fmistate : Instantiated, canReject : false,
        inputs : (< "valveState" : Input | value : <0>, type : integer, time : 0,
                                    contract : noContract, status : Undef >),
        outputs : (< "waterlevel" : Output | value : <0>, type : integer, time : 0,
                                    status : Undef, dependsOn : empty >) >

< "ctrl" : SU | parameters : (("high" |-> <5>) , ("low" |-> <0>)), canReject : false,
        localState : ("valve" |-> <false>), fmistate : Instantiated, time : 0,
        inputs : (< "waterlevel" : Input | value : <0>, type : integer, time : 0,
                                    contract : noContract, status : Undef >),
        outputs : (< "valveState" : Output | value : <0>, type : integer, time : 0,
                                    status : Undef, dependsOn : empty >) > .
```

We use an operator `findInstr` to instrument such partially instrumented scenarios, so that the state `findInstr`(*scenario*) becomes a fully instrumented scenario when all ports have been instrumented (rule `remove-findInstr`). The rules `instr-delayed` and `instr-reactive` set uninstrumented input ports to be either `delayed` or `reactive`:

```
rl [instr-delayed]:
 findInstr(< SU1 : SU | inputs : < I : Input | contract : noContract > IS > C)
=> findInstr(< SU1 : SU | inputs : < I : Input | contract : delayed > IS > C) .

rl [instr-reactive]:
 findInstr(< SU1 : SU | inputs : < I : Input | contract : noContract > IS > C)
=> findInstr(< SU1 : SU | inputs : < I : Input | contract : reactive > IS > C) .

crl [remove-findInstr]: findInstr(CONF) => CONF if instrumented(CONF) .
```

The different instrumentations of a partially instrumented scenario are found and explored using the following rule:

```
crl [findInstrumentation]: findContracts(INIT) => CONF
  if findInstr(INIT) => CONF
  /\ empty == tarjan(CONF)              --- no algebraic loops
  /\ runAnyAlgorithm CONF => run: ORC on: FINAL with: SIMDATA
  /\ simulationFinished(ORC)
  /\ desiredProperty(FINAL) .
```

This rule generates an instrumented scenario `CONF` from the partially instrumented scenario `INIT`. `CONF` is then simulated (in the rewrite condition), leading
to a final state `FINAL`. The instrumentation can be restricted by giving properties that the instrumented scenario `CONF` and/or the simulation result `FINAL`
must satisfy. For example, the condition `empty == tarjan(CONF)` says that the
instrumentation should not lead to algebraic loops, and the last conjunct in the
condition says that the simulation result `FINAL` must satisfy *desiredProperty*.

*Example 6.* We define *desiredProperty* to be that the water level of the tank is in
a desired range. The following Maude command then finds all instrumentations
which lead to simulations which end in a desired water level:

```
Maude> search findContracts(waterTankNotInstrumented) =>! C:Configuration .
```

This command returns the three instrumentations (with parts replaced by '...')

```
Solution 1
C:Configuration --> ...
< "ctrl" : SU | inputs : < "waterlevel" : Input | contract : delayed > ... >
< "tank" : SU | inputs : < "valveState" : Input | contract : reactive > ... >

Solution 2
C:Configuration --> ...
< "ctrl" : SU | inputs : < "waterlevel" : Input | contract : reactive > ... >
< "tank" : SU | inputs : < "valveState" : Input | contract : delayed > ... >

Solution 3
C:Configuration --> ...
< "ctrl" : SU | inputs : < "waterlevel" : Input | contract : delayed > ... >
< "tank" : SU | inputs : < "valveState" : Input | contract : delayed > ... >
```

## 5.2 Synthesizing SU Parameters

An SU may have different parameters. In our framework, the user can specify a
finite set of possible values for a parameter using a `choose` operator, and we can
then synthesize those parameter values that result in desired simulations.

*Example 7.* We want to synthesize the value of the parameter `flow` of the water
tank such that the water level is above 10 in the final simulation state. The
following predicate defines the desired water level:

```
op above10 : Configuration -> Bool .
eq above10(CONF < "tank" : SU | localState : "waterlevel" |-> <V>>) = V > 10.
```

To synthesize a `flow` value from the set $\{1, 2, 30\}$ we initialize `flow` accordingly:

```
< "tank" : SU | parameters : "flow" |-> choose(< 1 >, < 2 >, < 30 >), ... >
```

We use the following rule to synthesize parameter values that result in a
simulations that satisfy `above10`:

```
crl [getParamValues] : selectParams(UNITIALIZEDCONF) => CONF
  if UNITIALIZEDCONF => CONF
  /\ runAnyAlgorithm CONF =>
      run: < ALG : AlgData | Initialization : emptyList,
                            CosimStep : emptyList , Termination : emptyList >
      on: FINALSTATE with: SIMULATIONDATA
  /\ above10(FINALSTATE) .
```

The following Maude command gives all initialized scenarios which lead to desired simulations:

```
Maude> search selectParams(parametricWaterTank) =>! C:Configuration .

Solution 1
C:Configuration --> ... < "tank" : SU | parameters : "flow" |-> <30>, ... >

No more solutions.
```

We can also *simultaneously* synthesize both desired instrumentations *and* parameter values by having noContract ports and choose(...) values.

## 6 Related Work

A number of papers, e.g. [4,12,16,17], synthesize co-simulation algorithms for *fixed* scenarios. In contrast to our paper, this body of work does not provide *formal models* of co-simulation, and therefore no formal analysis. We exploit Maude's formal analysis features to synthesize suitable instrumentations and SU parameters, which is not addressed by the mentioned related work.

Design space exploration of SU parameters is described in [8,9]. This work uses genetic algorithms to find optimal parameters values. However, it does not consider how different instrumentations can affect the simulation result.

Another example of DSE of a CPS using Maude can be found in [20], where Maude is used to validate and analyze drone/unmanned aerial vehicle flight strategies to find the optimal flight strategy using an external simulation engine. In contrast, we use Maude's capabilities to validate co-simulation algorithms and formalize the co-simulation semantics instead of evaluating flight strategies.

Formal methods have been used for co-simulation, e.g., [1,5,18,25,26]. Thule et al. [25] formalize a given scenario and two given co-simulation algorithms for that scenario in PROMELA and use the SPIN model checker to compare the two simulation algorithms, e.g., in terms of reachability. In contrast, we provide a general formal framework for co-simulation, synthesize co-simulation algorithms for a given scenario, synthesize instrumentations and parameter values, and capture a broader class of co-simulation scenarios (e.g., including scenarios with algebraic loops and step rejection) than those in the case study in [25].

Cavalcanti et al. [5] provide the first behavioral semantics of FMI. They show how to prove essential properties of co-simulation algorithms using CSP, and also show that the co-simulation algorithm provided in the FMI standard is not consistent. We cover an extension of FMI scenarios, and also include feed-through, step rejection, and input port instrumentation. Furthermore, as already mentioned, we also synthesize co-simulation algorithms and parameters.

Amálio et al. [1] show how formal tools can detect algebraic loops in a scenario. We not only detect such loops, but also solve them to synthesize co-simulation algorithms. Zeyda et al. [26] formalize a co-simulation scenario in Isabelle/UTP, and prove different properties–including behavioral properties–about the scenario. In contrast, we use automatic model checking methods to both synthesize and analyze co-simulation algorithms, and also cover complex scenarios (algebraic loops, step rejection, etc.) not covered in [26].

On the Maude side, Mason et al. [20] use Maude and statistical model checking to analyze a system of UAVs ("drones"). The key point is that they integrate a quite realistic "external" UAV simulator, Ardupilot/SITL, into their Maude simulations. Maude and the simulator communicate by message passing. This work does not formalize co-simulation in our FMI sense, but shows that Maude can execute together with, and coordinate, external simulators for CPSs.

# 7   Concluding Remarks

We have presented a formal model of co-simulation in Maude for complex scenarios with algebraic loops and step negotiation. Using rewrite conditions, we have used Maude to generate and execute co-simulation algorithms, and to synthesize port instrumentations and parameter values (albeit from a finite set of possible values), such that the resulting co-simulation satisfies desired properties.

In future work we should validate our framework on larger applications. We should also explore how Maude's symbolic analysis methods can be used to synthesize algorithms and parameter values from symbolic initial states which represent infinitely many concrete states. Although users can define complex behaviors of their SUs, connecting Maude to a solver for real numbers such as dReal [10] could support defining the continuous dynamics of SUs using differential equations. Finally, we should exploit Maude's support for external objects to execute the synthesized algorithms on real systems.

**Acknowledgments.** We thank Claudio Gomes, Jaco van de Pol, José Meseguer, and Stefan Hallerstede for valuable discussions and feedback. We also thank the anonymous reviewers for useful comments on a previous version of this paper.

# References

1. Amálio, N., Payne, R.J., Cavalcanti, A., Woodcock, J.: Checking SysML models for co-simulation. In: Ogata, K., Lawford, M., Liu, S. (eds.) ICFEM 2016. LNCS, vol. 10009, pp. 450–465. Springer, Cham (2016). https://doi.org/10.1007/978-3-319-47846-3_28

2. Arnold, M., Clauß, C., Schierz, T.: Error analysis and error estimates for co-simulation in FMI for model exchange and co-simulation v2.0. In: Schöps, S., Bartel, A., Günther, M., ter Maten, E.J.W., Müller, P.C. (eds.) Progress in Differential-Algebraic Equations. DEF, pp. 107–125. Springer, Heidelberg (2014). https://doi.org/10.1007/978-3-662-44926-4_6

3. Blockwitz, T., et al.: Functional Mockup Interface 2.0: the standard for tool independent exchange of simulation models. In: Proceedings of 9th International Modelica Conference, pp. 173–184. Linköping University Electronic Press (2012)

4. Broman, D., et al.: Determinate composition of FMUs for co-simulation. In: Ernst, R., Sokolsky, O. (eds.) Proceedings of EMSOFT 2013. IEEE (2013)

5. Cavalcanti, A., Woodcock, J., Amálio, N.: Behavioural models for FMI co-simulations. In: Sampaio, A., Wang, F. (eds.) ICTAC 2016. LNCS, vol. 9965, pp. 255–273. Springer, Cham (2016). https://doi.org/10.1007/978-3-319-46750-4_15

6. Clavel, M., et al. (eds.): All About Maude. LNCS, vol. 4350. Springer, Heidelberg (2007). https://doi.org/10.1007/978-3-540-71999-1

7. FMI: Functional mock-up interface tools (2014). https://fmi-standard.org/tools/

8. Gamble, C.: Design space exploration in the INTO-CPS platform: integrated tool chain for model-based design of cyber physical systems (2016). https://projects.au.dk/fileadmin/D5.2d_DSE_in_the_INTO-CPS_Platform.pdf

9. Gamble, C., Pierce, K.: Design space exploration for embedded systems using co-simulation. In: Fitzgerald, J., Larsen, P.G., Verhoef, M. (eds.) Collaborative Design for Embedded Systems: Co-modelling and Co-simulation, pp. 199–222. Springer, Heidelberg (2014). https://doi.org/10.1007/978-3-642-54118-6_10

10. Gao, S., Kong, S., Clarke, E.M.: dReal: an SMT solver for nonlinear theories over the reals. In: Bonacina, M.P. (ed.) CADE 2013. LNCS (LNAI), vol. 7898, pp. 208–214. Springer, Heidelberg (2013). https://doi.org/10.1007/978-3-642-38574-2_14

11. Gomes, C., Broman, D., Vangheluwe, H., Thule, C., Larsen, P.G.: Co-simulation: a survey. ACM Comput. Surv. **51**(3), 1–33 (2018)

12. Gomes, C., Lucio, L., Vangheluwe, H.: Semantics of co-simulation algorithms with simulator contracts. In: Proceedings of the ACM/IEEE MODELS 2019. IEEE (2019)

13. Gomes, C., et al.: Semantic adaptation for FMI co-simulation with hierarchical simulators. J. Simul. **95**(3), 1–29 (2019)

14. Gomes, C., et al.: HintCO - hint-based configuration of co-simulations. In: Proceedings of Simultech 2019. SciTePress (2019)

15. Gomes, C., Thule, C., Lausdahl, K., Larsen, P.G., Vangheluwe, H.: Demo: stabilization technique in INTO-CPS. In: Mazzara, M., Ober, I., Salaün, G. (eds.) STAF 2018. LNCS, vol. 11176, pp. 45–51. Springer, Cham (2018). https://doi.org/10.1007/978-3-030-04771-9_4

16. Gomes, C., Thule, C., Lúcio, L., Vangheluwe, H., Larsen, P.G.: Generation of co-simulation algorithms subject to simulator contracts. In: Camara, J., Steffen, M. (eds.) SEFM 2019. LNCS, vol. 12226, pp. 34–49. Springer, Cham (2020). https://doi.org/10.1007/978-3-030-57506-9_4

17. Hansen, S.T., Gomes, C., Larsen, P.G., van de Pol, J.: Synthesizing co-simulation algorithms with step negotiation and algebraic loop handling. In: Martin, C.R., Blas, M.J., Inostrosa-Psijas, A. (eds.) Proceedings of Annual Modeling and Simulation Conference (ANNSIM 2021). IEEE (2021)

18. Hansen, S.T., Gomes, C., Palmieri, M., Thule, C., van de Pol, J., Woodcock, J.: Verification of co-simulation algorithms subject to algebraic loops and adaptive steps. In: Lluch Lafuente, A., Mavridou, A. (eds.) FMICS 2021. LNCS, vol. 12863, pp. 3–20. Springer, Cham (2021). https://doi.org/10.1007/978-3-030-85248-1_1

19. Kübler, R., Schiehlen, W.: Two methods of simulator coupling. Math. Comput. Model. Dyn. Syst. **6**(2), 93–113 (2000)
20. Mason, I.A., Nigam, V., Talcott, C., Brito, A.: A framework for analyzing adaptive autonomous aerial vehicles. In: Cerone, A., Roveri, M. (eds.) SEFM 2017. LNCS, vol. 10729, pp. 406–422. Springer, Cham (2018). https://doi.org/10.1007/978-3-319-74781-1_28
21. Meseguer, J.: Conditional rewriting logic as a unified model of concurrency. Theor. Comput. Sci. **96**(1), 73–155 (1992)
22. Oakes, B.J., Gomes, C., Holzinger, F.R., Benedikt, M., Denil, J., Vangheluwe, H.: Hint-based configuration of co-simulations with algebraic loops. In: Obaidat, M.S., Ören, T., Szczerbicka, H. (eds.) SIMULTECH 2019. AISC, vol. 1260, pp. 1–28. Springer, Cham (2021). https://doi.org/10.1007/978-3-030-55867-3_1
23. Paris, T., Wiart, J., Netter, D., Chevrier, V.: Teaching co-simulation basics through practice. In: Durak, U. (ed.) Proceedings of SummerSim 2019. ACM (2019)
24. Schweizer, B., Li, P., Lu, D.: Explicit and implicit cosimulation methods: stability and convergence analysis for different solver coupling approaches. J. Comput. Nonlinear Dyn. **10**(5), 051007 (2015)
25. Thule, C., Gomes, C., DeAntoni, J., Larsen, P.G., Brauer, J., Vangheluwe, H.: Towards the verification of hybrid co-simulation algorithms. In: Mazzara, M., Ober, I., Salaün, G. (eds.) STAF 2018. LNCS, vol. 11176, pp. 5–20. Springer, Cham (2018). https://doi.org/10.1007/978-3-030-04771-9_1
26. Zeyda, F., Ouy, J., Foster, S., Cavalcanti, A.: Formalising cosimulation models. In: Cerone, A., Roveri, M. (eds.) SEFM 2017. LNCS, vol. 10729, pp. 453–468. Springer, Cham (2018). https://doi.org/10.1007/978-3-319-74781-1_31

# An Efficient Canonical Narrowing Implementation for Protocol Analysis

Raúl López-Rueda[1]($\boxtimes$), Santiago Escobar[1], and José Meseguer[2]

[1] VRAIN, Universitat Politècnica de València, Valencia, Spain
{rloprue,sescobar}@upv.es
[2] University of Illinois at Urbana-Champaign, Urbana, IL, USA
meseguer@illinois.edu

**Abstract.** This work improves the *canonical narrowing* previously implemented using Maude 2.7.1 by taking advantage of the new functionalities that Maude 3.2 offers. In order to perform more faithful comparisons between algorithms, we have reimplemented Maude's built-in narrowing using Maude's metalevel. We compare these two metalevel implementations with Maude's built-in narrowing, implemented at the C++ level, through a function that collects all the solutions, since the original command only returns one at a time. The results of these experiments are relevant for narrowing-based protocol analysis tools, as well as for improving the analysis of many other narrowing-based applications such as logical model checking, theorem proving or partial evaluation.

**Keywords:** Canonical narrowing · Reachability analysis · Maude · Narrowing modulo · Security protocols

## 1 Introduction

Since verification of protocol security properties modulo the algebraic properties of a protocol's cryptographic functions for an arbitrary number of sessions is generally undecidable, and the state space is infinite, *symbolic* techniques such as unification and narrowing modulo a protocol's algebraic properties, as well as SMT solving, are particularly well suited to support symbolic model checking and theorem proving verification methods.

The Maude-NPA [7] is a symbolic model checker for cryptographic protocol analysis based on the above-mentioned symbolic techniques, which are efficiently supported by the underlying Maude language [4]. These Maude-based symbolic techniques are also used by other protocol analysis tools such as Tamarin [12] and AKISS [2].

This work has been partially supported by the EC H2020-EU grant agreement No. 952215 (TAILOR), by the grant RTI2018-094403-B-C32 funded by MCIN/AEI/10.13039/501100011033 and ERDF "A way of making Europe", by the grant PROMETEO/2019/098 funded by Generalitat Valenciana, and by the grant PCI2020-120708-2 funded by MICIN/AEI/10.13039/501100011033 and by the European Union NextGenerationEU/PRTR.

© Springer Nature Switzerland AG 2022
K. Bae (Ed.): WRLA 2022, LNCS 13252, pp. 151–170, 2022.
https://doi.org/10.1007/978-3-031-12441-9_8

State explosion is a significant challenge in this kind of symbolic model checking analysis modulo algebraic properties, particularly because unification modulo algebraic properties can generate large numbers of unifiers when computing symbolic transitions. Although Maude-NPA has quite effective state space reduction techniques [6], further state space reduction gains can be obtained by more sophisticated equational narrowing techniques such as *canonical narrowing* [8], whose state space reduction advantages were experimentally validated using Maude 2.7. The main motivation for the present work comes from the fact that the new unification and narrowing features supported by the current Maude 3.2, as well as its meta-level features, make possible a new implementation of canonical narrowing that we show can achieve additional computational and performance improvements. Throughout this work, we consider several experimental examples in order to demonstrate the effectiveness of the new implementation in Maude 3.2. Below we briefly explain these examples together with their Maude specification.

The first defined module, Example 5 below, is a classic in the Maude system. It is the coffee and apple vending machine, in which dollars and quarters are inserted to buy combinations of those products. The second defined module, Example 6 below, goes one step further at the level of complexity. In this case we implement a Maude specification of a very simple protocol using exclusive-or. Likewise, the third module, Example 7 below, is a very simple module with just one transition rule where symbolic reasoning takes place modulo the theory of an abelian group.

A fourth example explores the advantages of canonical narrowing modulo an equational theory that includes the idempotence property. The reason why we have chosen this property is because it is highly problematic in automated reasoning (even for matching and rewriting). It makes easier the representation of sets, in contrast to multisets, and it is useful when dealing with processes or agents. If we have several processes working at the same time, and it turns out that two of them are the same, the idempotence property allows us to eliminate one of them to avoid redundancy and reduce the use of computational resources.

*Example 1.* We can modify the equational theory of the vending machine to add some equations that express idempotence:

```
mod IDEMPOTENCE-VENDING-MACHINE is
    sorts Coin Item Marking Money State . subsort Coin < Money . subsort Money Item < Marking .

    op empty : -> Money .
    op __ : Money Money -> Money [assoc comm id: empty] .
    op __ : Marking Marking -> Marking [assoc comm id: empty] .
    op <_> : Marking -> State . ops $ q : -> Coin . ops c a : -> Item .
    var M : Marking .

    rl [buy-c] : < M $ > => < M c > [narrowing] .      eq [idem-dollar] : $ $ M = $ M [variant] .
    rl [buy-a] : < M $ > => < M a q > [narrowing] .     eq [idem-item-a] : a a M = a M [variant] .
    eq [change] : q q q q M = $ M [variant] .           eq [idem-item-c] : c c M = c M [variant] .
endm
```

Note that idempotence is not specified for quarters (q), but only for dollars ($), apples (a) and cups of coffee (c). This is because there is already an equation that reduces the repetition of four quarters to a dollar, so that adding idempotence for quarters would create a conflict.

If we consider an initial term < M1 > that only contains a variable of type Money, we would obtain several traces by using the narrowing algorithm. In each one of the observed narrowing states, it is necessary to unify with the left-hand side of the rules to determine the new narrowing steps that can be taken. Each of those possible steps results in a new branch in the reachability tree. One of these traces takes us to the term < $ a c q q M4 >, which also contains a variable M4 of type Money. The narrowing sequence associated to this term is as follows:

$$< \text{M1} > \rightsquigarrow_{\sigma_1} < \$ \text{a q M2} > \rightsquigarrow_{\sigma_2} < \text{a c q M3} > \rightsquigarrow_{\sigma_3} < \$ \text{a c q q M4} >$$

where M2 and M3 are also variables of type Money and the computed substitutions are $\sigma_1 = \{M \mapsto \$ \text{M2}\}$, $\sigma_2 = \{M2 \mapsto \$ \text{M3}\}$, and $\sigma_3 = \{M3 \mapsto \$ \text{M4}\}$. Note that in the first narrowing step, the substitution applied to the left-hand side of the rule buy-a is $\rho_1 = \{W1 \mapsto \$ \text{M2}\}$ for W1 the variable of a renamed version of rule buy-a. For the second narrowing step, the substitution applied to the left-hand side of the rule buy-c is $\rho_2 = \{W2 \mapsto \text{a q M3}\}$ for W2 the variable of a renamed version of rule buy-c. For the third narrowing step, the substitution applied to the left-hand side of the rule buy-a is $\rho_3 = \{W3 \mapsto \$ \text{a c q M4}\}$ for W3 the variable of a renamed version of rule buy-a. Note that extra $ are introduced by $\rho_1$ and $\rho_3$ due to equational unification using the variant equations and the axioms.

As we will see later, the use of *canonical narrowing* will allow us to introduce *irreducibility constraints* in the algorithm, which in many cases will significantly reduce the number of branches in the narrowing reachability tree.

The remaining of this paper is organized as follows. Section 2 provides some preliminaries on rewriting logic and narrowing. Section 3 gives a detailed presentation of canonical narrowing. Section 4 describes our new implementation of canonical narrowing in Maude 3.2. Section 5 presents the experiments carried out using (i) the standard built-in narrowing, (ii) our implementation of standard narrowing, and (iii) our implementation of canonical narrowing. Finally, Sect. 6 summarizes the paper and presents some future work. All of the Maude modules and experiments are available at https://github.com/ralorueda/canonical-narrowing.

## 2   Preliminaries

We follow the classical notation and terminology from [17] for term rewriting, and from [13] for rewriting logic and order-sorted notions.

We assume an order-sorted signature $\Sigma$ with a poset of sorts $(S, \leq)$. The poset $(S, \leq)$ of sorts for $\Sigma$ is partitioned into equivalence classes, called *connected components*, by the equivalence relation $(\leq \cup \geq)^+$. We assume that each connected component [s] has a *top element* under $\leq$, denoted $\top_{[s]}$ and called the *top sort* of [s]. This involves no real loss of generality, since if [s] lacks a top sort, it can be easily added.

We assume an S-sorted family $\mathcal{X} = \{\mathcal{X}_s\}_{s \in \mathsf{S}}$ of disjoint variable sets with each $\mathcal{X}_s$ countably infinite. $\mathcal{T}_\Sigma(\mathcal{X})_s$ is the set of terms of sort s, and $\mathcal{T}_{\Sigma,s}$ is the set of ground terms of sort s. We write $\mathcal{T}_\Sigma(\mathcal{X})$ and $\mathcal{T}_\Sigma$ for the corresponding order-sorted term algebras. Given a term $t$, $Var(t)$ denotes the set of variables in $t$.

A *substitution* $\sigma \in \mathcal{S}ubst(\Sigma, \mathcal{X})$ is a sorted mapping from a finite subset of $\mathcal{X}$ to $\mathcal{T}_\Sigma(\mathcal{X})$. Substitutions are written as $\sigma = \{X_1 \mapsto t_1, \ldots, X_n \mapsto t_n\}$ where the domain of $\sigma$ is $Dom(\sigma) = \{X_1, \ldots, X_n\}$ and the set of variables introduced by terms $t_1, \ldots, t_n$ is written $Ran(\sigma)$. The identity substitution is *id*. Substitutions are homomorphically extended to $\mathcal{T}_\Sigma(\mathcal{X})$. The application of substitution $\sigma$ to a term $t$ is denoted by $t\sigma$ or $\sigma(t)$.

A *$\Sigma$-equation* is an unoriented pair $t = t'$, where $t, t' \in \mathcal{T}_\Sigma(\mathcal{X})_s$ for some sort $s \in \mathsf{S}$. Given $\Sigma$ and a set $E$ of $\Sigma$-equations, order-sorted equational logic induces a congruence relation $=_E$ on terms $t, t' \in \mathcal{T}_\Sigma(\mathcal{X})$ (see [14]). Throughout this paper we assume that $\mathcal{T}_{\Sigma,s} \neq \emptyset$ for every sort s, because this affords a simpler deduction system. We write $\mathcal{T}_{\Sigma/E}(\mathcal{X})$ and $\mathcal{T}_{\Sigma/E}$ for the corresponding order-sorted term algebras modulo the congruence closure $=_E$, denoting the equivalence class of a term $t \in \mathcal{T}_\Sigma(\mathcal{X})$ as $[t]_E \in \mathcal{T}_{\Sigma/E}(\mathcal{X})$.

An *equational theory* $(\Sigma, E)$ is a pair with $\Sigma$ an order-sorted signature and $E$ a set of $\Sigma$-equations. An equational theory $(\Sigma, E)$ is *regular* if for each $t = t'$ in $E$, we have $Var(t) \doteq Var(t')$. An equational theory $(\Sigma, E)$ is *linear* if for each $t = t'$ in $E$, each variable occurs only once in $t$ and in $t'$. An equational theory $(\Sigma, E)$ is *sort-preserving* if for each $t = t'$ in $E$, each sort s, and each substitution $\sigma$, we have $t\sigma \in \mathcal{T}_\Sigma(\mathcal{X})_s$ iff $t'\sigma \in \mathcal{T}_\Sigma(\mathcal{X})_s$. An equational theory $(\Sigma, E)$ is *defined using top sorts* if for each equation $t = t'$ in $E$, all variables in $Var(t)$ and $Var(t')$ have a top sort.

An *E-unifier* for a $\Sigma$-equation $t = t'$ is a substitution $\sigma$ such that $t\sigma =_E t'\sigma$. For $Var(t) \cup Var(t') \subseteq W$, a set of substitutions $CSU_E^W(t = t')$ is said to be a *complete* set of unifiers for the equality $t = t'$ modulo $E$ away from $W$ iff: (i) each $\sigma \in CSU_E^W(t = t')$ is an E-unifier of $t = t'$; (ii) for any E-unifier $\rho$ of $t = t'$ there is a $\sigma \in CSU_E^W(t = t')$ such that $\sigma|_W \sqsupseteq_E \rho|_W$ (i.e., there is a substitution $\eta$ such that $(\sigma\eta)|_W =_E \rho|_W$); and (iii) for all $\sigma \in CSU_E^W(t = t')$, $Dom(\sigma) \subseteq (Var(t) \cup Var(t'))$ and $Ran(\sigma) \cap W = \emptyset$.

A *rewrite rule* is an oriented pair $l \to r$, where $l \notin \mathcal{X}$ and $l, r \in \mathcal{T}_\Sigma(\mathcal{X})_s$ for some sort $s \in \mathsf{S}$. An *(unconditional) order-sorted rewrite theory* is a triple $(\Sigma, E, R)$ with $\Sigma$ an order-sorted signature, $E$ a set of $\Sigma$-equations, and $R$ a set of rewrite rules. The set $R$ of rules is *sort-decreasing* if for each $t \to t'$ in $R$, each $s \in \mathsf{S}$, and each substitution $\sigma$, $t'\sigma \in \mathcal{T}_\Sigma(\mathcal{X})_s$ implies $t\sigma \in \mathcal{T}_\Sigma(\mathcal{X})_s$.

The rewriting relation on $\mathcal{T}_\Sigma(\mathcal{X})$, written $t \to_R t'$ or $t \to_{p,R} t'$ holds between $t$ and $t'$ iff there exist $p \in Pos_\Sigma(t)$, $l \to r \in R$ and a substitution $\sigma$, such that $t|_p = l\sigma$, and $t' = t[r\sigma]_p$. The relation $\to_{R/E}$ on $\mathcal{T}_\Sigma(\mathcal{X})$ is $=_E; \to_R; =_E$. The transitive (resp. transitive and reflexive) closure of $\to_{R/E}$ is denoted $\to_{R/E}^+$ (resp. $\to_{R/E}^*$). A term $t$ is called $\to_{R/E}$-irreducible (or just $R/E$-irreducible) if there is no term $t'$ such that $t \to_{R/E} t'$. For $\to_{R/E}$ confluent and terminating, the irreducible version of a term $t$ is denoted by $t\downarrow_{R/E}$.

A relation $\rightarrow_{R,E}$ on $\mathcal{T}_\Sigma(\mathcal{X})$ is defined as: $t \rightarrow_{p,R,E} t'$ (or just $t \rightarrow_{R,E} t'$) iff there are a non-variable position $p \in Pos_\Sigma(t)$, a rule $l \rightarrow r$ in $R$, and a substitution $\sigma$ such that $t|_p =_E l\sigma$ and $t' = t[r\sigma]_p$. Reducibility of $\rightarrow_{R/E}$ is undecidable in general since $E$-congruence classes can be arbitrarily large. Therefore, $R/E$-rewriting is usually implemented [11] by $R, E$-rewriting under some conditions on $R$ and $E$ such as confluence, termination, and coherence.

We call $(\Sigma, B, E)$ a *decomposition* of an order-sorted equational theory $(\Sigma, E \cup B)$ if $B$ is regular, linear, sort-preserving, defined using top sorts, and has a finitary and complete unification algorithm, which implies that $B$-matching is decidable, and the equations $E$ oriented into rewrite rules $\overrightarrow{E}$ are *convergent*, i.e., confluent, terminating, and strictly coherent [15] modulo $B$, and sort-decreasing.

Given a decomposition $(\Sigma, B, E)$ of an equational theory, $(t', \theta)$ is an $E, B$-*variant* [3,10] (or just a variant) of term $t$ if $t\theta{\downarrow}_{E,B} =_E t'$ and $\theta{\downarrow}_{E,B} =_E \theta$. A *complete set of $E, B$-variants* [10] (up to renaming) of a term $t$ is a subset, denoted by $[\![t]\!]_{E,B}$, of the set of all $E, B$-variants of $t$ such that, for each $E, B$-variant $(t', \sigma)$ of $t$, there is an $E, B$-variant $(t'', \theta) \in [\![t]\!]_{E,B}$ such that $(t'', \theta) \sqsupseteq_{E,B} (t', \sigma)$, i.e., there is a substitution $\rho$ such that $t' =_B t''\rho$ and $\sigma|_{Var(t)} =_B (\theta\rho)|_{Var(t)}$. A decomposition $(\Sigma, B, E)$ has the *finite variant property* (FVP) [10] (also called a *finite variant decomposition*) iff for each $\Sigma$-term $t$, a complete set $[\![t]\!]_{E,B}$ of its most general variants is finite.

In what follows, the set $G$ of equations will in practice be $G = E \uplus B$ and will have a decomposition $(\Sigma, B, E)$.

**Definition 1 (Reachability goal).** [16] *Given an order-sorted rewrite theory* $(\Sigma, G, R)$, *a reachability goal is defined as a pair* $t \stackrel{?}{\rightarrow}^*_{R/G} t'$, *where* $t, t' \in \mathcal{T}_\Sigma(\mathcal{X})_\mathsf{s}$. *It is abbreviated as* $t \stackrel{?}{\rightarrow}^* t'$ *when the theory is clear from the context; $t$ is the source of the goal and $t'$ is the* target. *A substitution $\sigma$ is a $R/G$-solution of the reachability goal (or just a solution for short) iff there is a sequence* $\sigma(t) \rightarrow_{R/G} \sigma(u_1) \rightarrow_{R/G} \cdots \rightarrow_{R/G} \sigma(u_{k-1}) \rightarrow_{R/G} \sigma(t')$.

*A set $\Gamma$ of substitutions is said to be a* complete set of solutions *of* $t \stackrel{?}{\rightarrow}^*_{R/G} t'$ *iff (i) every substitution $\sigma \in \Gamma$ is a solution of* $t \stackrel{?}{\rightarrow}^*_{R/G} t'$, *and (ii) for any solution $\rho$ of* $t \stackrel{?}{\rightarrow}^*_{R/G} t'$, *there is a substitution $\sigma \in \Gamma$ more general than $\rho$ modulo $G$, i.e.,* $\sigma|_{Var(t) \cup Var(t')} \sqsupseteq_G \rho|_{Var(t) \cup Var(t')}$.

This provides a tool-independent semantic framework for symbolic reachability analysis of protocols under algebraic properties. Note that we have removed the condition $Var(r) \subseteq Var(l)$ for rewrite rules $l \rightarrow r \in R$ and thus a solution of a reachability goal must be applied to all terms in the rewrite sequence. If the terms $t$ and $t'$ in a goal $t \stackrel{?}{\rightarrow}^*_{T/G} t'$ are ground and rules have no extra variables in their right-hand sides, then goal solving becomes a standard rewriting reachability problem. However, since we allow terms $t, t'$ with variables, we need a mechanism more general than standard rewriting to find solutions of reachability goals. *Narrowing* with $R$ modulo $G$ generalizes rewriting by performing *unification* at non-variable positions instead of the usual matching modulo $G$. Specifically, narrowing instantiates the variables in a term by a $G$-unifier that enables a rewrite modulo $G$ with a given rule of $R$ and a term position.

**Definition 2 (Narrowing modulo $G$).** [16] *Given an order-sorted rewrite theory* $(\Sigma, G, R)$, *the narrowing relation on* $T_\Sigma(\mathcal{X})$ *modulo* $G$ *is defined as* $t \leadsto_{\sigma, R, G} t'$ *(or* $\overset{\sigma}{\leadsto}$ *if* $R, G$ *is understood) iff there is* $p \in Pos_\Sigma(t)$, *a rule* $l \to r$ *in* $R$ *such that* $Var(t) \cap (Var(l) \cup Var(r)) = \emptyset$, *and* $\sigma \in CSU_G^V(t|_p = l)$ *for a set* $V$ *of variables containing* $Var(t)$, $Var(l)$, *and* $Var(r)$, *such that* $t' = \sigma(t[r]_p)$.

*The reflexive and transitive closure of narrowing is defined as* $t \leadsto_{\sigma, R, G}^* t'$ *iff either* $t = t'$ *and* $\sigma = id$, *or there are terms* $u_1, \ldots, u_n$, $n \geq 1$, *and substitutions* $\sigma_1, \ldots, \sigma_{n+1}$ *s.t.* $t \leadsto_{\sigma_1, R, G} u_1 \leadsto_{\sigma_2, R, G} u_2 \cdots u_n \leadsto_{\sigma_{n+1}, R, G} t'$ *and* $\sigma = \sigma_1 \cdots \sigma_{n+1}$.

Soundness and completeness of narrowing with rules $R$ modulo the equational theory $G$ for solving reachability goals are proved in [11,16] for order-sorted *topmost* rewrite theories, i.e., rewrite theories were all the rewrite steps happen at the top of the term.

## 3   Canonical Narrowing

This section gives an overview of the canonical narrowing strategy of [8]. The canonical narrowing relation $\leadsto_{R/E, B}$ includes irreducibility constraints only for the left-hand sides of the rules.

**Definition 3 (Canonical Constrained Narrowing).** [8] *Given an order-sorted rewrite theory* $(\Sigma, E \cup B, R)$ *such that* $(\Sigma, B, E)$ *is a decomposition of* $(\Sigma, E \cup B)$, *the* canonical narrowing relation with irreducibility constraints *holds between* $\langle t, \Pi \rangle$ *and* $\langle t', \Pi' \rangle$, *denoted*

$$\langle t, \Pi \rangle \leadsto_{\alpha, R/E, B} \langle t', \Pi' \rangle$$

*iff there exists* $l \to r \in R$, *which we always assume renamed, so that* $Var(\langle t, \Pi \rangle) \cap (Var(r) \cup Var(l)) = \emptyset$, *and a unifier* $\alpha \in CSU_{E \cup B}^W(t = l)$, *where* $W = Var(\langle t, \Pi \rangle) \cup Var(r) \cup Var(l)$, *and*

*1.* $\langle t', \Pi' \rangle = \langle r\alpha, \Pi\alpha \cup \{(l\alpha)\!\downarrow_{\overrightarrow{E}, B}\} \rangle$, *and*

*2.* $\Pi\alpha \cup \{(l\alpha)\!\downarrow_{\overrightarrow{E}, B}\}$ *are* $\overrightarrow{E}, B$-*irreducible.*

Soundness and completeness of canonical narrowing with rules $R$ modulo the equational theory $E \cup B$ w.r.t. canonical rewriting for solving reachability goals are proved in [8].

Note that we do not require a narrowing step to compute $CSU_{E \cup B}(t = l)$ anymore, we perform regular equational unification but impose an irreducibility constraint on the normal form of the instantiated left-hand side, which can be handled in Maude by using asymmetric unification [5].

The irreducibility constraints are computed by using the normalized left-hand side of the rules that are used in the narrowing steps. Each trace will carry a different set of irreducibility constraints, although several of the conditions are shared by having common predecessor nodes. In each new narrowing step, the list of irreducibility constraints computed previously in that sequence must be taken into account, so that if it is necessary to reduce one of the terms appearing in the list to compute a new step, it will be discarded. In this way, we eliminate redundancy as well as branches of the reachability tree, which will be less and less wide than the tree resulting from using standard narrowing. In some cases, we will even get infinite reachability trees to become finite, ensuring termination.

*Example 2.* If we look at the module of Example 1, we can define an equational unification problem of the form $t = t'$. Specifically, if we consider the narrowing trace shown in that example, we can place ourselves in the third term, just before taking the last step. To compute the next possible steps from that term, it is necessary to try to unify it with the left-hand side of each of the defined rules. In this case, we will focus on the rule buy-a, which is also used to take the first step of the trace. The specification of the unification problem would then be $t = $ < a c q M3 > and $t' = $ < W3 \$ >, where W3 is a variable of type Marking (money, items, or combinations of them) corresponding to the variable of a renamed version of rule buy-a. If we run the unification problem using Maude's command, we will get 5 unifiers as a solution:

```
Maude> variant unify < a c q M3:Money > =? < W3:Marking $ > .
```

```
            Unifier #1                          Unifier #2
            M3:Money --> $ %1:Money             M3:Money --> q q q #1:Money
            W3:Marking --> q c a %1:Money       W3:Marking --> c a #1:Money


            Unifier #3                          Unifier #4
            M3:Money --> $ #1:Money             M3:Money --> $ q q q %1:Money
            W3:Marking --> $ q c a #1:Money     W3:Marking --> c a %1:Money

                        Unifier #5
                        M3:Money --> q q q %1:Money
                        W3:Marking --> $ c a %1:Money
```

Note that $\rho_3$ of Example 1 corresponds to the third unifier. But of those 5 unifiers, there are 3 that could be ignored, since the accumulated substitution makes the left-hand side of the buy-a rule used at the first narrowing step reducible. Canonical narrowing would have computed irreducibility constraints that come from normalizing the instantiated left-hand side of the rules when taking the first and second step. That is, the terms < M3 \$ > (i.e., < W1 \$ >$\rho_1\downarrow_{E,B} = $ < \$ \$ M2 >$\downarrow_{E,B} = $ < \$ M2 > and < \$ M2 >$\sigma_2\downarrow_{E,B} = $ < M3 \$ >) and < a q M3 > (i.e., < W2 \$ >$\rho_2\downarrow_{E,B} = $ < a q M3 >) are assumed to be irreducible when we want to take the last step of the trace. Maude's unification command allows us to indicate this irreducibility constraint using such that M3 \$ irreducible at the end command. If we run it now, we can see how the number of unifiers found is reduced to 2, since the first, third and fourth unifiers from the previous command are discarded:

```
Maude> variant unify < a c q M3:Money > =? < W3:Marking $ >
>       such that M3:Money $ irreducible .
```

```
Unifier #1                              Unifier #2
M3:Money --> q q q #1:Money             M3:Money --> q q q %1:Money
W3:Marking --> c a #1:Money             W3:Marking --> $ c a %1:Money
```

As can be seen, the use of irreducibility constraints manages to reduce the number of unifiers. By applying them to the narrowing algorithm, as canonical narrowing does, then this implies the reduction of possible steps (branches in the reachability tree) from the term in which we were, since for each one of the unifiers found between the term and the right part of a rule, we will have a new narrowing step.

## 4    Implementation

Our approach has been to create a meta-level command in which one of the input parameters allows us to choose between the standard narrowing algorithm or the canonical narrowing algorithm.

### 4.1    Using the Meta-level

To achieve the implementation of the command it is necessary to use some calls to the Maude meta-level available in Maude 3.2. Thanks to this, we can reuse functionalities that are integrated at the native level in C++, achieving much better performance than if we implemented them from scratch; as it happened in the previous implementation in Maude 2.7.

Each user command in Maude is represented by a corresponding command at the meta-level, allowing us greater control and management of their outputs. For example, the variant unify command that we saw in Example 2 corresponds to the metaVariantUnify command at the meta level. It is precisely this command that we use to carry out the unification step in our implementation, since it allows us to perform equational unification modulo variant equations and axioms. The operator that defines the command is the following:

```
op metaVariantUnify :
    Module UnificationProblem TermList Qid VariantOptionSet Nat ~> UnificationPair? .    .
```

The command receives six parameters and returns a structure of type UnificationPair?, an error or a pair consisting of a substitution and an identifier of the family of variables used. The first command received is the module that defines the rewriting theory to work on. The second is the unification problem to which solutions are sought. The third is a list of irreducibility terms, which is of vital importance in the canonical narrowing algorithm. The fourth corresponds to the identifier of the family of variables to avoid (the one used for the variables of the unification problem). The fifth is a parameter used to indicate if we want to filter the returned unifiers. Finally, a natural number parameter is received in which the unifier to be searched is indicated. We show an execution of this command in Example 3, using in turn the module of the vending machine with idempotence as a rewriting system (see Example 1).

*Example 3.* Considering the module from Example 1 again as a rewriting system, we can use the `metaVariantUnify` command to find the unifiers seen in Example 2. We simply indicate the same equational unification problem, and by means of the last argument of the command we can select each of the unifiers to obtain. Additionally, we can use an irreducibility condition to reduce the number of unifiers just like we have seen before. For example, by using the same irreducibility condition, we can obtain one of the unifiers as follows:

```
Maude> reduce in META-LEVEL :
>       metaVariantUnify(upModule('IDEMPOTENCE-VENDING-MACHINE, true),
>       '<_>['__['a.Item,'c.Item,'q.Coin,'M3:Money]] =? '<_>['__['$.Coin,'W3:Marking]],
>       '<_>['__['$.Coin,'M3:Money]], '@, none, 0) .
result UnificationPair: {
  'M3:Money <- '__['q.Coin,'q.Coin,'q.Coin,'#1:Money] ;
  'W3:Marking <- '__['a.Item,'c.Item,'#1:Money],'#}}
```

Another meta-level functionality that has been necessary to use is the `metaNarrowingApply` command. It performs a narrowing step, using the arguments shown in its definition below. Thanks to this command and the `metaVariantUnify` one, we can abstract from the unification processes, which are the most costly at the computational level. By invoking meta-level commands to do so, execution is done natively in C++ code, which turns out to be much faster and more efficient. The operator that defines the command is the following:

```
op metaNarrowingApply :
    Module Term TermList Qid VariantOptionSet Nat -> NarrowingApplyResult? .
```

In this case, the command receives as the first parameter, again, the module that represents the rewrite theory to be used. The second parameter represents the term from which to perform the narrowing step. The third parameter is a list of irreducibility terms, important for canonical narrowing. The fourth parameter is the identifier of the family of variables to avoid. The fifth parameter is used to indicate if we want to filter the returned unifiers in order to get only the most general unifiers. Finally, the sixth parameter is the step that you want to take, that is, the "branch" of the tree that you want to generate from the given term. The result will be of type `NarrowingApplyResult?`, a data structure that contains either an error, or the necessary information from the narrowing step performed.

*Example 4.* We use again the module from Example 1. As an initial term we consider the same that we will use later for the experiments whose results are shown in Table 4. The `metaNarrowingApply` command allows us to give (among others) the first step of narrowing from that term:

```
Maude> reduce in META-LEVEL :
>       metaNarrowingApply(upModule('IDEMPOTENCE-VENDING-MACHINE, true),
>                   '<_>['M1:Money], empty, '@, none, 0) .
result NarrowingApplyResult: { '<_>['__['a.Item,'q.Coin,'%1:Money]],'State,
  [], 'buy-a, 'M1:Money <- '__['$.Coin,'%1:Money], 'M:Marking <- '%1:Money, '% }
```

The output returned by Maude shows how the rule labeled as buy-a has been used to perform the narrowing step, resulting in two different assignments. On the one hand, a dollar is assigned together with a fresh variable to the variable M1 of type Money. On the other hand, the same fresh variable is assigned to the variable M2 of type Marking (Note that in this case, this is possible only because Money is a subsort of Marking).

There is also a metaNarrowingSearch command that performs the entire instead of only one step, but we have not used it since we need to perform intermediate operations between each narrowing step to implement the canonical narrowing algorithm.

## 4.2    Data Structures and the narrowing Command

All narrowing algorithms perform one-step transitions from one symbolic state to another—the narrowing steps—using the rewrite rules of the given specification. We use a tree as a data structure, in which each of these narrowing steps gives rise to a new node, with its associated term. Thus, the root node of the tree will have as its associated term the initial term (reduced to normal form) indicated by the user. At the same time, each of the nodes is itself a data structure, in which we not only find the associated term, but also some extra information that allows us to locate the node and generate new terms from it.

Our implementation is built in such a way that ten parameters are requested from the user to invoke the command, as follows:

```
narrowing(Module, Term, SearchArrow, Term, Algorithm, VariantOptionSet, TermList, Qid,
          Bound, Bound)
```

The first argument receives the rewrite theory to perform the unification and narrowing steps. The second and fourth arguments are used to indicate the initial term and the target term respectively. The third argument corresponds to the search arrow that we want to use, so that solutions are included or discarded depending on the rewriting steps performed to achieve them. This argument may take values to indicate that only solutions that involve a single rewrite step, one or more steps, or any number of steps can be considered. The combination of the fifth and sixth parameters will indicate the type of algorithm to use. Combinations indicating the use of standard narrowing or canonical narrowing are currently accepted. The seventh argument is used to indicate a list of initial irreducibility terms to consider. This argument will be taken into account in all unification calls and in each narrowing step, allowing the value empty to indicate that we do not want to use irreducibility constraints. The eighth argument receives the identifier used to name the variables in the initial and target terms, to avoid later clashes. Finally, the ninth and tenth arguments are used to impose bounds on the algorithm, being able to indicate a maximum depth to expand the search tree or a maximum of solutions to search.

## 4.3    Search for Solutions

When we receive the parameters from the user, the first necessary step is to verify that the value of the depth limits and solutions are admissible. If they are, the strategy to follow will be determined according to the indicated search arrow.

Once all the above is prepared, the first nodes of the search tree are generated from the root, that is, from the initial term. The tree will be generated by levels, so that children of any node belonging to the next level will not be generated until that level is completely generated. Each node contains its associated term plus some extra information. Specifically, for each node we need a unique identifier, a reference to its parent node, the branch of the tree to which it belongs, the depth to which it is located and, in the case of the use of canonical narrowing, a list of the irreducibility terms calculated so far in that branch.

Each time a new node is generated, an attempt is made to unify its associated term with the target term indicated by the user. If unifiers exist, a solution will be built for each of the unifiers found. To do this, it is necessary to go backwards through the branch to which the node belongs, combining the substitutions made to compute the accumulated substitution. If we are using canonical narrowing, when a new node is generated it will also be necessary to modify the list of irreducibility terms, adding the irreducibility term that is calculated from the normalized left-hand side of the rule used to reach the node (see Definition 3).

## 4.4  Avoiding Variable Clashes

For the generation of new nodes, some calls are made to internal commands of the Maude meta-level. These commands only allow the indication of a variable identifier to avoid (which must be the one used previously), preventing possible variable clashes. Each of the calls to these internal commands will result in a random use of the rest of the variable identifiers handled by Maude. This gives rise to the possibility that variables can be repeated in different nodes, which is not an a priori problem, but it can't be assumed when it is required to calculate the cumulative substitution of a reachability solution.

To avoid this problem, we have chosen the strategy of renaming each of the fresh variables that Maude generates on the fly, using a new identifier, the $ symbol. That is why in the final result returned to the user, all the fresh variables that contain the narrowing solutions will be identified with that symbol, thus ensuring that none of them clashes with the rest.

## 4.5  Algorithm Performance Improvement

Due to the nature of the algorithm and the uses for which it is intended, performance of the algorithm plays a very important role. To improve this characteristic, different aspects have been taken into account regarding the sequence in which the algorithm acts and the data structures it handles.

Regarding the operators and equations in the code, they have been divided into three main parts, which correspond to the main steps of the algorithm at a theoretical level: (i) the generation of nodes (terms) in the reachability tree, (ii) the attempt to unify each new term with the target term, and (iii) the computation of solutions in case the unification is successful. Likewise, each of these parts is divided into subparts that facilitate not only the understanding of the code, but also a structured scheme to add new functionalities easily. Thanks to this, once we reimplemented the standard narrowing algorithm, it was relatively easy to add the new functionalities that modified it to achieve the canonical narrowing algorithm.

We can also consider the way in which the algorithm handles the data structures it works with. A priori, it could be thought that the nodes that are generated can go to a set of nodes that is subsequently processed. However, our strategy is to use an ordered list in which the nodes are processed taking into account an order similar to that of a recursion queue. In the same way, the nodes that are being processed in that list (that is, those in which the children have been generated) go to another list. This second list is used for the computation of the accumulated substitutions in the solutions. There is also another list in which the found unifiers are stored. It is also ordered to facilitate working with it recursively and calculating the solutions from the unifiers.

In addition to all this, extra parameters are dragged in the main data structure and also locally in each of the nodes. These parameters will later help to perform certain operations more quickly and efficiently. For example, each node has a reference to its parent node identifier, making it easy to go backwards on its branch if a cumulative substitution needs to be calculated.

## 5    Experiments

To test the operation and efficiency of the new command, as well as to check the performance differences between the different algorithms, we have used the modules mentioned in the introduction. That is, we used for the experiments the module of the vending machine (Example 5 below), the module of a protocol using the exclusive-or property (Example 6 below), the module of a process counter using the properties of an abelian group, the module of the vending machine with idempotence (Example 7 below), and Example 1. These modules allow us to check how the narrowing algorithms behave in those cases, subjecting the command to executions of different complexity for various applications.

The reimplementation of both the standard narrowing and canonical narrowing in the same command presented in Sect. 4 allows us to perform more faithful comparisons between the algorithms, independently of the standard built-in narrowing algorithm provided by Maude at the C++ level. However, since the built-in narrowing returns only one solution when executed via its meta-level function, we have also built a command that iteratively obtains all solutions. In this way, in the tables below we include (i) the standard built-in narrowing, (ii) our implementation of standard narrowing, and (iii) our implementation of canonical narrowing. It can be noted how, in certain cases, our canonical narrowing algorithm manages to surpass even the built-in narrowing command.

**Table 1.** Experiments using the vending machine module.

| Algorithm | Depth limit | Execution time | Solutions found |
|-----------|-------------|----------------|-----------------|
| Native    | 4 | 32 ms     | 163  |
| Standard  | 4 | 75 ms     | 163  |
| Canonical | 4 | 60 ms     | 137  |
| Native    | 5 | 112 ms    | 550  |
| Standard  | 5 | 496 ms    | 550  |
| Canonical | 5 | 324 ms    | 119  |
| Native    | 6 | 460 ms    | 1850 |
| Standard  | 6 | 6384 ms   | 1850 |
| Canonical | 6 | 2724 ms   | 1213 |
| Native    | 7 | 3092 ms   | 6216 |
| Standard  | 7 | 166828 ms | 6216 |
| Canonical | 7 | 45808 ms  | 3559 |

## 5.1   Experiments with the Vending Machine

*Example 5.* This Maude's system module is a classic in the Maude community. It is the coffee and apple vending machine, in which dollars and quarters are inserted to buy combinations of those products. To do this, we specify that each coffee costs one dollar and each apple three-quarters of a dollar. Two rules handle state transitions for those specifications. Furthermore, an equation is used to specify the change of four-quarters of a dollar to one dollar. Note the addition of a variable M of type Marking to make the rules and equations ACU-coherent.

```
mod NARROWING-VENDING-MACHINE is
    sorts Coin Item Marking Money State .
    subsort Coin < Money .
    op empty : -> Money .
    op __ : Money Money -> Money [assoc comm id: empty] .
    subsort Item < Marking .
    op __ : Marking Marking -> Marking [assoc comm id: empty] .
    op <_> : Marking -> State .
    ops $ q : -> Coin .
    ops c a : -> Item .
    var M : Marking .
    rl [buy-c] : < M $ > => < M c > [narrowing] .
    rl [buy-a] : < M $ > => < M a q > [narrowing] .
    eq [change] : q q q q M = $ M [variant] .
endm
```

We use the reachability problem $< M1 > \leadsto^*_{\alpha, R/E, B} St$ where M1 is a variable of type Money and St is a variable of type State. That is, we are asking for all the states that can be reached from an initial state containing only quarters and dollars. It is a fairly generic problem that allows us to see the number of nodes that are being generated in the reachability tree. Table 1 shows the results of running the command with this reachability problem.

These initial experiments use a simple rewrite theory and a simple reachability problem. As a consequence, the narrowing included natively in Maude turns out to be faster than either of our two algorithms, thanks to its coding in C++. However, we can see that even in these cases, if we compare our standard narrowing implementation with our canonical narrowing implementation, the latter has always a better performance. This leads us to think that a natively programmed canonical narrowing would be able to outperform Maude's standard narrowing even using these simple parameters. To strengthen this idea, we can look at the number of solutions (which in this case represent the number of states in the tree) found. For example, for depth level 7, canonical narrowing is capable of reducing the number of states generated by almost half regarding standard narrowing. If it was implemented natively in Maude, its execution time would obviously be much less, since it has to go through far fewer rewriting steps. In addition, the decrease in solutions represents in itself a relevant improvement, since those that come from redundancy in the rewriting traces are being discarded.

## 5.2  Experiments with a Protocol Using the Exclusive-Or Property

*Example 6.* The equational theory used in the protocol below corresponds to the XOR property. Note the addition of the second equation for AC-coherence.

```
fmod EXCLUSIVE-OR is
  sort XOR .
  op mt : -> XOR .
  op _*_ : XOR XOR -> XOR [assoc comm] .
  vars X Y Z U V : [XOR] .
  eq [idem] :      X * X = mt      [variant] .
  eq [idem-Coh] : X * X * Z = Z [variant] .
  eq [id] :        X * mt = X      [variant] .
endfm
```

In the XOR-PROTOCOL module, the equation theory is imported and the rest of the protocol is implemented. The main structure is a state that stores the set of messages that have been sent and the new messages to be sent. The exchange of messages is done between two users for the protocol to take place. The – and + symbols are used as operators to distinguish between the messages to be received or sent respectively. The *Nonces* generation is included in the protocol, as well as data structures that specify the knowledge that an intruder might have.

```
mod XOR-PROTOCOL is protecting EXCLUSIVE-OR .
  sorts Name Nonce Fresh Msg . subsort Name Nonce XOR < Msg . subsort Nonce < XOR .
  ops a b c : -> Name . op n : Name Fresh -> Nonce .
  op pk : Name Msg -> Msg . ops r1 r2 r3 : -> Fresh .
  sort SMsg . sort SMsgList . subsort SMsg < SMsgList .
  ops + - : Msg -> SMsg .
  op nil : -> SMsgList .
  op _',_ : SMsgList SMsgList -> SMsgList [assoc] .

  sort Strand . sort StrandSet . subsort Strand < StrandSet .
  op '[_|_'] : SMsgList SMsgList -> Strand .
  op mt : -> StrandSet .
  op _&_ : StrandSet StrandSet -> StrandSet [assoc comm id: mt] .

  sort IntruderKnowledge .
  op mt : -> IntruderKnowledge .
  op inI : Msg -> IntruderKnowledge .
  op nI : Msg -> IntruderKnowledge .
  op _',_ : IntruderKnowledge IntruderKnowledge -> IntruderKnowledge [assoc comm id: mt] .
```

**Table 2.** Experiments using the XOR-protocol module.

| Algorithm | Execution time | Solutions found |
|-----------|----------------|-----------------|
| Native    | 1660 ms        | 84              |
| Standard  | 16124 ms       | 84              |
| Canonical | 2300 ms        | 1               |

```
sort State .
op Sta : -> State .
op '{_'{_'}'} : StrandSet IntruderKnowledge -> State .

vars IK IK1 IK2 : IntruderKnowledge .  vars A B : Name .
vars X Y Z U V : [XOR] .  vars SS SS1 SS2 : StrandSet .
var M : Msg .  vars L1 L2 : SMsgList .  vars NA NB : Nonce .

rl [r1] : { (SS & [ ( L1 , -(M)) | L2 ]) { (inI(M) , IK) } } =>
          { (SS & [ L1 | (-(M) , L2) ]) { (inI(M) , IK) } }
          [narrowing] .
rl [r2] : { (SS & [ (L1 , +(M)) | L2 ]) { (inI(M) , IK) } } =>
          { (SS & [ L1 | (+(M) , L2) ]) { (nI(M) , IK) } }
          [narrowing] .
endm
```

We can define a reachability problem by using a basic message exchange between users. To do this, we consider a backwards execution, so that the target term will be the initial state of the message stack, while the initial term will be the final state. If a solution is found, it means that execution trace exists, so it could occur in the protocol. The Maude-NPA [7] tool works in a similar way to this.

Considering X and Y as variables of type Msg, the reachability problem that we have defined for the experiments is the following:

$$
\begin{aligned}
&\{[nil, +(pk(a, n(b, r1))), -(pk(b, Y)), +(Y * n(b, r1)) \mid nil] \\
&\& [nil, -(pk(a, X)), +(pk(b, n(a, r2))), -(X * n(a, r2)) \mid nil] \\
&\quad \{inI(X * n(a, r2)), inI(pk(a, X)), inI(pk(b, Y))\}\} \\
&\qquad\qquad \overset{?}{\leadsto}{}^{*} \\
&\{[nil \mid + (pk(a, n(b, r1))), -(pk(b, Y)), +(Y * n(b, r1)), nil] \\
&\& [nil \mid - (pk(a, X)), +(pk(b, n(a, r2))), -(X * n(a, r2)), nil] \\
&\quad \{nI(X * n(a, r2)), nI(pk(a, X)), nI(pk(b, Y))\}\}
\end{aligned}
$$

In this context we are working with a finite search space, so we can ignore the limit of solutions and the depth limit (although all the solutions are found in depth 6, so we could also use that depth limit). Results are shown in Table 2.

In this case, the native standard narrowing in Maude again manages to be faster than our two algorithms, although the difference is less than before. The usual impact that canonical narrowing has on the number of returned solutions is further noticeable. As we mentioned before, thanks to carrying out a "pruning" of the tree by discarding those redundant traces, canonical narrowing is able to reduce the 84 initial solutions to only 1.

It is important to note here the usefulness of canonical narrowing in the field of security protocols, and specifically for tools relying on unification and/or narrowing, such as the Maude-NPA tool [7], Tamarin [12] and AKISS [2]. By managing to rule out redundancies when generating the branches of the reachability tree, as seen in Example 2, when we analyze a protocol with canonical narrowing we achieve higher performance with less numerous reachable states but still complete results.

## 5.3  Experiments Using the Properties of an Abelian Group

We have already seen that canonical narrowing is useful even when—due to the prototype nature of its present implementation—it cannot be faster than the C++ based native standard narrowing in Maude. We have also concluded that if it were also integrated natively, it could be substantially faster and generate fewer states than standard narrowing in many cases. But if we also carry out experiments with systems in which there are many variants, these claims will be further reinforced.

We can see the real potential of canonical narrowing by resorting to a module using only one simple transition rule and the equations of an abelian group.

*Example 7.* We first implement a simple module defining the properties of an abelian group. That will be the equational theory used.

```
fmod ABELIAN-GROUP is
   sort Int .
   ops 0 1 : -> Int [ctor] .
   op _+_ : Int Int -> Int [assoc comm prec 30] .
   op -_ : Int -> Int .
   vars X Y Z : Int .

   eq X + 0 = X [variant] .
   eq X + (- X) = 0 [variant] .
   eq X + (- X) + Y = Y [variant] .
   eq - (- X) = X [variant] .
   eq - 0 = 0 [variant] .
   eq (- X) + (- Y) = -(X + Y) [variant] .
   eq -(X + Y) + Y = - X [variant] .
   eq -(- X + Y) = X + (- Y) [variant] .
   eq (- X) + (- Y) + Z = -(X + Y) + Z [variant] .
   eq - (X + Y) + Y + Z = (- X) + Z [variant] .
endfm
```

A rewrite theory which consists of a pair of integers that function as process counters is defined. The first integer represents the processes that are running, while the second represents those that have finished their execution. The transition rule represents the termination of a process that was in running, so that the value of the first integer of the pair is decreased by one, and at the same time the value of the second integer is increased by one. The transition rule allows for narrowing, while the abelian group equations allow for the generation of variants. Combining everything, we get a system of transitions that, despite looking simple, turns out to be quite complex, due to the large number of variants that any term will normally have.

```
mod PROC-COUNTER is protecting ABELIAN-GROUP .
   sort State .
   op <_,_> : Int Int -> State [ctor] .
   vars X Y Z : Int .

   rl [finish-proc] : < (X + 1),Y > => < ((X + 1) + (- 1)),(Y + 1) > [narrowing] .
endm
```

**Table 3.** Experiments using the process counter module.

| Algorithm | Depth limit | Execution time | Solutions found |
|-----------|-------------|----------------|-----------------|
| Native    | 1           | 424 ms         | 184             |
| Standard  | 1           | 420 ms         | 184             |
| Canonical | 1           | 440 ms         | 184             |
| Native    | 2           | > 8 h          | –               |
| Standard  | 2           | > 8 h          | –               |
| Canonical | 2           | 2752 ms        | 719             |
| Native    | 3           | > 8 h          | –               |
| Standard  | 3           | > 8 h          | –               |
| Canonical | 3           | 12548 ms       | 2033            |
| Native    | 4           | > 8 h          | –               |
| Standard  | 4           | > 8 h          | –               |
| Canonical | 4           | 73070 ms       | 4969            |

Considering X and Y are variables of type Int, we use a common initial term: $< 0, 1 + X >$. The target term will vary slightly allowing us to fix the depth at which we want to find the solution. For depth one, it will be $< -1, Y >$. For depth two, it will be $< -(1 + 1), Y >$. For depth three, it will be $< -(1 + 1 + 1), Y >$, and for depth four, it will be $< -(1 + 1 + 1 + 1), Y >$.

Apparently, the initial term and the target terms are very simple in this example, but due to the large number of variants that an abelian group contains, the computation becomes very complex, since the reachability tree will grow very quickly in width. Table 3 shows the results of executing the above problems using different depth limits.

In this case, we can see how the execution time for the first level (i.e., first reachability problem) is very similar in any of the three algorithms. Furthermore, it is striking that the solutions returned are the same. This is normal, since canonical narrowing does not have any kind of impact on the first level, because it has not yet calculated (see Definition 3) irreducibility constraints (unless we specify them as part of the initial call). However, we can see how from depth 2, our standard narrowing algorithm does not even manage to finish in a reasonable execution time. The built-in narrowing doesn't do it either. In contrast, the canonical narrowing algorithm does terminate, returning a large number of solutions in a relatively short time. The deeper we go into the tree, the more solutions are found. At the same time, the execution time grows, but it is still acceptable.

This is a clear example of the enormous improvement that canonical narrowing can bring over standard narrowing in many cases, even using the one found natively in Maude. Obviously, if we put canonical narrowing at the same

**Table 4.** Experiments using the vending machine with idempotence of Example 1

| Algorithm | Depth limit | Execution time | Solutions found |
|-----------|-------------|----------------|-----------------|
| Native    | 3           | 2268 ms        | 3804            |
| Standard  | 3           | 6528 ms        | 3804            |
| Canonical | 3           | 708 ms         | 856             |
| Native    | 4           | $\sim$ 2 min   | 40284           |
| Standard  | 4           | $\sim$ 37 min  | 40284           |
| Canonical | 4           | 11264 ms       | 4284            |
| Native    | 5           | > 8 h          | –               |
| Standard  | 5           | > 8 h          | –               |
| Canonical | 5           | $\sim$ 5 min   | 18963           |

level, that is, included in Maude natively, the performance difference would be extremely large in favor of canonical narrowing, especially in this type of cases.

### 5.4    Experiments with the Vending Machine Using Idempotence

As we mentioned earlier in the introduction, idempotence is a very important property in computing, since, for example, set data types enjoy it. Canonical narrowing seems to behave very well managing this property when compared to standard narrowing (even better than the experiments with an abelian group). We have done some experiments in which this property is used to corroborate this. We use the vending machine module with idempotence on items and dollars (see Example 1). The reachability problem defined in this case is $< M1 > \leadsto^{*}_{\alpha,R/E,B} < \$\, a\, c\, M2 >$, where M1 is a variable of type `Money` and M2 is a variable of type `Marking`.

Table 4 shows the results obtained when using the reachability problem. We must bear in mind that in this case, once again, the growth of the tree in width is very large, due to the large number of variants of the system.

In this case we can see that the executions with a lower depth limit are extremely fast. However, even in those cases the difference is obvious, with the canonical narrowing being faster and returning fewer solutions. As we increase the depth limit, the difference in performance becomes more and more noticeable.

Just by looking at the experiments with depth limit 4, we can see that canonical narrowing achieves a performance at the computational level about 10 times better than Maude's native standard narrowing. And if we instead look at our Maude reimplementation of standard narrowing, for a fair comparison, the difference is huge. More than half an hour of execution is reduced to just 11 s. The number of solutions returned is also reduced to one tenth, showing that the large percentage of those calculated by standard narrowing were unnecessary. For depth limit 5, the execution times of the standard narrowing are no longer reasonable, while the canonical narrowing manages to finish in just 5 min.

# 6   Conclusions

In this work, we have presented a new efficient implementation of canonical narrowing in Maude. The algorithm uses irreducibility constraints to reduce the width of the reachability tree without losing completeness. The experiments that we have presented demonstrate the improvements that canonical narrowing offers over standard narrowing, both in terms of performance and solutions. Typically, the greater the number of variants calculated for each unification problem, the greater the improvement. Furthermore, the deeper the reachability tree to be generated, the greater the performance relationship between both algorithms. The results are important for tools such as Maude-NPA or others that are used to analyze protocols. They are also relevant in many other areas of computing, such as when performing symbolic model checking verification of concurrent systems, theorem proving or partial evaluation.

The most obvious next step is to include the command at the user level, making the outputs returned by it more readable and understandable. Another interesting step forward consists of integrating canonical narrowing with the computation of most general unifiers [1,9]. This would involve combining an improvement of the standard narrowing algorithm with an improvement in unification with equations and axioms. If formalized and implemented correctly, this should result in an even better algorithm than the current canonical narrowing.

# References

1. Aparicio-Sánchez, D., Escobar, S., Sapiña, J.: Variant-based equational unification under constructor symbols. In: Ricca, F., et al. (eds.) Proceedings 36th International Conference on Logic Programming (Technical Communications), ICLP Technical Communications 2020 (Technical Communications) UNICAL, Rende (CS), Italy, 18–24th September 2020, EPTCS, vol. 325, pp. 38–51 (2020)
2. Chadha, R., Cheval, V., Ciobâcă, Ş, Kremer, S.: Automated verification of equivalence properties of cryptographic protocols. ACM Trans. Comput. Log. **17**(4), 23:1-23:32 (2016)
3. Comon-Lundh, H., Delaune, S.: The finite variant property: how to get rid of some algebraic properties. In: Giesl, J. (ed.) RTA 2005. LNCS, vol. 3467, pp. 294–307. Springer, Heidelberg (2005). https://doi.org/10.1007/978-3-540-32033-3_22
4. Durán, F., et al.: Programming and symbolic computation in Maude. J. Log. Algebraic Methods Program. **110**, 100497 (2020)
5. Erbatur, S., et al.: Asymmetric unification: a new unification paradigm for cryptographic protocol analysis. In: Bonacina, M.P. (ed.) CADE 2013. LNCS (LNAI), vol. 7898, pp. 231–248. Springer, Heidelberg (2013). https://doi.org/10.1007/978-3-642-38574-2_16
6. Escobar, S., Meadows, C., Meseguer, J., Santiago, S.: State space reduction in the Maude-NRL protocol analyzer. Inf. Comput. **238**, 157–186 (2014)
7. Escobar, S., Meadows, C., Meseguer, J.: Maude-NPA: cryptographic protocol analysis modulo equational properties. In: Aldini, A., Barthe, G., Gorrieri, R. (eds.) FOSAD 2007-2009. LNCS, vol. 5705, pp. 1–50. Springer, Heidelberg (2009). https://doi.org/10.1007/978-3-642-03829-7_1

8. Escobar, S., Meseguer, J.: Canonical narrowing with irreducibility constraints as a symbolic protocol analysis method. In: Guttman, J.D., Landwehr, C.E., Meseguer, J., Pavlovic, D. (eds.) Foundations of Security, Protocols, and Equational Reasoning. LNCS, vol. 11565, pp. 15–38. Springer, Cham (2019). https://doi.org/10.1007/978-3-030-19052-1_4

9. Escobar, S., Sapiña, J.: Most general variant unifiers. In: Bogaerts, B., et al. (eds.) Proceedings 35th International Conference on Logic Programming (Technical Communications), ICLP 2019 Technical Communications, Las Cruces, NM, USA, 20–25 September 2019, EPTCS, vol. 306, pp. 154–167 (2019)

10. Escobar, S., Sasse, R., Meseguer, J.: Folding variant narrowing and optimal variant termination. J. Log. Algebr. Program. 81(7–8), 898–928 (2012)

11. Jouannaud, J.-P., Kirchner, H.: Completion of a set of rules modulo a set of equations. SIAM J. Comput. 15(4), 1155–1194 (1986)

12. Meier, S., Schmidt, B., Cremers, C., Basin, D.: The TAMARIN prover for the symbolic analysis of security protocols. In: Sharygina, N., Veith, H. (eds.) CAV 2013. LNCS, vol. 8044, pp. 696–701. Springer, Heidelberg (2013). https://doi.org/10.1007/978-3-642-39799-8_48

13. Meseguer, J.: Conditioned rewriting logic as a united model of concurrency. Theor. Comput. Sci. 96(1), 73–155 (1992)

14. Meseguer, J.: Membership algebra as a logical framework for equational specification. In: Presicce, F.P. (ed.) WADT 1997. LNCS, vol. 1376, pp. 18–61. Springer, Heidelberg (1998). https://doi.org/10.1007/3-540-64299-4_26

15. Meseguer, J.: Strict coherence of conditional rewriting modulo axioms. Theor. Comput. Sci. 672, 1–35 (2017)

16. Meseguer, J., Thati, P.: Symbolic reachability analysis using narrowing and its application to verification of cryptographic protocols. High. Order Symb. Comput. 20(1–2), 123–160 (2007)

17. TeReSe (ed.): Term Rewriting Systems. Cambridge University Press, Cambridge (2003)

# Checking Sufficient Completeness
# by Inductive Theorem Proving

José Meseguer(✉)

Department of Computer Science, University of Illinois at Urbana-Champaign,
Urbana, USA
`meseguer@illinois.edu`

**Abstract.** Sufficient completeness of an equational program ensures
that each input can be fully evaluated to a data result. Checking this fundamental property for programs in expressive equational languages supporting conditional equations, types and subtypes, and rewriting modulo
structural axioms is a challenging problem for which few methods currently exist. This work presents a new method that reduces sufficient
completeness verification to a standard inductive theorem proving problem for a wide class of conditional equational programs in such languages.

## 1 Introduction

Sufficient completeness of an equational program is a fundamental property. It
means that the recursive equations defining each of its functions cover all the
required cases, so that any concrete input evaluates to actual *data*. Lack of
sufficient completeness is a common programming error, similar to bugs missing or getting wrong some conditions in an imperative program. Besides being
an essential correctness requirement, sufficient completeness is also *an essential
condition in program verification*. For example, in inductive theorem proving,
induction is typically based on *data constructors*, so that the correctness of an
inductive proof essentially depends on the specification's sufficient completeness.

As equational programming languages become more expressive, their very
richness and power makes achieving sufficient completeness more challenging for
several reasons: (i) the greater expressiveness of *conditional* equations can lead to
missing some conditions needed for evaluation; (ii) the power and expressiveness
of rewriting modulo *structural axioms* such as associativity and/or commutativity and/or identity allows recursive function definitions that use a rich variety of
patterns in its recursive equations; but this greater power, as all power, has to be
used wisely; (iii) *types and subtypes* naturally support case analysis and, together
with structural axioms, allow definition of very sophisticated data structures; but
this again requires greater care when defining functions.

Although methods for proving sufficient completeness have been studied since
the 1970s s to the present, e.g., [1, 2, 6, 16–22, 25, 30–33], in general this is an undecidable property [21] whose verification can only be fully automated in restricted
cases. For example, sufficient completeness of *unconditional* equational programs

© Springer Nature Switzerland AG 2022
K. Bae (Ed.): WRLA 2022, LNCS 13252, pp. 171–190, 2022.
https://doi.org/10.1007/978-3-031-12441-9_9

such that the lefthand sides of its recursive equations do not have repeated variables can be automatically checked by either: (i) *tree automata* techniques, e.g., [6], which, using *equational tree automata*, have been generalized to support structural axioms, types and subtypes, and context-sensitive rewriting in [18,19]; or (ii) *Boolean operations on term patterns* [23,30]; in particular, [30] supports the general *order-sorted* case, involving types and subtypes. For more general and expressive classes of programs, verifying completeness becomes harder. As further discussed in Sect. 6, for expressive equational languages supporting conditional equations, structural axioms and order-sorted typing such as OBJ [15], CafeOBJ [13] and Maude [5], few sufficient completeness verification methods currently exist for general programs.

This work proposes a new method of proving sufficient completeness of programs in expressive equational languages such as the ones mentioned above. It offers four main advantages: (1) it applies to a class of order-sorted conditional equational programs modulo structural axioms that is quite general in actual practice; (2) it does *not* require developing a new sufficient completeness verification tool, because it *reduces* such verification to standard inductive theorem proving: any such prover supporting conditional order-sorted specifications modulo axioms can be used; (3) it supports a *hierarchical* proof methodology allowing simpler automated methods, such as those in [19], to be applied to subprograms, so that only more complex function definitions need to be dealt with; and (4) such a hierarchical proof methodology is shared with that in the companion paper [25], which supports verification of both ground confluence, and of sufficient completeness by methods that nicely complement the one proposed in this work. Proofs of all theorems are relegated to Appendix B.

## 2    Preliminaries

I assume familiarity with the notions of an order-sorted signature $\Sigma$ on a poset of sorts $(S, \leqslant)$, an order-sorted $\Sigma$-algebra $A$, and the term $\Sigma$-algebras $T_\Sigma$ and $T_\Sigma(X)$ for $X$ an $S$-sorted set of variables. I also assume familiarity with the notions of: (i) $\Sigma$-homomorphism $h : A \rightarrow B$ between $\Sigma$-algebras $A$ and $B$, so that $\Sigma$-algebras and $\Sigma$-homomorphisms form a category $\mathbf{OSAlg}_\Sigma$; (ii) order-sorted (i.e., sort-preserving) substitution $\theta$, its domain $dom(\theta)$ and range $ran(\theta)$, and its application $t\theta$ to a term $t$; (iii) *preregular* order-sorted signature $\Sigma$, i.e., a signature such that each term $t$ has a least sort, denoted $ls(t)$; (iv) the set $\widehat{S} = S/(\geqslant \cup \leqslant)^+$ of *connected components* of a poset $(S, \leqslant)$ viewed as a DAG; and (v) for $A$ a $\Sigma$-algebra, the set $A_s$ of it elements of sort $s \in S$, and the set $A_{[s]} = \bigcup_{s' \in [s]} A_{s'}$ of all elements in a connected component $[s] \in \widehat{S}$. We furthermore assume that all signatures $\Sigma$ have *non-empty sorts*, i.e., $T_{\Sigma,s} \neq \varnothing$ for each $s \in S$. $[A \rightarrow B]$ denotes the $S$-sorted functions from $A$ to $B$. All these notions are explained in detail in [14,28]. The material below is adapted from [25,26,29].

**Order-Sorted Algebra and $E$-Unification.** An OS *equational theory* is a pair $T = (\Sigma, E)$, with $E$ a set of (possibly conditional) $\Sigma$-equations. $\mathbf{OSAlg}_{(\Sigma,E)}$

denotes the full subcategory of $\mathbf{OSAlg}_{\Sigma}$ with objects those $A \in \mathbf{OSAlg}_{\Sigma}$ such that $A \models E$, called the $(\Sigma, E)$-*algebras*. $\mathbf{OSAlg}_{(\Sigma, E)}$ has an *initial algebra* $T_{\Sigma/E}$ [28]. If $\mathcal{E} = (\Sigma, E)$, $T_{\mathcal{E}}$ abbreviates $T_{\Sigma/E}$. $Form(\Sigma)$ denotes the set of first-order $\Sigma$-formulas. For $T = (\Sigma, E)$ and $\varphi \in Form(\Sigma)$, we call $\varphi$ $T$-*valid*, written $E \models \varphi$, iff $A \models \varphi$ for all $A \in \mathbf{OSAlg}_{(\Sigma, E)}$. The inference system in [28] is *sound and complete* for OS equational deduction, i.e., for any OS equational theory $(\Sigma, E)$, and $\Sigma$-equation $u = v$ we have an equivalence $E \vdash u = v \iff E \models u = v$. Deducibility $E \vdash u = v$ is abbreviated as $u =_E v$, called $E$-*equality*. An $E$-*unifier* of a system of $\Sigma$-equations, i.e., of a conjunction $\phi = u_1 = v_1 \wedge \ldots \wedge u_n = v_n$ of $\Sigma$-equations, is a substitution $\sigma$ such that $u_i\sigma =_E v_i\sigma$, $1 \leqslant i \leqslant n$. An $E$-*unification algorithm* for $(\Sigma, E)$ is an algorithm generating a *complete set* of $E$-unifiers $Unif_E(\phi)$ for any system of $\Sigma$ equations $\phi$, where "complete" means that for any $E$-unifier $\sigma$ of $\phi$ there is a $\tau \in Unif_E(\phi)$ and a substitution $\rho$ such that $\sigma =_E (\tau\rho)|_{dom(\sigma) \cup dom(\tau)}$, where $=_E$ here means that for any variable $x$ we have $x\sigma =_E x(\tau\rho)|_{dom(\sigma) \cup dom(\tau)}$. The algorithm is *finitary* if it always terminates with a *finite set* $Unif_E(\phi)$ for any $\phi$. Given a set of equations $B$ used for deduction modulo $B$, a preregular OS signature $\Sigma$ is called $B$-*preregular*[1] iff for each $u = v \in B$ and substitutions $\rho$, $ls(u\rho) = ls(v\rho)$.

**Convergent Theories and Constructors.** Given an order-sorted equational theory $\mathcal{E} = (\Sigma, E \cup B)$, where $B$ is a collection of associativity and/or commutativity and/or identity axioms and $\Sigma$ is $B$-preregular, we can associate to it a corresponding *rewrite theory* [27] $\vec{\mathcal{E}} = (\Sigma, B, \vec{E})$ by orienting the equations $E$ as left-to-right rewrite rules. That is, each $(u = v) \in E$ is transformed into a rewrite rule $u \rightarrow v$. For simplicity we recall here the case of unconditional equations; for how conditional equations (whose conditions are conjunctions of equalities) are likewise transformed into conditional rewrite rules see, e.g., [24]. The main purpose of the rewrite theory $\vec{\mathcal{E}}$ is to reduce the complex bidirectional reasoning with equations to the much simpler unidirectional reasoning with rules under suitable assumptions. We assume familiarity with the notion of subterm $t|_p$ of $t$ at a term position $p$ and of term replacement $t[w]_p$ of $t|_p$ by $w$ at position $p$ (see, e.g., [8]). The rewrite relation $t \rightarrow_{\vec{E}, B} t'$ holds iff there is a subterm $t|_p$ of $t$, a rule $(u \rightarrow v) \in \vec{E}$ and a substitution $\theta$ such that $u\theta =_B t|_p$, and $t' = t[v\theta]_p$. We denote by $\rightarrow^*_{\vec{E}, B}$ the reflexive-transitive closure of $\rightarrow_{\vec{E}, B}$. The requirements on $\vec{\mathcal{E}}$ allowing us to reduce equational reasoning to rewriting are the following: (i) $vars(v) \subseteq vars(u)$; (ii) *sort-decreasingness*: for each substitution $\theta$ we have $ls(u\theta) \geqslant ls(v\theta)$; (iii) *strict $B$-coherence*: if $t_1 \rightarrow_{\vec{E}, B} t_1'$ and $t_1 =_B t_2$ then there exists $t_2 \rightarrow_{\vec{E}, B} t_2'$ with $t_1' =_B t_2'$; (iv) *confluence* (resp. *ground confluence*) modulo $B$: for each term $t$ (resp. ground term $t$) if $t \rightarrow^*_{\vec{E}, B} v_1$ and $t \rightarrow^*_{\vec{E}, B} v_2$, then there exist rewrite sequences $v_1 \rightarrow^*_{\vec{E}, B} w_1$ and $v_2 \rightarrow^*_{\vec{E}, B} w_2$

---

[1] If $B = B_0 \uplus U$, with $B_0$ associativity and/or commutativity axioms, and $U$ identity axioms, the $B$-preregularity notion can be *broadened* by requiring only that: (i) $\Sigma$ is $B_0$-preregular in the standard sense that $ls(u\rho) = ls(v\rho)$ for all $u = v \in B_0$ and substitutions $\rho$; and (ii) the axioms $U$ oriented as rules $\vec{U}$ are *sort-decreasing* in the sense explained below.

such that $w_1 =_B w_2$; (v) *termination*: the relation $\to_{\vec{E},B}$ is well-founded (for $\vec{E}$ conditional, we require *operational termination* [24]). If $\vec{\mathcal{E}}$ satisfies conditions (i)–(v) (resp. the same, but (iv) weakened to ground confluence modulo $B$), then it is called *convergent* (resp. *ground convergent*). The key point is that then, given a term (resp. ground term) $t$, all terminating rewrite sequences $t \to^*_{\vec{E},B} w$ end in a term $w$, denoted $t!_{\vec{\mathcal{E}}}$, that is unique up to $B$-equality, and its called $t$'s *canonical form*. Three major results then follow for the ground convergent case: (1) for any ground terms $t, t'$ we have $t =_{E \cup B} t'$ iff $t!_{\vec{\mathcal{E}}} =_B t'!_{\vec{\mathcal{E}}}$, (2) the $B$-equivalence classes of canonical forms are the elements of the *canonical term algebra* $C_{\Sigma/E,B}$, where for each $f : s_1 \ldots s_n \to s$ in $\Sigma$ and $B$-equivalence classes of canonical terms $[t_1], \ldots, [t_n]$ with $ls(t_i) \leqslant s_i$ the operation $f_{C_{\Sigma/E,B}}$ is defined by the identity: $f_{C_{\Sigma/E,B}}([t_1] \ldots [t_n]) = [f(t_1 \ldots t_n)!_{\vec{\mathcal{E}}}]$, and (3) we have an isomorphism $T_{\mathcal{E}} \cong C_{\Sigma/E,B}$.

A ground convergent rewrite theory $\vec{\mathcal{E}} = (\Sigma, B, \vec{E})$ is called *sufficiently complete* with respect to a subsignature $\Omega$, whose operators are then called *constructors*, iff for each ground $\Sigma$-term $t$, $t!_{\vec{\mathcal{E}}} \in T_\Omega$. Furthermore, for $\vec{\mathcal{E}} = (\Sigma, B, \vec{E})$ sufficiently complete w.r.t. $\Omega$, a ground convergent rewrite subtheory $(\Omega, B_\Omega, \vec{E}_\Omega) \subseteq (\Sigma, B, \vec{E})$ is called a *constructor subspecification* iff $T_{\mathcal{E}}|_\Omega \cong T_{\Omega/E_\Omega \cup B_\Omega}$. If $E_\Omega = \varnothing$, then $\Omega$ is called a signature of *free constructors modulo axioms $B_\Omega$*. Note that $\vec{\mathcal{E}} = (\Sigma, B, \vec{E})$ is sufficiently complete with respect to $\Omega$ iff each ground $\Sigma$-term $f(u_1, \ldots, u_n)$ with $f \in \Sigma \backslash \Omega$ and $u_i \in T_\Omega$, $1 \leqslant i \leqslant n$, is $\vec{E}, B$-reducible, i.e., $f(u_1, \ldots, u_n) \to_{\vec{E},B} t$ for some $t \in T_\Sigma$.

**Generator Sets** generalize standard structural induction on the constructors of a sort. They are particularly useful for inductive reasoning when constructors obey structural axioms $B$ like associativity or associativity-commutativity for which structural induction may be ill-suited. A *generator set* for a sort $s$ is a set of constructor terms of sort $s$ or smaller such that, up to $B$-equality, any ground constructor term of sort $s$ is a ground substitution instance of one of the patterns in the generator set. Here is the general definition (identity axioms are not needed thanks to the theory transformation[2] $\vec{\mathcal{E}} \mapsto \vec{\mathcal{E}}_U$ in [9]):

**Definition 1.** *For $\Omega$ an order-sorted signature of constructors which may satisfy axioms $B$ of associativity and/or commutativity, and $s$ a sort in $\Omega$, a $B$-generator set for sort $s$ is a finite set of terms $\{u_1, \ldots, u_k\}$, with $u_1, \ldots, u_k \in T_\Omega(X)_s$ and such that*

$$T_{\Omega/B,s} = \bigcup_{1 \leqslant i \leqslant k} \{[u_i \, \rho] \in T_{\Omega/B,s} \mid \rho \in [X \to T_\Omega]\}.$$

**Checking the Correctness of Generator Sets.** How do we *know* that a proposed generator set $\{u_1, \ldots, u_k\}$ it truly one modulo axioms $B$ for a given

---

[2] If a theory $\vec{\mathcal{G}}$ has axioms $B \uplus U$, with $B$ associative and/or commutative axioms and $U$ unit element axioms, then the axioms $U$ can be eliminated by turning them into rules $\vec{U}$ by means of the semantics-preserving theory transformation $\vec{\mathcal{G}} \mapsto \vec{\mathcal{G}}_U$ defined in [9], so that the axioms of the semantically equivalent $\vec{\mathcal{G}}_U$ are just $B$.

sort $s$ and constructors $\Omega$? Assuming that the terms $u_1, \ldots, u_k$ are all *linear*, i.e., have no repeated variables—which is the usual case for generator sets— this check can be reduced to an automatic *sufficient completeness check* with Maude's Sufficient Completeness Checker (SCC) tool [19], which is based on tree automata decision procedures modulo axioms $B$. The reduction is extremely simple: define a new unary predicate $s : s \to Bool$ with equations $s(u_i) = true$, $1 \leqslant i \leqslant k$. Then, $\{u_1, \ldots, u_k\}$ is a correct generator set for sort $s$ modulo $B$ for the constructor signature $\Omega$ iff the predicate $s$ is sufficiently complete, which can be automatically checked by the SCC tool. Furthermore, if $\{u_1, \ldots, u_k\}$ is *not* a generator set for sort $s$, the SCC tool will output a useful counterexample.

**Inductive Theorem Proving.** An inductive theorem prover implements a sound inference system to prove *inductive theorems* $\varphi$ in a given equational theory $\mathcal{E}$, i.e., formulas $\varphi$ such that $T_{\mathcal{E}} \models \varphi$. Although the methods I present for proving sufficient completeness by inductive theorem proving do not depend on the given inductive inference system, the examples presented in Sect. 5 will use an extended version of the inductive inference system for order-sorted equational specifications presented in [26].

# 3    A Hierarchical Methodology

I present a hierarchical methodology to prove sufficient completeness by inductive theorem proving. This methodology uses the same assumptions about the input theory and about theory hierarchies as those in a similar hierarchical method for proving ground confluence and sufficient completeness presented in the companion paper [25], which is not based on the use of a standard inductive theorem prover. Thus, the two methods help each other and can profitably be used in combination. Since this section focuses on the assumptions and infrastructure common to both methods, it follows closely the presentation in [25].

**Basic Assumptions About $\mathcal{E}$.** We assume throughout a, possibly conditional, equational theory $\mathcal{E} = (\Sigma, E \cup U \cup B)$ such that: (i) $\Sigma$ decomposes as a disjoint union $\Sigma = \Delta \uplus \Omega$, where $\Omega$ are the—intended but not yet proved to be— constructor symbols that furthermore are free modulo $B$ and $\Delta$ are the intended defined symbols; (ii) $B$ is any combination of associativity and/or commutativity axioms[3], but any binary $f \in \Delta$ may not satisfy any axioms except commutativity[4]; (iii) $\vec{U}$ are sort-decreasing unit axiom rules of the form $c(e, x) \to x$ or

---

[3]  For any $f$ that is commutative we always assume a top typing $f : s\,s \to s_0$ with all other typings of the form $f : s'\,s' \to s_0'$, with $s \leqslant s'$, $s_0 \leqslant s_0'$. Regarding the absence of unit element axioms, they are precisely the equations $U$, that will be used as rules $\vec{U}$ (see, e.g., Example 4, and Footnote 2). I.e., our results apply as well to theories $\mathcal{G}$ with axioms $B \uplus U$ such that $\mathcal{G}_U$ has the properties (i)–(vi) listed in what follows.

[4]  Since axioms $B$ are primarily used to specify constructor data structures, in actual practice, limiting axioms for defined symbols is a mild restriction. Furthermore, as explained in [25, Footnote 3], this restriction on axioms can be lifted *a posteriori* by further inductive theorem proving.

$c(x, e) \rightarrow x$, where $c$ is a constructor name and $e$ is an $\Omega$-term. However, $c(x, e)$ is not an $\Omega$-term because some of $c$'s type declarations do not belong to $\Omega$ (see Examples 4 and 5 in Sect. 5). This makes it possible for constructors to be free modulo $B$ in spite of such unit rules; (iv) $E = \bigcup_{f \in \Delta} E_f$, where for each $f \in \Delta$, its associated rewrite rules $\vec{E}_f$ are sort-decreasing and have the form: $\vec{E}_f = \{[i] : f(\vec{u_i}) \rightarrow r_i \;\; if \;\; \Gamma_i\}_{i \in I}$ such that: (a) the $\vec{u_i}$ are $\Omega$-terms; and (b) for each $i \in I$, $\Gamma_i = \bigwedge_{j \in J} w_j = w'_j$ and $vars(f(\vec{u_i})) \supseteq vars(r_i) \cup vars(\Gamma_i)$. (v) There is a $B$-compatible recursive path order (RPO) $>$ (see [8]) such that for each $i \in I$, $f(\vec{u_i}) > r_i$ and, for $j \in J$, $f(\vec{u_i}) > w_j$ and $f(\vec{u_i}) > w'_j$, which makes the rules $\vec{E} \cup \vec{U}$ operationally terminating modulo $B$. (vi) The rules $\vec{E} \cup \vec{U}$ are strictly $B$-coherent.

The main goal of this paper is to develop a hierarchical method based on inductive theorem proving to prove that a theory $\vec{\mathcal{E}}$ enjoying properties (i)–(vi) above is sufficiently complete with respect to a constructor signature $\Omega$. We first need to consider theory hierarchies based on the "call graph" of $\vec{\mathcal{E}}$.

**Call Graph and Theory Hierarchies**. We assume that all function symbols in $\Delta$ are *subsort-overloaded*, i.e., for any $f : s_1 \ldots s_n \rightarrow s$ and $f : s'_1 \ldots s'_n \rightarrow s'$ we have $[s] = [s']$, and $[s_i] = [s'_i]$, $1 \leqslant i \leqslant n$. This can always be achieved by renaming any "ad-hoc overloaded" symbols—i.e., symbols $f$ with typings $f : s_1 \ldots s_n \rightarrow s$ and $f : s'_1 \ldots s'_n \rightarrow s'$ failing the above condition—that might exist in $\Delta$. Let $F_\Delta$ be the set of *names* for the function symbols in $\Delta$, disregarding their typing. The *calling relation* is a binary relation $C$ on $F_\Delta$, where for each $f, g \in F_\Delta$, $(f, g) \in C$ iff there exists a rule $f(\vec{u_i}) \rightarrow r_i \;\; if \;\; \Gamma_i$ in $\vec{E}_f$ such that the function symbol $g$ occurs in either $r_i$ or in $\Gamma_i$. Let $C^*$ denote the reflexive-transitive closure of $C$, and $\equiv_C$ the equivalence relation on $F_\Delta$ defined by the equivalence: $f \equiv_C g$ iff $f C^* g$ and $g C^* f$. Then, the quotient set $F_\Delta / \equiv_C$ has an associated partial order defined by the equivalence $[f] \geqslant [g] \Leftrightarrow f C^* g$.

The hierarchical method we propose is based on a *hierarchy of theory inclusions* chosen as follows. Given our theory $\vec{\mathcal{E}}$ we: (i) identity a subtheory $\vec{\mathcal{E}}_0$ having subsignature $\Delta_0 \uplus \Omega$ containing all the subsort-overloaded typings of any $f \in \Delta_0$ and having rules $\vec{U} \cup \vec{E}_0$, with $\vec{E}_0 = \bigcup_{f \in \Delta_0} \vec{E}_f$, and axioms $B_0 = B_{\Delta_0 \uplus \Omega} = \bigcup_{f \in \Delta_0 \uplus \Omega} B_f$, where $B_f$ are the associative and/or commutative axioms, if any, for $f$ in $B$, and such that $\vec{\mathcal{E}}_0 \cup \vec{U}$ is sufficiently complete with respect to $\Omega$ and ground convergent. Of course, we should choose $\vec{\mathcal{E}}_0$ as big as possible: in the worse case we may have $\vec{E}_0 = \varnothing$ and keep only $\vec{U}$. We furthermore assume that we can find a sequence of theory inclusions:

$$\vec{\mathcal{E}}_0 \subset \vec{\mathcal{E}}_1 \subset \ldots \vec{\mathcal{E}}_{n-1} \subset \vec{\mathcal{E}}_n$$

such that: (a) $\vec{\mathcal{E}}_n \doteq \vec{\mathcal{E}}$, (b) each $\vec{\mathcal{E}}_k$ has signature $\Delta_k \uplus \Omega$ containing all subsort-overloaded typings of any $f \in \Delta_k$ and having axioms $B_k = B_{\Delta_k \uplus \Omega}$ and, besides $\vec{U}$, rules $\vec{E}_k = \bigcup_{f \in \Delta_k} \vec{E}_f$, where for each $k \geqslant 1$ and each rule $[i] : f(\vec{u_i}) \rightarrow r_i \;\; if \;\; \Gamma_i$ in $\vec{E}_f$, the condition $\Gamma_i$ is a conjunction of $\vec{\mathcal{E}}_{k-1}$-equalities, (c) for each $0 \leqslant k < k + 1 \leqslant n$, there exists a function symbol $g \in F_{\Delta_{k+1}} \backslash F_{\Delta_k}$ such that

$F_{\Delta_{k+1}} = F_{\Delta_k} \uplus [g]$; that is, we add all the symbols in a new $\equiv_C$-equivalence class $[g]$ to climb up each step in the theory hierarchy.

**The Hierarchical Proof Method.** All now boils down to finding proof methods to *climb up the theory hierarchy one step at a time*. We then repeat this method $n$ times, with $n$ the length of the chain of theory inclusions. That is, we focus on a single theory inclusion $\vec{\mathcal{E}}_0 \subset \vec{\mathcal{E}}$, where $\vec{\mathcal{E}}_0$ has already been proved ground convergent and sufficiently complete with respect to the constructor signature $\Omega$, and then prove that $\vec{\mathcal{E}}$ is also ground convergent and sufficiently complete as follows. First of all, we define a new theory $\vec{\mathcal{E}}^\Delta$, with the same rules $\vec{E} \cup \vec{U}$ as in $\vec{\mathcal{E}}$, and having also a theory inclusion $\vec{\mathcal{E}}_0 \subset \vec{\mathcal{E}}^\Delta$, but where, if $\Sigma_0$ and $\Sigma$ are the respective signatures of $\vec{\mathcal{E}}_0$ and $\vec{\mathcal{E}}$, and $\Delta = \Sigma \backslash \Sigma_0$, then $\vec{\mathcal{E}}^\Delta$ has a signature $\Sigma^\Delta$ that extends $\Sigma_0$ by: (i) adding to each connected component of the poset of sorts $(S, \leqslant)$ of $\Sigma_0$ the kind $[s]$ as a new top sort, i.e., $\forall s' \in [s]$, $s' \leqslant [s]$, and (ii) lifting to the kind levels all $f \in \Sigma$. That is, we extend the function symbols of $\Sigma_0$ by adding for each $f : s_1 \ldots s_n \to s$, $n \geqslant 1$, in $\Sigma$ a function symbol $f : [s_1] \ldots [s_n] \to [s]$ to $\Sigma^\Delta$. In $\vec{\mathcal{E}}^\Delta$ the axioms $B$ are lifted to kinds. Note that $\Sigma^\Delta$ *adds no new terms* to the original sorts $s \in S$, i.e., $T_{\Sigma^\Delta}(X)_s = T_{\Sigma_0}(X)_s$. For three concrete examples of the $\vec{\mathcal{E}}^\Delta$ construction, see modules OE-DELTA, NAT-PRESBURGER-DELTA and and MULTISET-ALGEBRA-DELTA in Sect. 5. The hierarchical proof methodology then proceeds as follows:

1. We first prove that $\vec{\mathcal{E}}^\Delta$ is ground convergent.
2. We then prove that for any $f \in \Delta$, maximal typing $f : s_1, \ldots, s_n \to s$ and ground constructor substitution $\rho$, the term $f(x_1, \ldots, x_n)\rho$, with $x_i$ of sort $s_i$, $1 \leqslant i \leqslant n$, can be rewritten with some rule in $\vec{\mathcal{E}}^\Delta$.
3. (1) and (2) actually *prove* that $\vec{\mathcal{E}}$ is ground convergent and sufficiently complete with respect to $\Omega$.

Methods for proving (1), as well as a proof that (1) and (2) imply (3) can be found in [25]. In this paper we focus on a new method that reduces proving (2) to a proof by standard inductive theorem proving.

## 4  Proving Sufficient Completeness Inductively

The reduction of sufficient completeness proofs to inductive proofs is based on a new general theory transformation $\vec{\mathcal{E}} \mapsto \vec{\mathcal{E}}$: defined below.

**The $\vec{\mathcal{E}} \mapsto \vec{\mathcal{E}}$: Transformation.** We assume a ground convergent and possibly conditional theory (with rules having no extra variables in their righthand side or condition) $\vec{\mathcal{E}} = (\Sigma, B, \vec{E})$ such that: (i) $B$ are associativity and/or commutativity axioms and $\Sigma$ is $B$-preregular; (ii) all its function symbols are subsort-overloaded; (iii) its poset of sorts $(S, <)$ is such that each connected component $[s]$ of $S$ has a top sort, which we denote $\top_{[s]} \in [s]$; (iv) any $f : s_1 \ldots s_n \to s$, $n \geqslant 1$, in $\Sigma$ has also a typing $f : \top_{[s_1]} \ldots \top_{[s_n]} \to \top_{[s]}$; (v) there is a $B$-compatible RPO $>$ making $\vec{\mathcal{E}}$ operationally terminating. In what follows $\vec{\mathcal{E}}$ will always be a theory of the form $\vec{\mathcal{E}}^\Delta$ as defined in the methodology of Sect. 3. The transformation $\vec{\mathcal{E}} \mapsto \vec{\mathcal{E}}$: maps $\vec{\mathcal{E}} = (\Sigma, B, \vec{E})$ to $\vec{\mathcal{E}} := (\Sigma:, B, \vec{E} \cup \vec{E}_\Sigma)$, where:

- $\Sigma_:$ extends $\Sigma$ by adding: (i) a fresh new sort *Pred* of predicates with a constant *tt* not related to any other sort in the subsort order; and (ii) for each connected component $[s]$ of the sort poset $(S, <)$ in $\Sigma$ and each $s' \in [s]$ a unary function symbol $\_ : s' : \top_{[s]} \to Pred$ called a *sort predicate*.
- the set $\vec{E}_\Sigma$ of rewrite rules contains: (i) for each $s \in S$ a rule,

$$x : s \to tt$$

where $x$ is a variable of sort $s$; (ii) for each $f : s_1 \ldots s_n \to s$ in $\Sigma$, $n \geqslant 0$, a rule,

$$f(x_1, \ldots, x_n) : s \to tt \quad if \quad x_1 : s_1 = tt \wedge \ldots \wedge x_n : s_n = tt$$

where $x_i$ has sort $\top_{[s_i]}$, $1 \leqslant i \leqslant n$; (iii) for each $s < s'$ in $(S, <)$ a rule,

$$x : s' \to tt \quad if \quad x : s = tt.$$

where $x$ has sort $\top_{[s]}$.

Two key results about the $\vec{\mathcal{E}} \mapsto \vec{\mathcal{E}}_:$ transformation include:

**Theorem 1.** *Under the assumptions on $\vec{\mathcal{E}}$, $\vec{\mathcal{E}}_:$ is operationally terminating.*

**Theorem 2.** *Under the assumptions on $\vec{\mathcal{E}}$, $\vec{\mathcal{E}}_:$ is ground convergent. Furthermore, for each $s \in S$ and $t \in T_{\Sigma, \top_{[s]}}$ we have the equivalences:*

$$(t:s)!_{\vec{\mathcal{E}}_:} = tt \Leftrightarrow s \geqslant ls(t!_{\vec{\mathcal{E}}}) \quad and \quad (t:s)!_{\vec{\mathcal{E}}_:} = t!_{\vec{\mathcal{E}}} : s \Leftrightarrow s \not\geqslant ls(t!_{\vec{\mathcal{E}}}).$$

**Reducing Sufficient Completeness Checking to Inductive Reasoning.** As already mentioned our focus of interest is in the transformation: $\vec{\mathcal{E}}^\Delta \mapsto \vec{\mathcal{E}}_:^\Delta$, which will give us the desired reduction of sufficient completeness checking to inductive theorem proving. Here is the main theorem:

**Theorem 3.** *Let $\vec{\mathcal{E}}_0 \subset \vec{\mathcal{E}}$ and $\vec{\mathcal{E}}_0 \subset \vec{\mathcal{E}}^\Delta$ be theory inclusions satisfying the assumptions in Sect. 3, where $\vec{\mathcal{E}}_0$ is sufficiently complete with respect to $\Omega$ and $\vec{\mathcal{E}}^\Delta$ has already been proved ground convergent. Then, for all $f \in \Delta$, maximal typing $f : s_1, \ldots, s_n \to s$ of $f$, and all ground constructor substitutions $\rho$, the term $f(x_1, \ldots, x_n)\rho$, with $x_i$ of sort $s_i$, $1 \leqslant i \leqslant n$, can be rewritten with some rule in $\vec{\mathcal{E}}^\Delta$ if for $F_\Delta = \{f_1, \ldots f_k\}$, and each maximal typing $f_j : s_1^i \ldots s_{n_j}^i \to s_i$, $i \in I_j$, for each $f_j$, $1 \leqslant j \leqslant k$, we have,*

$$T_{\vec{\mathcal{E}}:\Delta} \models \bigwedge_{1 \leqslant j \leqslant k, i \in I_j} f_j(x_1^i, \ldots, x_{n_j}^i) : s_i = tt$$

*where $x_m^i$ has sort $s_m^i$, $1 \leqslant m \leqslant n_j$.*

The reason for proving the above conjunction as an inductive theorem, as opposed to proving each conjunct $f_j(x_1^i, \ldots, x_{n_j}^i) : s_i = tt$ separately, is that, since $F_\Delta$ is one of the nodes in the calling graph of $\vec{\mathcal{E}}$, the symbols in $F_\Delta$ call each other, typically due to some mutual recursion. Therefore, it may be considerably easier to prove the entire conjunction than proving each conjunct separately. Let us see a simple example.

*Example 1.* (Odd and Even). Let $\mathcal{OE}$ be the theory of the Peano Naturals with sort *Nat*, which imports the Booleans, has constructors 0 and $s : Nat \to Nat$, and with $\Delta$ the predicates *odd* : $Nat \to Bool$ and *even* : $Nat \to Bool$, which are defined by rules: $odd(0) \to false$, $even(0) \to true$, $odd(s(n)) \to \neg even(n)$, and $even(s(n)) \to \neg odd(n)$, so that the calling graph has the single equivalence class node $\{odd, even\}$. Then, to prove sufficient completeness with the above method we just need to prove in $\mathcal{OE}{:}^{\Delta}$ the inductive theorem:

$$odd(n){:}Nat \wedge even(n){:}Nat$$

Theorem 3 has a quite useful corollary for finite sorts.

**Corollary 1.** *Under the assumptions and notation in Theorem 3, if a maximal typing in* $\Delta$, *say,* $f_j : s_1^i \ldots s_{n_j}^i \to s_i$ *is such that* $T_{\Omega/B,s_i}$ *is a finite set, say,* $T_{\Omega/B,s_j} = \{[u_1], \ldots, [u_m]\}$, *then Theorem 3 still holds replacing the conjunct* $f_j(x_1^i, \ldots, x_{n_j}^i){:}s_i = tt$ *by the disjunction:*

$$\bigvee_{1 \leqslant k \leqslant m} f_j(x_1^i, \ldots, x_{n_j}^i) = u_k.$$

Theorem 3 and Corollary 1 provide the desired reduction of sufficient completeness checking for $\vec{\mathcal{E}}$ to inductive theorem proving in $\vec{\mathcal{E}}{:}^{\Delta}$. In fact, if all operators in $\Delta$ have finite sorts, the conjunction of disjunctions in Corollary 1 can be inductively proved just in $\vec{\mathcal{E}}^{\Delta}$.

*Example 2.* (Odd and Even Revisited). From the above corollary it follows that to prove sufficient completeness of $\mathcal{OE}$ it is enough to prove that

$$(odd(n) = true \vee odd(n) = false) \wedge (even(n) = true \vee even(n) = false)$$

is an inductive theorem of $\mathcal{OE}^{\Delta}$.

## 5  Some Examples

As a warmup exercise, let us prove sufficient completeness of the $\mathcal{OE}$ theory in Example 1. Of course, since the equations are left-linear, a different, automatic proof based on tree automata can be given for $\mathcal{OE}$ using, for example, Maude's SCC tool [19]. But one can easily find similar mutually recursive function definitions—for example, conditional ones—outside the scope of automated tools.

*Example 3.* In Maude, the theory $\mathcal{OE}^{\Delta}$ can be specified as follows:

```
fmod OE-DELTA is protecting BOOL-OPS .
  sort Nat .
  op 0 : -> Nat [ctor] .
  op s : Nat -> Nat [ctor] .
  op odd : [Nat] -> [Bool] .
```

```
op even : [Nat] -> [Bool] .
var n : Nat .

eq odd(0) = false .
eq odd(s(n)) = not(even(n)) .
eq even(0) = true .
eq even(s(n)) = not(odd(n)) .
endfm
```

where [Nat] and [Bool] are the "kind" supersorts automatically added by Maude above, respectively, Nat and Bool, and therefore need not be declared. The theory BOOL-OPS can be trivially shown to be sufficiently complete by truth table inspection, so in this case the theory $\mathcal{OE}_0$ in the inclusion $\mathcal{OE}_0 \subset \mathcal{OE}^\Delta$ is just BOOL-OPS together with the constructors $\{0, s\}$. Also, OE-DELTA can easily be checked to be confluent and terminating. As pointed out in Example 2, we just need to prove that:

$$(odd(n) = true \lor odd(n) = false) \land (even(n) = true \lor even(n) = false)$$

is an inductive theorem of OE-DELTA. We can do so by standard induction on $n$ (which is a special case of the **GSI** rule in[5] [26]). The **Base Case** can be proved automatically by the *Equality Predicate Simplification* rule (**EPS**) in [26]. For the **Induction Step** we have induction hypotheses: $odd(\overline{k}) = true \lor odd(\overline{k}) = false$ and $even(\overline{k}) = true \lor even(\overline{k}) = false$, and need to prove the conjunction:

$$(odd(s(\overline{k})) = true \lor odd(s(\overline{k})) = false) \land (even(s(\overline{k})) = true \lor even(s(\overline{k})) = false)$$

where $\overline{k}$ is a fresh constant of sort *Nat*. Using the **EPS** simplification rule this goal reduces to:

$$(\neg(even(\overline{k})) = true \lor \neg(even(\overline{k})) = false) \land (\neg(odd(\overline{k})) = true \lor \neg(odd(\overline{k})) = false)$$

Applying the *Split* rule ( **SP**) in [26] with the induction hypothesis disjunction $even(\overline{k}) = true \lor even(\overline{k}) = false$ we get subgoals:

$even(\overline{k}) = true \rightarrow$

$$(\neg(even(\overline{k})) = true \lor \neg(even(\overline{k})) = false) \land (\neg(odd(\overline{k})) = true \lor \neg(odd(\overline{k})) = false)$$

and

$even(\overline{k}) = false \rightarrow$

$$(\neg(even(\overline{k})) = true \lor \neg(even(\overline{k})) = false) \land (\neg(odd(\overline{k})) = true \lor \neg(odd(\overline{k})) = false)$$

---

[5] The inference rules in [26] have been extended in work submitted for publication. In the examples in the section some rules will be used in their extended form.

Applying the *Inductive Contextual Rewriting* (**ICC**) rule in [26], both of these subgoals automatically simplify to the goal:

$$\neg(odd(\overline{k})) = true \vee \neg(odd(\overline{k})) = false$$

After applying the **SP** rule to this goal with the disjunction hypothesis $odd(\overline{k}) = true \vee odd(\overline{k}) = false$ we again get two goals that can be automatically discharged by means of the **ICC** rule. This finishes the sufficient completeness proof for $\mathcal{OE}$.

I next present a Presburger arithmetic specification for which I am not aware of any proof of sufficient completeness by any other method than the one here.

*Example 4.* (Presburger Arithmetic) Consider the following specification of Presburger Arithmetic for the naturals:

```
fmod NAT-PRESBURGER is protecting TRUTH-VALUE .
  sorts NzNat Nat .
  subsort NzNat < Nat .
  op 0 : -> Nat [ctor] .
  op 1 : -> NzNat [ctor] .
  op _+_ : Nat Nat -> Nat [assoc comm] .
  op _+_ : NzNat NzNat -> NzNat [ctor assoc comm] .
  op _>_ :  Nat Nat -> Bool .

  vars n m : Nat .  var k : NzNat .

  eq n + 0 = n [variant] .

  eq k > 0 =  true [variant] .
  eq n + k > n = true [variant] .
  eq 0 > m = false [variant] .
  eq n > n = false [variant] .
  eq n > n + m = false [variant] .
endfm
```

The equations are size-decreasing and therefore clearly terminating. It is easy to check that they are also confluent using Maude's Church-Rosser Checker [10]. Furthermore, these equations enjoy the finite variant property [7,12]; this can be easily checked in Maude by the method proposed in [4]. Therefore, $E \cup B$-unification in this theory is finitary [12]. The [variant] attribute allows Maude to use this knowledge to compute $E \cup B$-unifiers. Note that the lefthand sides of the second, fourth and fifth rules for $>$ are non-linear. This places the sufficient completeness checking problem outside the scope of equational tree automata techniques such as those used in [19]. That the equations defining $>$ are sufficiently complete is intuitively clear; but giving a formal proof in a suitable inference system is a different matter. Here is where the new inductive methodology is helpful. Of course, calling $\vec{\mathcal{E}}$ the entire theory, and $\vec{\mathcal{E}_0}$ the theory obtained by dropping the $>$ operator, we get a subtheory $\vec{\mathcal{E}_0}$ that is convergent and sufficiently complete. This is intuitively clear, since any term of sort Nat is either

0, and thus a constructor, or is of the form $1 + .^n. +1$, and therefore a constructor term. This can be automatically checked with Maude's SCC tool [19]. Our theory $\vec{\mathcal{E}}^{\Delta}$ is then:

```
fmod NAT-PRESBURGER-DELTA is protecting TRUTH-VALUE .
  sorts NzNat Nat .
  subsort NzNat < Nat .
  op 0 : -> Nat [ctor] .
  op 1 : -> NzNat [ctor] .
  op _+_ : Nat Nat -> Nat [assoc comm] .
  op _+_ : NzNat NzNat -> NzNat [ctor assoc comm] .
  op _+_ : [Nat] [Nat] -> [Nat] [assoc comm] .
  op _>_ :  [Nat] [Nat] -> [Bool] .

  vars n m : Nat .   var k : NzNat .

  eq n + 0 = n [variant] .

  eq k > 0 =  true [variant] .
  eq n + k > n = true [variant] .
  eq 0 > m = false [variant] .
  eq n > n = false [variant] .
  eq n > n + m = false [variant] .
endfm
```

where $\vec{\mathcal{E}}^{\Delta}$ can easily be checked to be terminating and convergent. Since the sort Bool is finite, we can use Corollary 1 to reduce proving the sufficient completeness of $\vec{\mathcal{E}}$ to proving in $\vec{\mathcal{E}}^{\Delta}$ the inductive theorem:

$$x > y = true \vee x > y = false$$

with $x, y$ of sort Nat. Let us prove this theorem using the inference system proposed in [26]. Using the generator set $\{0, 1, 1 + k\}$ for sort Nat, where $k$ has sort NzNat, we can induct on, say, $x$ and apply the *Generator Set Induction* rule (**GSI**) to get three sub goals: (1) $0 > y = true \vee 0 > y = false$, (2) $1 > y = true \vee 1 > y = false$, and (3) $\overline{k} + 1 > y = true \vee \overline{k} + 1 > y = false$ with induction hypothesis $\overline{k} > y = true \vee \overline{k} > y = false$, where $\overline{k}$ is a fresh constant of sort NzNat.

By applying the *Case* rule (**CAS**) (which is similar to **GSI** but does not add induction hypotheses) to $y$ with the same generator set for Nat, both goals (1) and (2) split into three subgoals each. For example, goal (1) generates subgoals: (1.1) $0 > 0 = true \vee 0 > 0 = false$, (1.2) $0 > 1 = true \vee 0 > 1 = false$, (1.3) $0 > k' + 1 = true \vee 0 > k' + 1 = false$, with $k'$ a fresh variable of sort NzNat. All such subgoals can be automatically discharged by the *Equality Predicate Simplification* rule (**EPS**). For example, the above three subgoals (1.1)–(1.3) can all be discharged by the **EPS** rule using the third rule for $>$. Regarding goal (3), we can use the induction hypothesis $\overline{k} > y = true \vee \overline{k} > y = false$ to apply the *Split* rule (**SP**) with predicate term $\overline{k} > y$ to get subgoals:

$$(3.1) \quad \overline{k} > y = true \rightarrow (\overline{k} + 1 > y = true \vee \overline{k} + 1 > y = false)$$

$$(3.2) \quad \overline{k} > y = \textit{false} \rightarrow (\overline{k} + 1 > y = \textit{true} \lor \overline{k} + 1 > y = \textit{false})$$

The key point to observe is that both goals are clauses whose respective conditions are equations between terms in a theory $\vec{\mathcal{E}}^\Delta$ enjoying the finite variant property, and therefore having a finitary $\vec{\mathcal{E}}^\Delta$-unification algorithm. The "only" problem is that, to perform such a variant unification, we would like to replace the fresh constant $\overline{k}$ by fresh variables $k$ of the same sort. In fact, using an extended version of the *Constructor Variant Unification Left* rule (**CVUL**)— which allows and justifies this replacement of fresh constants by fresh variables— we can do just that.

Applying **CVUL** to (3.1) involves solving the premise equation $k > y = \textit{true}$, which yields two constructor variant unifiers: $\alpha_1 = \{y \mapsto 0\}$, $\alpha_2 = \{k \mapsto p + q, y \mapsto p\}$, where $p, q$ have of sort NzNat. Since the "magic" of the **CVUL** rule involves converting the original and resulting variables that correspond to constants (in this case, $k$, $p$ and $q$) back into fresh constants, we then get the following instantiated conclusions as subgoals:

$$(3.1.1) \quad \overline{k} + 1 > 0 = \textit{true} \lor \overline{k} + 1 > 0 = \textit{false}$$

$$(3.1.2) \quad \overline{p} + \overline{q} > \overline{p} = \textit{true} \lor \overline{p} + \overline{q} > \overline{p} = \textit{false}$$

(3.1.1) (resp. (3.1.2)) can be immediately discharged by **EPS** simplification using the first equation (resp. the second equation) for $>$.

We can likewise apply **CVUL** to (3.2) by solving the premise equation $k > y = \textit{false}$, which yields the single constructor variant unifier $\beta = \{y \mapsto k + z\}$, where $z$ has sort Nat. We then get the instantiated conclusion:

$$(3.2.1) \quad \overline{k} + 1 > \overline{k} + z = \textit{true} \lor \overline{k} + 1 > \overline{k} + z = \textit{false}$$

which can be discharged by applying the **CAS** rule to variable $z$—with the same generator set for sort Nat—followed by **EPS** simplification.

This finishes the proof of sufficient completeness for the above Presburger arithmetic specification. This formal proof is important because, together with the fact that the constructors are free modulo associativity and commutativity, it ensures that we can use this specification to decide quantifier-free Presburger arithmetic formulas using the *variant satisfiability* algorithm in [29].

The previous example has illustrated the fact that proving the sufficient completeness of *unconditional* theories when lefthand sides of rules are *nonlinear* can be nontrivial. The case of *conditional* theories can be even harder. The following example, besides illustrating the application of these method to conditional theories for a function whose result sort is infinite, does also illustrate that the hierarchical framework used here is the same as the one used in the companion paper [25], where a theory of multisets of numbers was used as a running example. The example below illustrates that the method for proving sufficient completeness in [25]—based on *constrained patterns*– and the one presented here *complement each other*: one can use either method. In [25], sufficient completeness of the intersection function $\_\cap\_$, which is specified by *conditional* rules, was

184    J. Meseguer

proved. Here I prove it instead by induction, counting on the ground convergence of the corresponding theory $\vec{\mathcal{E}}^{\Delta}$ already proved in [25]. The Maude specification for $\vec{\mathcal{E}}^{\Delta}$ is given below. That for the original theory $\vec{\mathcal{E}}$ is given in Appendix A.

*Example 5.* (Multisets). The theory $\vec{\mathcal{E}}^{\Delta}$ for multisets is as follows:

```
fmod MULTISET-ALGEBRA-DELTA is
  protecting TRUTH-VALUE .
  sorts Nat NeMult Mult .
  subsort Nat < NeMult < Mult .
  op 0 : -> Nat [ctor] .
  op s : Nat -> Nat [ctor] .
  op mt : -> Mult [ctor] .                       *** empty multiset
  op _,_ : Mult Mult -> Mult [assoc comm] .       *** multiset union
  op _,_ : NeMult NeMult -> NeMult [ctor assoc comm] .  *** multiset union
  op _.=._ : Nat Nat -> Bool [comm] .             *** nats equality
  op _in_ : Nat Mult -> Bool .                    *** membership
  op _\_  : Mult Mult -> Mult .                   *** difference

  op _/\_ : [Mult] [Mult] -> [Mult] .             *** intersection
  op s : [Mult] -> [Mult] [ctor] .
  op _,_ : [Mult] [Mult] -> [Mult] [assoc comm] .
  op _.=._ : [Mult] [Mult] -> [Bool] [comm] .
  op _in_ : [Mult] [Mult] -> [Bool] .
  op _\_  : [Mult] [Mult] -> [Mult] .

  vars n m k : Nat .  vars U V W : Mult .

  eq U,mt = U .                                   *** unit equation

  eq n .=. n = true .
  eq 0 .=. s(n) = false .
  ceq s(n) .=. s(m) = false if n .=. m = false .

  eq n in mt = false .
  eq n in n = true .
  ceq n in m = false if (n .=. m) = false .
  eq n in (n,U) = true .
  ceq n in (m,U) = false
          if (n .=. m) = false /\ (n in U) = false .

  eq mt \ U = mt .
  eq U \ mt = U .
  eq m \ m = mt .
  ceq m \ n = m if n .=. m = false .
  eq (m,U) \ m =  U .
  ceq (m,U) \ n = m,(U \ n) if n .=. m = false .
  eq U \ (n,V) = (U \ n) \ V .
```

```
  eq mt /\ V = mt .
  ceq n /\ V = n if (n in V) = true .
  ceq n /\ V = mt if (n in V) = false .
  ceq (n,U) /\ V = n,(U /\ (V \ n)) if (n in V) = true .
  ceq  (n,U) /\ V = U /\ V  if (n in V) = false .
endfm
```

In this case, the subtheory $\mathcal{E}_0$ is everything except the kinds, the operator $\_ \cap \_$ and its defining equations. To show that $\_ \cap \_$ is sufficiently complete, we need to prove in $\vec{\mathcal{E}}{:}^{\Delta}$ the inductive theorem:

$$U \cap V : Mult = tt$$

We apply the **GSI** rule to variable $U$ with generator set $\{\varnothing, x, (y, Q)\}$ with $x, y$ of sort Nat and $Q$ of sort NeMult, and get subgoals: (1) $\varnothing \cap V : Mult = tt$, which is automatically discharged by **EPS** simplification, (2) $x \cap V : Mult = tt$ and (3) $(\overline{y}, \overline{Q}) \cap V : Mult = tt$, with induction hypotheses $\overline{y} \cap V : Mult = tt$ and $\overline{Q} \cap V : Mult = tt$. We can now apply to goal (2) the *Split* rule (**SP**) with predicate term $x \in V$ to get subgoals:

$$(2.1) \quad x \in V = true \rightarrow x \cap V : Mult = tt$$

$$(2.2) \quad x \in V = false \rightarrow x \cap V : Mult = tt$$

which can both be automatically discharged using the **ICC** simplification rule. Likewise, we can apply **SP** to goal (3) with predicate term $\overline{y} \in V$ to get subgoals:

$$(3.1) \quad \overline{y} \in V = true \rightarrow (\overline{y}, \overline{Q}) \cap V : Mult = tt$$

$$(3.2) \quad \overline{y} \in V = false \rightarrow (\overline{y}, \overline{Q}) \cap V : Mult = tt$$

which can both be automatically discharged using the **ICC** simplification rule. The reason why this is so is worth pointing out. The **ICC** rule can use the next-to-last conditional equation defining $\_ \cap \_$ to simplify the consequent of (3.1) to $\overline{y}, (\overline{Q}) \cap (V \backslash \overline{y}) : Mult = tt$; but then it can use the induction hypothesis $\overline{Q} \cap V : Mult = tt$ and the rules $X, Y : Mult \rightarrow tt$ *if* $X : Mult = tt \wedge X : Mult = tt$, $X : Mult \rightarrow tt$ *if* $X : Nat = tt$, and $x : Nat \rightarrow tt$ in $\vec{\mathcal{E}}{:}^{\Delta}$, where $x$ has sort Nat and $X, Y$ have sort [Mult], to further simplify the conclusion to $tt = tt$, thus discharging the goal. In the case of subgoal (3.2), the **ICC** rule just uses the last conditional equation defining $\_ \cap \_$ and then the induction hypothesis $\overline{Q} \cap V : Mult = tt$. This finishes the proof of the sufficient completeness of $\_ \cap \_$.

## 6    Related Work and Conclusions

Research on sufficient completeness goes back to Guttag's thesis in the 1970's and includes, e.g., [1,2,6,16–22,25,30–33]. Four papers most closely related to this work, because all of them deal with order-sorted theories, are [31], [1], [17] and [25]. The work in [31] provides some useful methods for proving sufficient completeness of order-sorted CafeOBJ specifications and shares with this work

the use of module hierarchies; however, the methods used in [31] do not seem
to support rewriting modulo axioms. The work in [1] has several relevant similarities with the present work: (i) it supports conditional order-sorted theories;
and (ii) it emphasizes that proofs of sufficient completeness and of ground confluence help each other. However, [1] does not support rewriting modulo axioms.
The paper [17] did not support rewriting modulo axioms either; but it has two
relevant similarities with the present work: (i) it could analyze specifications in
membership equational logic [3,28], which is also supported by Maude and is
more general than order-sorted equational logic; and (ii) it used a previous version of the Maude Inductive Theorem prover to discharge verification conditions.
In comparison with [17], the implicit similarity is that the $\vec{\mathcal{E}} \mapsto \vec{\mathcal{E}}$: transformation
in this work endows $\vec{\mathcal{E}}$ with membership equational logic reasoning capabilities,
while remaining within the simpler order-sorted framework. The relation with
the companion paper [25] has already been discussed in the body of the paper:
they complement each other.

In conclusion, I have presented a new hierarchical methodology to verify
sufficient completeness by inductive theorem proving and have illustrated it
with three examples. Since order-sorted specifications contain many-sorted and
unsorted ones as special cases, the sufficient completeness proof methods presented here apply in particular to many-sorted and unsorted specifications in
any equational language supporting them. In fact, Example 1 is many-sorted. It
would be highly desirable to combine the inference systems for proving ground
convergence in [11] with those in [25] and in [11] within a tool that would also
support verification of sufficient completeness by the hierarchical methods in [25]
and in this paper. Such a tool would use as a backend the new Maude Inductive Theorem Prover under construction—which supports an extension of the
inference sytem in [26] illustrated in the examples of Sect. 5.

**Acknowledgements.** I cordially thank the anonymous referees for their excellent
suggestions for improving the paper.

## A    Multiset Theory

```
fmod MULTISET-ALGEBRA is
  protecting TRUTH-VALUE .
  sorts Nat NeMult Mult .
  subsort Nat < NeMult < Mult .
  op 0 : -> Nat [ctor] .
  op s : Nat -> Nat [ctor] .
  op mt : -> Mult [ctor] .                         *** empty multiset
  op _,_ : Mult Mult -> Mult [assoc comm] .        *** multiset union
  op _,_ : NeMult NeMult -> NeMult [ctor assoc comm] .  *** multiset union
  op _.=._ : Nat Nat -> Bool [comm] .              *** nats equality
  op _in_ : Nat Mult -> Bool .                     *** membership
  op _\_ : Mult Mult -> Mult .                     *** difference
  op _/\_ : Mult Mult -> Mult .                    *** intersection
```

```
vars n m k : Nat .   vars U V W : Mult .

eq U,mt = U .                                    *** unit equation

eq n .=. n = true .
eq 0 .=. s(n) = false .
ceq s(n) .=. s(m) = false if n .=. m = false .

eq n in mt = false .
eq n in n = true .
ceq n in m = false if (n .=. m) = false .
eq n in (n,U) = true .
ceq n in (m,U) = false
        if (n .=. m) = false /\ (n in U) = false .

eq mt \ U = mt .
eq U \ mt = U .
eq m \ m = mt .
ceq m \ n = m if n .=. m = false .
eq (m,U) \ m = U .
ceq (m,U) \ n = m,(U \ n) if n .=. m = false .
eq U \ (n,V) = (U \ n) \ V .

eq mt /\ V = mt .
ceq n /\ V = n if (n in V) = true .
ceq n /\ V = mt if (n in V) = false .
ceq (n,U) /\ V = n,(U /\ (V \ n)) if (n in V) = true .
ceq (n,U) /\ V = U /\ V if (n in V) = false .
endfm
```

# B   Proofs

**Proof of Theorem** 1. We can extend the RPO order on function symbols making $\vec{\mathcal{E}}$ operationally terminating modulo $B$ by adding, for $(S, <)$ the sort poset in $\vec{\mathcal{E}}$, the new ordered pairs $\_ : s > tt$ for each sort $s \in S$, as well as the pairs $\_ : s > \_ : s' > tt$ for each $s, s' \in S$ such that $s > s'$. It is easy to check that this order makes the new added rules in $\vec{E}_\Sigma$ operationally terminating. Note that this crucially depends on $B$ containing only associativity and/or commutativity axioms by assumption, since any axioms for an identity element for a binary function symbol $f$ could make the rule $f(x_1, x_2) : s \to tt$ *if* $x_1 : s_1 = tt \wedge x_2 : s_2 = tt$ non-terminating.                □

**Proof of Theorem** 2. All terms in $\vec{\mathcal{E}}$ can be rewritten with $\to_{\vec{\mathcal{E}}:}$ iff they can be rewritten with $\to_{\vec{\mathcal{E}}}$, so we only need to prove ground convergence for ground terms of sort *Pred* in $\vec{\mathcal{E}}:$, which are either $tt$, which is in canonical form, or ground terms of the form $t : s$ for some sort $s \in S$. Any rewrite sequence $t : s \to^*_{\vec{\mathcal{E}}:} w$ must

be either of the form: (i) $t:s \to^*_{\vec{\mathcal{E}}} t':s$, or (ii) of the form $t:s \to^*_{\vec{\mathcal{E}}} t':s \to_{\vec{E}_{\Sigma},B} tt$.
In particular, this applies to terminating sequences, so that all canonical forms
of ground terms of sort *Pred* must be of the from (a) $tt$, or of the form (b) $t!_{\vec{\mathcal{E}}}:s$.
$\vec{\mathcal{E}}$: will be ground convergent if we can prove that any ground predicate term $t:s$
has a *unique* canonical form. This follows from the following lemma, whose easy
proof by structural induction on the term structure of $t$ is left to the reader:

**Lemma 1.** *For any ground term $t$ in $\vec{\mathcal{E}}$, $t:s \to_{\vec{E}_{\Sigma},B} tt$ iff $ls(t) \leqslant s$.*

Uniqueness (up to $B$-equality) of the canonical normal form of a ground predicate
term $t:s$ then follows easily from the fact that, since $\vec{\mathcal{E}}$ is ground convergent, then
it is also sort-decreasing, so that for any rewrite sequence $t:s \to^*_{\vec{\mathcal{E}}} t':s \to^*_{\vec{\mathcal{E}}} t!_{\vec{\mathcal{E}}}:s$
we must have $ls(t') \geqslant ls(t!_{\vec{\mathcal{E}}})$. Therefore, either (a) $ls(t!_{\vec{\mathcal{E}}}) \leqslant s$ and the canonical
form of $t:s$ is $tt$, or (b) $ls(t!_{\vec{\mathcal{E}}}) \not\leqslant s$ and the canonical form of $t:s$ must be $t!_{\vec{\mathcal{E}}}:s$,
so that the two equivalences stated in the theorem hold.    □

**Proof of Theorem 3.** For any $f \in \Delta$, maximal typing $f : s_1, \ldots, s_n \to s$ and
ground constructor substitution $\rho$, we need to show that the term $f(x_1, \ldots, x_n)\rho$,
with $x_i$ of sort $s_i$, $1 \leqslant i \leqslant n$, is $\vec{\mathcal{E}}^{\Delta}$-reducible assuming that for $F_{\Delta} = \{f_1, \ldots f_k\}$
and each family $\{f_j : s_1^j \ldots s_{n_j}^j \to s_j\}_{1 \leqslant j \leqslant k}$ of maximal typings in $\Sigma \backslash \Sigma_0$, we
have,

$$T_{\vec{\mathcal{E}}:\Delta} \models \bigwedge_{1 \leqslant j \leqslant k} f_j(x_1^j, \ldots, x_{n_j}^j):s_j = tt$$

where $x_i^j$ has sort $s_i^j$. In particular, we have that $T_{\vec{\mathcal{E}}:\Delta} \models f(x_1, \ldots, x_n):s = tt$.
Suppose that for some ground constructor substitution $\rho$ the term $f(x_1, \ldots, x_n)\rho$
is $\vec{\mathcal{E}}^{\Delta}$-irreducible. Since $\vec{\mathcal{E}}:^{\Delta}$ is ground confluent and $tt$ is irreducible, we then
must have $f(x_1, \ldots, x_n)\rho:s!_{\vec{\mathcal{E}}:\Delta} = tt$. But since $f(x_1, \ldots, x_n)\rho$ is irreducible, this
can only happen if $f(x_1, \ldots, x_n)\rho:s \to_{\vec{E}_{\Sigma}} tt$. But since $ls(f(x_1, \ldots, x_n)\rho) = [s]$
and $[s] > s$, this is impossible by Lemma 1.    □

**Proof of Corollary 1.** Since $\vec{\mathcal{E}}^{\Delta}$ is assumed ground convergent, if a maximal
typing in $\Delta$, say, $f : s_1 \ldots s_n \to s$ is such that $T_{\Omega/B,s}$ is a finite set, say, $T_{\Omega/B,s} = \{[u_1], \ldots, [u_m]\}$, we need to show that if

$$(\ddagger) \quad T_{\vec{\mathcal{E}}\Delta} \models \bigvee_{1 \leqslant i \leqslant m} f(x_1, \ldots, x_n) = u_i.$$

with $x_j$ of sort $s_j$, $1 \leqslant j \leqslant n$, holds, then, replacing the conjunct $f(x_1, \ldots, x_n):$
$s = tt$ by the disjunction $\bigvee_{1 \leqslant i \leqslant m} f(x_1, \ldots, x_n) = u_i$ in the conjunction of Theorem 3, the theorem still holds. We reason by contradiction assuming that the
resulting conjunction after this replacement is an inductive theorem of $T_{\vec{\mathcal{E}}:\Delta}$ but
Theorem 3 fails. In particular, all other conjuncts for the remaining maximal typ-
ings of $\Delta$ are inductive theorems of $T_{\vec{\mathcal{E}}:\Delta}$, so that, by the above proof of Theorem
3, all corresponding ground instances for those operators and their typings are
reducible. Therefore, the theorem can only fail if there is a ground constructor
substitution $\rho$ such that the term $f(x_1, \ldots, x_n)\rho$ is $\vec{\mathcal{E}}^{\Delta}$-irreducible. But since $(\ddagger)$

holds, then by the ground convergence of $\vec{\mathcal{E}}^{\Delta}$ there must be $[u_k] \in \{[u_1], \ldots, [u_m]\}$ such that $(f(x_1, \ldots, x_n)\rho)!_{\vec{\mathcal{E}}^{\Delta}} =_B u_k$. But since $f(x_1, \ldots, x_n)\rho$ is $\vec{\mathcal{E}}^{\Delta}$-irreducible, this is impossible. $\qquad\square$

# References

1. Bouhoula, A.: Simultaneous checking of completeness and ground confluence for algebraic specifications. ACM Trans. Comput. Log. **10**(3), 20:1–20:33 (2009)
2. Bouhoula, A., Jouannaud, J.P.: Automata-driven automated induction. Inf. Comput. **169**(1), 1–22 (2001)
3. Bouhoula, A., Jouannaud, J.P., Meseguer, J.: Specification and proof in membership equational logic. Theor. Comput. Sci. **236**, 35–132 (2000)
4. Cholewa, A., Meseguer, J., Escobar, S.: Variants of variants and the finite variant property. Technical report, CS Department University of Illinois at Urbana-Champaign (2014). http://hdl.handle.net/2142/47117
5. Clavel, M., et al.: All About Maude-A High-Performance Logical Framework. LNCS, vol. 4350. Springer, Heidelberg (2007). https://doi.org/10.1007/978-3-540-71999-1
6. Comon, H.: Sufficient completeness, term rewriting systems and "anti-unification". In: Siekmann, J. (ed.) CADE 1986. LNCS, vol. 230, pp. 128–140. Springer, Heidelberg (1986). https://doi.org/10.1007/3-540-16780-3_85
7. Comon-Lundh, H., Delaune, S.: The finite variant property: how to get rid of some algebraic properties. In: Giesl, J. (ed.) RTA 2005. LNCS, vol. 3467, pp. 294–307. Springer, Heidelberg (2005). https://doi.org/10.1007/978-3-540-32033-3_22
8. Dershowitz, N., Jouannaud, J.P.: Rewrite systems. In: van Leeuwen, J. (ed.) Handbook of Theoretical Computer Science, Vol. B, pp. 243–320. North-Holland (1990)
9. Durán, F., Lucas, S., Meseguer, J.: Termination modulo combinations of equational theories. In: Ghilardi, S., Sebastiani, R. (eds.) FroCoS 2009. LNCS (LNAI), vol. 5749, pp. 246–262. Springer, Heidelberg (2009). https://doi.org/10.1007/978-3-642-04222-5_15
10. Durán, F., Meseguer, J.: On the Church-Rosser and coherence properties of conditional order-sorted rewrite theories. J. Algebraic Logic Program. **81**, 816–850 (2012)
11. Durán, F., Meseguer, J., Rocha, C.: Ground confluence of order-sorted conditional specifications modulo axioms. J. Log. Algebraic Methods Program. **111**, 100513 (2020)
12. Escobar, S., Sasse, R., Meseguer, J.: Folding variant narrowing and optimal variant termination. J. Algebraic Logic Program. **81**, 898–928 (2012)
13. Futatsugi, K., Diaconescu, R.: CafeOBJ Report. World Scientific (1998)
14. Goguen, J., Meseguer, J.: Order-sorted algebra I: equational deduction for multiple inheritance, overloading, exceptions and partial operations. Theor. Comput. Sci. **105**, 217–273 (1992)
15. Goguen, J.A., Winkler, T., Meseguer, J., Futatsugi, K., Jouannaud, JP.: Introducing OBJ. In: Goguen, J., Malcolm, G. (eds.) Software Engineering with OBJ. Advances in Formal Methods, vol 2. Springer, Boston (2000). https://doi.org/10.1007/978-1-4757-6541-0_1
16. Guttag, J.V., Horning, J.J.: The algebraic specification of abstract data types. Acta Informatica **10**, 27–52 (1978). https://doi.org/10.1007/BF00260922
17. Hendrix, J., Clavel, M., Meseguer, J.: A sufficient completeness reasoning tool for partial specifications. In: Giesl, J. (ed.) RTA 2005. LNCS, vol. 3467, pp. 165–174. Springer, Heidelberg (2005). https://doi.org/10.1007/978-3-540-32033-3_13

18. Hendrix, J., Meseguer, J.: On the completeness of context-sensitive order-sorted specifications. In: Baader, F. (ed.) RTA 2007. LNCS, vol. 4533, pp. 229–245. Springer, Heidelberg (2007). https://doi.org/10.1007/978-3-540-73449-9_18

19. Hendrix, J., Meseguer, J., Ohsaki, H.: A sufficient completeness checker for linear order-sorted specifications modulo axioms. In: Furbach, U., Shankar, N. (eds.) IJCAR 2006. LNCS (LNAI), vol. 4130, pp. 151–155. Springer, Heidelberg (2006). https://doi.org/10.1007/11814771_14

20. Jouannaud, J.P., Kounalis, E.: Automatic proofs by induction in theories without constructors. Inf. Comput. **82**(1), 1–33 (1989)

21. Kapur, D., Narendran, P., Rosenkrantz, D.J., Zhang, H.: Sufficient-completeness, ground-reducibility and their complexity. Int. J. Biometeorol. **36**(4), 311–350 (1991). https://doi.org/10.1007/BF01212959

22. Kikuchi, K., Aoto, T.: Simple derivation systems for proving sufficient completeness of non-terminating term rewriting systems. In: 41st IARCS Annual Conference on Foundations of Software Technology and Theoretical Computer Science, FSTTCS 2021. LIPIcs, vol. 213, pp. 49:1–49:15. Schloss Dagstuhl - Leibniz-Zentrum für Informatik (2021)

23. Lassez, J.L., Marriott, K.: Explicit representation of terms defined by counter examples. J. Autom. Reasoning **3**(3), 301–317 (1987). https://doi.org/10.1007/bf00243794

24. Lucas, S., Meseguer, J.: Normal forms and normal theories in conditional rewriting. J. Log. Algebr. Meth. Program. **85**(1), 67–97 (2016)

25. Meseguer, J., Skeirik, S.: On ground convergence and completeness of conditional equational program hierarchies. In: Bae, K. (ed.) WRLA 2022. LNCS, vol. 13252, pp. 191–211. Springer, Cham (2022)

26. Meseguer, J., Skeirik, S.: Inductive reasoning with equality predicates, contextual rewriting and variant-based simplification. In: Escobar, S., Martí-Oliet, N. (eds.) WRLA 2020. LNCS, vol. 12328, pp. 114–135. Springer, Cham (2020). https://doi.org/10.1007/978-3-030-63595-4_7

27. Meseguer, J.: Conditional rewriting logic as a unified model of concurrency. Theor. Comput. Sci. **96**(1), 73–155 (1992)

28. Meseguer, J.: Membership algebra as a logical framework for equational specification. In: Presicce, F.P. (ed.) WADT 1997. LNCS, vol. 1376, pp. 18–61. Springer, Heidelberg (1998). https://doi.org/10.1007/3-540-64299-4_26

29. Meseguer, J.: Variant-based satisfiability in initial algebras. Sci. Comput. Program. **154**, 3–41 (2018)

30. Meseguer, J., Skeirik, S.: Equational formulas and pattern operations in initial order-sorted algebras. Formal Aspects Comput. **29**(3), 423–452 (2017). https://doi.org/10.1007/s00165-017-0415-5

31. Nakamura, M., Ogata, K., Futatsugi, K.: Incremental proofs of termination, confluence and sufficient completeness of OBJ specifications. In: Iida, S., Meseguer, J., Ogata, K. (eds.) Specification, Algebra, and Software. LNCS, vol. 8373, pp. 92–109. Springer, Heidelberg (2014). https://doi.org/10.1007/978-3-642-54624-2_5

32. Shiraishi, T., Kikuchi, K., Aoto, T.: A proof method for local sufficient completeness of term rewriting systems. In: Cerone, A., Ölveczky, P.C. (eds.) ICTAC 2021. LNCS, vol. 12819, pp. 386–404. Springer, Cham (2021). https://doi.org/10.1007/978-3-030-85315-0_22

33. Thiel, J.J.: Stop losing sleep over incomplete data type specification. In: Kennedy, K. (ed.) Proceedings, Eleventh Symposium on Principles of Programming Languages. Association for Computing Machinery (1984)

# On Ground Convergence
# and Completeness of Conditional
# Equational Program Hierarchies

José Meseguer[⊠] and Stephen Skeirik

Department of Computer Science, University of Illinois at Urbana-Champaign,
Urbana, USA
{meseguer,skeirik2}@illinois.edu

**Abstract.** Both complete definition of functions by equations and determinism (i.e., evaluation to a unique result), are fundamental correctness properties of equational programs. But for expressive functional languages supporting conditional equations, types and subtypes and rewriting modulo axioms, proof methods for verifying such properties under general conditions are currently quite limited. This work proposes a hierarchical proof methodology where both properties are simultaneously verified in a hierarchical manner under termination assumptions.

## 1  Introduction

Equational programs define functions by means of equations. Such programs may be faulty in various ways. A common problem is *lack of complete definition*, that is, for a specific concrete input the program cannot be evaluated to concrete data. This is called lack of *sufficient completeness*. Maude [7] color codes unevaluated function symbols in the result of an evaluation to alert the user when this happens. A more subtle problem is *lack of determinism*. For any input, a terminating equational program should evaluate to a *unique result*. Lack of determinism may not be detected at runtime, since a functional expression will evaluate to *some* result following a given evaluation strategy that *assumes* such determinism.

These two properties, completeness and determinism, are fundamental for program correctness and are assumed when proving many other properties, for example by inductive theorem proving. For expressive equational languages supporting conditional equations, types and subtypes and rewriting modulo axioms like associativity and/or commutativity and/or identity such as OBJ [19], CafeOBJ [15] and Maude [7], decision procedures ensuring these properties only exist for restricted program classes of *unconditional*, terminating programs not involving associative but noncommutative axioms. For them: (i) joinability of critical pairs (see, e.g., [9]) ensures determinism; and (ii) equational tree automata methods can check sufficient completeness if lefthand sides of equational definitions are left-linear [25]. Even under those restrictions, checking (i)

© Springer Nature Switzerland AG 2022
K. Bae (Ed.): WRLA 2022, LNCS 13252, pp. 191–211, 2022.
https://doi.org/10.1007/978-3-031-12441-9_10

is only a *sufficient* condition for determinism. This is because only a weaker condition, namely, *ground joinability* of critical pairs is needed, since an equational program only executes *concrete* inputs, i.e., ground terms. There are many perfectly fine equational programs whose critical pairs are ground joinable but not joinable. For a good example, see the equational definition of finitary set theory in [12]. Alas, ground joinability of critical pairs is an *undecidable* property [28]. Checking completeness and determinism of function definitions becomes much harder for *conditional* specifications, where both checks are undecidable. But this does not make the need to ensure these fundamental properties any less pressing.

Ground joinability of critical pairs is called *ground confluence*, and, under the assumption of operational termination, *ground convergence*. As further discussed in Sect. 4, a lot of work has been done on proof methods for both sufficient completeness, e.g., [5,8,22–25,27,29,30,39,42,44] and ground confluence, e.g., [1,3,4,12,14,16,17,28,41,46]. However, for *equational order-sorted conditional programs modulo axioms*, the only prior work we are aware of is that in [39] and [12]. This paper develops new proof methods for both completeness and ground convergence of such programs. For ground convergence it further advances the ideas in [12]. Two key features of our proof methodology are that: (i) sufficient completeness and ground convergence proofs help and depend on each other; and (ii) the proof methods are *hierarchical* and therefore incremental: proofs are obtained by climbing up a tower of theory inclusions.

## 2    Preliminaries

We assume familiarity with the notions of an order-sorted signature $\Sigma$ on a poset of sorts $(S, \leqslant)$, an order-sorted $\Sigma$-algebra $A$, and the term $\Sigma$-algebras $T_\Sigma$ and $T_\Sigma(X)$ for $X$ an $S$-sorted set of variables. We also assume familiarity with the notions of: (i) $\Sigma$-homomorphism $h : A \to B$ between $\Sigma$-algebras $A$ and $B$, so that $\Sigma$-algebras and $\Sigma$-homomorphisms form a category **OSAlg**$_\Sigma$; (ii) order-sorted (i.e., sort-preserving) substitution $\theta$, its domain $dom(\theta)$ and range $ran(\theta)$, and its application $t\theta$ to a term $t$; (iii) *preregular* order-sorted signature $\Sigma$, i.e., a signature such that each term $t$ has a least sort, denoted $ls(t)$; (iv) the set $\widehat{S} = S/(\geqslant \cup \leqslant)^+$ of *connected components* of a poset $(S, \leqslant)$ viewed as a DAG; and (v) for $A$ a $\Sigma$-algebra, the set $A_s$ of its elements of sort $s \in S$, and the set $A_{[s]} = \bigcup_{s' \in [s]} A_{s'}$ of all elements in a connected component $[s] \in \widehat{S}$. We furthermore assume that all signatures $\Sigma$ have *non-empty sorts*, i.e., $T_{\Sigma,s} \neq \varnothing$ for each $s \in S$. $[A \to B]$ denotes the $S$-*sorted functions* from $A$ to $B$. These notions are explained in [18,36]. The material below is adapted from [34,37].

**Order-Sorted Algebra and $E$-Unification.** An OS *equational theory* is a pair $T = (\Sigma, E)$, with $E$ a set of (possibly conditional) $\Sigma$-equations. **OSAlg**$_{(\Sigma,E)}$ denotes the full subcategory of **OSAlg**$_\Sigma$ with objects those $A \in$ **OSAlg**$_\Sigma$ such that $A \models E$, called the $(\Sigma, E)$-*algebras*. **OSAlg**$_{(\Sigma,E)}$ has an *initial algebra* $T_{\Sigma/E}$ [36]. If $\mathcal{E} = (\Sigma, E)$, $T_\mathcal{E}$ abbreviates $T_{\Sigma/E}$. For $\Sigma$ an OS-signature, $Form(\Sigma)$

denotes the set of its first-order formulas, whose atoms are $\Sigma$-equations. Given $T = (\Sigma, E)$ and $\varphi \in Form(\Sigma)$, we call $\varphi$ *T-valid*, written $E \models \varphi$, iff $A \models \varphi$ for all $A \in \mathbf{OSAlg}_{(\Sigma,E)}$. We call $\varphi$ *T-satisfiable* iff there exists $A \in \mathbf{OSAlg}_{(\Sigma,E)}$ with $\varphi$ satisfiable in $A$; that is, there exists an *assignment* $a$, i.e., and $S$-sorted function $a \in [fv(\varphi) \to A]$ with $fv(\varphi)$ the free variables of $\varphi$, such that $A, a \models \varphi$. Note that $\varphi$ is *T-valid* iff $\neg\varphi$ is *T-unsatisfiable*. The inference system in [36] is *sound and complete* for OS equational deduction, i.e., for any OS equational theory $(\Sigma, E)$, and $\Sigma$-equation $u = v$ we have an equivalence $E \vdash u = v \Leftrightarrow E \models u = v$. Deducibility $E \vdash u = v$ is abbreviated as $u =_E v$, called *E-equality*. An *E-unifier* of a system of $\Sigma$-equations, i.e., of a conjunction $\phi = u_1 = v_1 \wedge \dots \wedge u_n = v_n$ of $\Sigma$-equations, is a substitution $\sigma$ such that $u_i\sigma =_E v_i\sigma$, $1 \leqslant i \leqslant n$. An *E-unification algorithm* for $(\Sigma, E)$ is an algorithm generating a *complete set* of *E-unifiers* $Unif_E(\phi)$ for any system of $\Sigma$ equations $\phi$, where "complete" means that for any *E-unifier* $\sigma$ of $\phi$ there is a $\tau \in Unif_E(\phi)$ and a substitution $\rho$ such that $\sigma =_E (\tau\rho)|_{dom(\sigma) \cup dom(\tau)}$, where $=_E$ here means that for any variable $x$ we have $x\sigma =_E x(\tau\rho)|_{dom(\sigma) \cup dom(\tau)}$. The algorithm is *finitary* if it always terminates with a *finite set* $Unif_E(\phi)$ for any $\phi$. Given a set of equations $B$ used for deduction modulo $B$, a preregular OS signature $\Sigma$ is called *B-preregular*[1] iff for each $u = v \in B$ and substitutions $\rho$, $ls(u\rho) = ls(v\rho)$.

**Convergent Theories and Sufficient Completeness.** Given an order-sorted equational theory $\mathcal{E} = (\Sigma, E \cup B)$, where $B$ is a collection of associativity and/or commutativity and/or identity axioms and $\Sigma$ is $B$-preregular, we can associate to it a corresponding *rewrite theory* [35] $\vec{\mathcal{E}} = (\Sigma, B, \vec{E})$ by orienting the equations $E$ as left-to right rewrite rules. That is, each $(u = v) \in E$ is transformed into a rewrite rule $u \to v$. For simplicity we recall here the case of unconditional equations. Since in this work we will consider *conditional* theories $\vec{\mathcal{E}}$, we refer to [32] for full details on the general definition of convergent theory. The main purpose of the rewrite theory $\vec{\mathcal{E}}$ is to reduce the complex bidirectional reasoning with equations to the much simpler unidirectional reasoning with rules under suitable assumptions. We assume familiarity with the notion of subterm $t|_p$ of $t$ at a term position $p$ and of term replacement $t[w]_p$ of $t|_p$ by $w$ at position $p$ (see, e.g., [9]). The rewrite relation $t \to_{\vec{E},B} t'$ holds iff there is a subterm $t|_p$ of $t$, a rule $(u \to v) \in \vec{E}$ and a substitution $\theta$ such that $u\theta =_B t|_p$, and $t' = t[v\theta]_p$. We denote by $\to_{\vec{E},B}^*$ the reflexive-transitive closure of $\to_{\vec{E},B}$. For $\vec{\mathcal{E}}$ unconditional, the convergence requirements are as follows (see [32] for $\vec{\mathcal{E}}$ conditional): (i) $vars(v) \subseteq vars(u)$; (ii) *sort-decreasingness*: for each substitution $\theta$, $ls(u\theta) \geqslant ls(v\theta)$; (iii) *strict B-coherence*: if $t_1 \to_{\vec{E},B} t'_1$ and $t_1 =_B t_2$ then there exists $t_2 \to_{\vec{E},B} t'_2$ with $t'_1 =_B t'_2$; (iv) *confluence* (resp. *ground confluence*) modulo $B$: for each term $t$ (resp. ground term $t$) if $t \to_{\vec{E},B}^* v_1$ and $t \to_{\vec{E},B}^* v_2$, then

---

[1] If $B = B_0 \uplus U$, with $B_0$ associativity and/or commutativity axioms, and $U$ identity axioms, the $B$-preregularity notion can be *broadened* by requiring only that: (i) $\Sigma$ is $B_0$-preregular in the standard sense that $ls(u\rho) = ls(v\rho)$ for all $u = v \in B_0$ and substitutions $\rho$; and (ii) the axioms $U$ oriented as rules $\vec{U}$ are *sort-decreasing* in the sense explained below.

there exist rewrite sequences $v_1 \to^*_{\vec{E},B} w_1$ and $v_2 \to^*_{\vec{E},B} w_2$ such that $w_1 =_B w_2$;

(v) *termination*: the relation $\to_{\vec{E},B}$ is well-founded (for $\vec{E}$ conditional, we require *operational termination* [32]). If $\vec{\mathcal{E}}$ satisfies conditions (i)–(v) (resp. the same, but (iv) weakened to ground confluence modulo $B$), then it is called *convergent* (resp. *ground convergent*). The key point is that then, given a term (resp. ground term) $t$, all terminating rewrite sequences $t \to^*_{\vec{E},B} w$ end in a term $w$, denoted $t!_{\vec{\mathcal{E}}}$, that is unique up to $B$-equality, and its called $t$'s *canonical form*. Ground convergence implies three major results: (1) for any ground terms $t, t'$ we have $t =_{E \cup B} t'$ iff $t!_{\vec{\mathcal{E}}} =_B t'!_{\vec{\mathcal{E}}}$, (2) the $B$-equivalence classes of canonical forms are the elements of the *canonical term algebra* $C_{\Sigma/E,B}$, where for each $f : s_1 \ldots s_n \to s$ in $\Sigma$ and $B$-equivalence classes of canonical terms $[t_1], \ldots, [t_n]$ with $ls(t_i) \leqslant s_i$ the operation $f_{C_{\Sigma/E,B}}$ is defined by the identity: $f_{C_{\Sigma/E,B}}([t_1] \ldots [t_n]) = [f(t_1 \ldots t_n)!_{\vec{\mathcal{E}}}]$, and (3) we have an isomorphism $T_{\mathcal{E}} \cong C_{\Sigma/E,B}$.

A ground convergent rewrite theory $\vec{\mathcal{E}} = (\Sigma, B, \vec{E})$ is called *sufficiently complete* with respect to a subsignature $\Omega$, whose operators are then called *constructors*, iff for each ground $\Sigma$-term $t$, $t!_{\vec{\mathcal{E}}} \in T_\Omega$. Furthermore, for $\vec{\mathcal{E}} = (\Sigma, B, \vec{E})$ sufficiently complete w.r.t. $\Omega$, a ground convergent rewrite subtheory $(\Omega, B_\Omega, \vec{E}_\Omega) \subseteq (\Sigma, B, \vec{E})$ is called a *constructor subspecification* iff $T_{\mathcal{E}}|_\Omega \cong T_{\Omega/E_\Omega \cup B_\Omega}$. If $E_\Omega = \varnothing$, then $\Omega$ is called a *signature* of *free constructors modulo axioms* $B_\Omega$. Note that $\vec{\mathcal{E}} = (\Sigma, B, \vec{E})$ is sufficiently complete with respect to $\Omega$ iff each ground $\Sigma$-term $f(u_1, \ldots, u_n)$ with $f \in \Sigma \setminus \Omega$ and $u_i \in T_\Omega$, $1 \leqslant i \leqslant n$, is $\vec{E}, B$-*reducible*, i.e., $f(u_1, \ldots, u_n) \to_{\vec{E},B} t$ for some $t \in T_\Sigma$.

**Generator Sets.** Generator sets generalize standard structural induction on the constructors of a sort. They are particularly useful for inductive reasoning when constructors obey structural axioms $B$ including associativity and/or associativity-commutativity for which structural induction may be ill-suited.

A *generator set* for a sort $s$ is a set of constructor terms of sort $s$ or smaller such that any ground constructor term of sort $s$ is a ground substitution instance of one of the patterns in the generator set. Here is the general definition (identity axioms are not needed thanks to the theory transformation $\vec{\mathcal{E}} \mapsto \vec{\mathcal{E}}_U$ in [10]):

**Definition 1.** *For an order-sorted signature of constructors $\Omega$—which may satisfy axioms $B$ of associativity and/or commutativity—and $s$ a sort in $\Omega$, a $B$-generator set for sort $s$ is a finite set of terms $\{u_1, \ldots, u_k\} \subseteq T_\Omega(X)_s$ such that*

$$T_{\Omega/B,s} = \bigcup_{1 \leqslant i \leqslant k} \{[u_i \, \rho] \in T_{\Omega/B,s} \mid \rho \in [X \to T_\Omega]\}.$$

**Checking the Correctness of Generator Sets.** How do we *know* that a proposed generator set $\{u_1, \ldots, u_k\}$ it truly one modulo axioms $B$ for a given sort $s$ and constructors $\Omega$? Assuming that the terms $u_1, \ldots, u_k$ are all *linear*, i.e., have no repeated variables—which is the usual case for generator sets—this check can be reduced to an automatic *sufficient completeness check* with Maude's Sufficient Completeness Checker (SCC) tool [25], which is based on tree automata decision procedures modulo axioms $B$. The reduction is extremely simple: define a new unary predicate $s : s \rightarrow Bool$ with equations $s(u_i) = true$, $1 \leqslant i \leqslant k$. Then, $\{u_1, \ldots, u_k\}$ is a correct generator set for sort $s$ modulo $B$ for the constructor signature $\Omega$ iff the predicate $s$ is sufficiently complete, which can be automatically checked by the SCC tool. Furthermore, if $\{u_1, \ldots, u_k\}$ is *not* a generator set for sort $s$, the SCC tool will output a useful counterexample.

## 3 Proving Ground Convergence And Sufficient Completeness Hierarchically

**Basic Assumptions About $\vec{\mathcal{E}}$.** We assume throughout a conditional equational theory $\mathcal{E} = (\Sigma, E \cup U \cup B)$ such that: (i) $\Sigma$ decomposes as a disjoint union $\Sigma = \Delta \uplus \Omega$, where $\Omega$ are the—intended but not yet proved to be—constructor symbols, that furthermore are free modulo $B$, and $\Delta$ are the intended defined symbols. (ii) $B$ is any combination of associativity and/or commutativity axioms,[2] but any binary $f \in \Delta$ may not satisfy any axioms except commutativity.[3] (iii) $\vec{U}$ are sort-decreasing unit axiom rules of the form $c(e, x) \rightarrow x$ or $c(x, e) \rightarrow x$, where $c$ is a constructor name and $e$ is an $\Omega$-term. However, $c(x, e)$ is not an $\Omega$-term because some of $c$'s type declarations do not belong to $\Omega$ (see Example 1). This makes it possible for constructors to be free modulo $B$ in spite of such unit rules. (iv) $E = \bigcup_{f \in \Delta} E_f$, where for each $f \in \Delta$, its associated rewrite rules $\vec{E}_f$ are sort-decreasing and have the form: $\vec{E}_f = \{[i] : f(\vec{u_i}) \rightarrow r_i \ if \ \Gamma_i\}_{i \in I}$

---

[2] Furthermore, for any $f$ that is commutative we always assume a top typing $f : s\ s \rightarrow s_0$ with all other typings of the form $f : s'\ s' \rightarrow s'_0$, with $s \leqslant s'$, $s_0 \leqslant s'_0$. Regarding the absence of unit element axioms, they are precisely the equations $U$, that will be used as rules $\vec{U}$ (see, e.g., Example 1). The point is that, for both confluence and termination purposes, if $\vec{\mathcal{G}}$ has axioms $B \uplus U$, with $B$ associative and/or commutative axioms and $U$ unit element axioms, then the axioms $U$ can be eliminated by turning them into rules $\vec{U}$ thanks to the semantics-preserving theory transformation $\vec{\mathcal{G}} \mapsto \vec{\mathcal{G}}_U$ defined in [10], so that the axioms of the semantically equivalent $\vec{\mathcal{G}}_U$ are just $B$. Therefore, Our results apply as well to theories $\vec{\mathcal{G}}$ with axioms $B \uplus U$ such that $\vec{\mathcal{G}}_U$ has the properties (i)–(vi) listed in what follows.

[3] Since axioms $B$ are primarily used to specify constructor data structures, in actual practice, limiting axioms for defined symbols to just commutativity is a mild restriction. Furthermore, this restriction can be removed *a posteriori* in the following sense. After $\vec{\mathcal{E}}$ has been shown ground convergent and sufficiently complete, if we can prove by inductive theorem proving that the initial algebra $T_\mathcal{E}$ does satisfy additional associativity and/or commutativity axioms for some binary $f \in \Delta$, then we can add to $\vec{\mathcal{E}}$: (a) those extra axioms for $f$, and (b) the $A$-, resp. $AC$-extensions (see [40]) of the rules $\vec{E}_f$ in the sense of (iv) below (to ensure $B$-coherence). One can then show that the theory thus extended is also ground convergent and sufficiently complete if its rules remain operationally terminating modulo the extended axioms. For example, in the MULTISET-ALGEBRA module of Example 1, we can prove the associativity and commutativity of the intersection operator $\_\cap\_$ as inductive theorems and then add those properties as axioms of $\_\cap\_$ (the $AC$-extensions of $\vec{E}_\cap$ do not need to be added explicitly: they are added automatically by Maude).

such that: (a) the $\vec{u_i}$ are $\Omega$-terms; and (b) for each $i \in I$, $\Gamma_i = \bigwedge_{j \in J} w_j = w'_j$ and $vars(f(\vec{u_i})) \supseteq vars(r_i) \cup vars(\Gamma_i)$. (v) There is a $B$-compatible recursive path order (RPO) $>$ (see [9]) such that for each $i \in I$, $f(\vec{u_i}) > r_i$, and $f(\vec{u_i}) > w_j, w'_j$, $j \in J$, which makes the rules $\vec{E} \cup \vec{U}$ operationally terminating modulo $B$. (vi) The rules $\vec{E} \cup \vec{U}$ are strictly $B$-coherent.

The main goal of this paper is to develop hierarchical methods to prove that a theory $\vec{\mathcal{E}}$ enjoying properties (i)–(vi) above is ground convergent. As it turns out, such hierarchical methods will also allow us to prove that $\vec{\mathcal{E}}$ is sufficiently complete with respect to its hypothesized constructor signature $\Omega$. As we shall see, hierarchical proofs of ground convergence and of sufficient completeness will help each other. Since the rules in $\vec{\mathcal{E}}$ can be conditional, they are generally outside the scope of equational tree-automata methods for checking sufficient completeness supported by tools like Maude's Sufficient Completeness Checker (SCC) [25], which assume unconditional and left-linear rules: new proof methods are needed. Sufficient completeness is important both for program correctness and because it allows constructor-based inductive reasoning.

**Calling Graph and Theory Hierarchies.** We assume that all function symbols in $\Delta$ are *subsort-overloaded*, i.e., for any $f : s_1 \ldots s_n \to s$ and $f : s'_1 \ldots s'_n \to s'$ we have $[s] = [s']$, and $[s_i] = [s'_i]$, $1 \leq i \leq n$. This can always be achieved by renaming $\Delta$. Let $F_\Delta$ be the set of *names* for the function symbols in $\Delta$, disregarding their typing. The *calling relation* is a binary relation $C$ on $F_\Delta$, where for each $f, g \in F_\Delta$, $(f, g) \in C$ iff there exists a rule $f(\vec{u_i}) \to r_i$ *if* $\Gamma_i$ in $\vec{E}_f$ such that the function symbol $g$ occurs in either $r_i$ or in $\Gamma_i$. Let $C^*$ denote the reflexive-transitive closure of $C$, and $\equiv_C$ the equivalence relation on $F_\Delta$ defined by the equivalence: $f \equiv_C g$ iff $f C^* g$ and $g C^* f$. Then, the quotient set $F_\Delta / \equiv_C$ has a partial order defined by the equivalence $[f] \geq [g] \Leftrightarrow f C^* g$.

The hierarchical method we propose is based on a *hierarchy of theory inclusions* chosen as follows. Given our theory $\vec{\mathcal{E}}$ we: (i) identity a subtheory $\vec{\mathcal{E}_0}$ having subsignature $\Delta_0 \uplus \Omega$ containing all the subsort-overloaded typings of any $f \in \Delta_0$ and having rules $\vec{U} \cup \vec{E_0}$, with $\vec{E_0} = \bigcup_{f \in \Delta_0} \vec{E}_f$, and axioms $B_0 = B_{\Delta_0 \uplus \Omega} = \bigcup_{f \in \Delta_0 \uplus \Omega} B_f$, where $B_f$ are the associative and/or commutative axioms, if any, for $f$ in $B$, and such that $\vec{\mathcal{E}_0} \cup \vec{U}$ is sufficiently complete with respect to $\Omega$ and ground convergent. Of course, we should choose $\vec{\mathcal{E}_0}$ as big as possible: in the worse case we may have $\vec{E_0} = \varnothing$ and keep only $\vec{U}$. We furthermore assume that we can find a sequence of theory inclusions:

$$\vec{\mathcal{E}_0} \subset \vec{\mathcal{E}_1} \subset \ldots \vec{\mathcal{E}_{n-1}} \subset \vec{\mathcal{E}_n}$$

such that: (a) $\vec{\mathcal{E}_n} = \vec{\mathcal{E}}$, (b) each $\vec{\mathcal{E}_k}$ has signature $\Delta_k \uplus \Omega$ containing all subsort-overloaded typings of any $f \in \Delta_k$ and having axioms $B_k = B_{\Delta_k \uplus \Omega}$ and, besides $\vec{U}$, rules $\vec{E_k} = \bigcup_{f \in \Delta_k} \vec{E}_f$, where for each $k \geq 1$ and each rule $[i] : f(\vec{u_i}) \to r_i$ *if* $\Gamma_i$ in $\vec{E}_f$, the condition $\Gamma_i$ is a conjunction of $\vec{\mathcal{E}_{k-1}}$-equalities, (c) for each $0 \leq k < k+1 \leq n$, there exists a function symbol $g \in F_{\Delta_{k+1}} \backslash F_{\Delta_k}$ such that $F_{\Delta_{k+1}} = F_{\Delta_k} \uplus [g]$; that is, we add all the symbols in a new $\equiv_C$-equivalence class $[g]$ to climb up each step in the theory hierarchy. Let us see an example.

*Example 1* (Multisets of Natural Numbers). As a running example we use a theory of *multisets of natural numbers* with number equality _.=._ and multiset membership _∈_ predicates, multiset difference _\_, intersection _∩_ and union _,_. Its Maude specification (with self-explanatory syntax for $(\Sigma, E \cup U \cup B)$) is:

```
fmod MULTISET-ALGEBRA is
  protecting TRUTH-VALUE .
  sorts Nat Mult .
  subsort Nat < NeMult < Mult .
  op 0 : -> Nat [ctor] .
  op s : Nat -> Nat [ctor] .
  op mt : -> Mult [ctor] .                            *** empty multiset
  op _,_ : Mult Mult -> Mult [assoc comm] .           *** multiset union
  op _,_ : NeMult NeMult -> NeMult [ctor assoc comm] . *** multiset union
  op _.=._ : Nat Nat -> Bool [comm] .                 *** nats equality
  op _in_ : Nat Mult -> Bool .                        *** membership
  op _\_  : Mult Mult -> Mult .                       *** difference
  op _/\_ : Mult Mult -> Mult .                       *** intersection

  vars n m k : Nat .  vars U V W : Mult .

  eq U,mt = U .                                       *** unit equation

  eq n .=. n = true .
  eq 0 .=. s(n) = false .
  ceq s(n) .=. s(m) = false if n .=. m = false .

  eq n in mt = false .
  eq n in n = true .
  ceq n in m = false if (n .=. m) = false .
  eq n in (n,U) = true .
  ceq n in (m,U) = false
          if (n .=. m) = false /\ (n in U) = false .

  eq mt \ U = mt .
  eq U \ mt = U .
  eq m \ m = mt .
  ceq m \ n = m if n .=. m = false .
  eq (m,U) \ m =  U .
  ceq (m,U) \ n = m,(U \ n) if n .=. m = false .
  eq U \ (n,V) = (U \ n) \ V .

  eq mt /\ V = mt .
  ceq n /\ V = n if (n in V) = true .
  ceq n /\ V = mt if (n in V) = false .
  ceq (n,U) /\ V = n,(U /\ (V \ n)) if (n in V) = true .
  ceq  (n,U) /\ V = U /\ V  if (n in V) = false .
endfm
```

Its calling graph is described in Fig. 1.

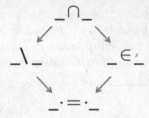

**Fig. 1.** Calling Graph of Multiset Operations

Let $\vec{\mathcal{E}}$ denote the above theory of multisets of natural numbers, with the defined operations $\_.=.\_$, $\_\in\_$, $\_\setminus\_$, and $\_\cap\_$, and the usual Boolean operations. Its *constructors* $\Omega$ are: $0$, $s$, $\top$ and $\bot$, $\varnothing$, and the union operator $\_,\_$ for non-empty multisets. $\vec{\mathcal{E}}$ satisfies properties (i)–(vi). Furthermore, we have a tower of theories:

$$\vec{\mathcal{E}}_{.=.,\in} \subset \vec{\mathcal{E}}_{.=.,\in,.\setminus.} \subset \vec{\mathcal{E}}$$

where each theory contains the Boolean values, plus the constructors, plus the mentioned operations, plus the rules defining such operations. $\vec{\mathcal{E}}_{.=.,\in}$ is both convergent and sufficiently complete, so we just have two steps to climb up the tower to prove $\vec{\mathcal{E}}$ both ground convergent and sufficiently complete.

**The Hierarchical Proof Method.** All now boils down to finding proof methods to *climb up the theory hierarchy one step at a time*. We then repeat this method $n$ times, with $n$ the length of the chain of theory inclusions. That is, we focus on a single theory inclusion $\vec{\mathcal{E}}_0 \subset \vec{\mathcal{E}}$, where $\vec{\mathcal{E}}_0$ has already been proved ground convergent and sufficiently complete with respect to the constructor signature $\Omega$, and then prove that $\vec{\mathcal{E}}$ is also ground convergent and sufficiently complete as follows. First of all, we define a new theory $\vec{\mathcal{E}}^\Delta$, with the same rules $E \cup U$ as in $\vec{\mathcal{E}}$, and having also a theory inclusion $\vec{\mathcal{E}}_0 \subset \vec{\mathcal{E}}^\Delta$, but where, if $\Sigma_0$ and $\Sigma$ are the respective signatures of $\vec{\mathcal{E}}_0$ and $\vec{\mathcal{E}}$, and $\Delta = \Sigma \setminus \Sigma_0$, then $\vec{\mathcal{E}}^\Delta$ has a signature $\Sigma^\Delta$ that extends $\Sigma_0$ by: (i) adding to each connected component $[s] \in \hat{S}$ of the poset of sorts $(S, \leqslant)$ of $\Sigma_0$ $[s]$ itself as a new "kind" top sort, i.e., $\forall s' \in [s]$, $s' < [s]$, and (ii) lifting to the kind levels the $f \in \Sigma$. That is, we extend $\Sigma_0$ by adding to $\Sigma^\Delta$ a function symbol $f : [s_1]\ldots[s_n] \to [s]$ for each $f : s_1\ldots s_n \to s$, $n \geqslant 1$, in $\Sigma$. In $\vec{\mathcal{E}}^\Delta$ the axioms $B$ are lifted to kinds. For example, for the theory inclusion $\vec{\mathcal{E}}_{.=.,\in,.\setminus.} \subset \vec{\mathcal{E}}$ in our running example, $\Sigma_0$, $\Sigma$ and $\Sigma^\Delta$ can be depicted as follows (subsort inclusions in vertical lines):

$\Sigma_0$  $\Sigma$  $\Sigma^{\Delta}$

Note that $\Sigma^{\Delta}$ *adds no new terms* to the original sorts $S$, i.e., $T_{\Sigma^{\Delta}}(X)_s = T_{\Sigma_0}(X)_s$, $s \in S$. The hierarchical proof methodology then proceeds as follows:

1. We first prove that $\vec{\mathcal{E}}^{\Delta}$ is ground convergent.
2. We then prove that for any $f \in \Delta$, typing $f : s_1, \ldots, s_n \to s$ maximal in the subsort order, and ground constructor substitution $\rho$, $f(x_1, \ldots, x_n)\rho$, with $x_i$ of sort $s_i$, $1 \leqslant i \leqslant n$, can be rewritten with some rule in $\vec{\mathcal{E}}^{\Delta}$.
3. By Theorem 1, (1) and (2) actually *prove* that $\vec{\mathcal{E}}$ is both ground convergent and sufficiently complete with respect to $\Omega$.

All we now need to do is to give a sound inference system that will allow us to carry out the proofs for (1) and (2), and then prove Theorem 1. The inference system for proving (1) and (2) works in the context of the theory inclusion $\vec{\mathcal{E}}_0 \subset \vec{\mathcal{E}}^{\Delta}$, where $\vec{\mathcal{E}}_0$ has equations $U \uplus E_0 \uplus B_0$. Properties are specified as *constrained properties* of the form $p \mid \varphi$, where $p$ is a *property* and $\varphi$ is a conjunction of $\Sigma_0$-equations. Semantically, $p \mid \varphi$ describes the set $[\![p \mid \varphi]\!]$ of *ground constructor instances* of property $p$ that satisfy $\varphi$. More precisely:

$$[\![p \mid \varphi]\!] = \{p\theta \mid \theta \in [X \to T_{\Omega}] \wedge \mathcal{E}_0 \vdash \varphi\theta\}.$$

The hierarchical inference system uses two kinds of constrained properties:

1. For proving that all instances of a term $f(x_1, \ldots, x_n)$ by a ground substitution $\rho$ are $\vec{E}_f, B$-*reducible*, where $f \in \Delta$, we use constrained properties of the form $red(f(u_1, \ldots, u_n)) \mid \varphi$, where the $u_i$ are $\Omega$-terms. By definition, $red(f(u_1, \ldots, u_n)) \mid \varphi$ *holds* iff for each ground constructor substitution $\rho$ such that $\varphi\rho$ holds in $\mathcal{E}_0$, the term $f(u_1, \ldots, u_n)\rho$ is $\vec{E}_f, B$-*reducible*.

2. For proving the *ground joinability* of a conditional critical pair[4] (CCP) of the form: $\varphi \Rightarrow t = t'$ for the theory $\vec{\mathcal{E}}^\Delta$, we represent this property as the constrained property $t \downarrow t' \mid \varphi$ and think of $\_ \downarrow \_$ as a binary predicate. We always assume that sorts of all variables in $t \downarrow t' \mid \varphi$ are in $S$ and $\varphi$ is a conjunction of $\mathcal{E}_0$-equalities, which will always be the case for any CCP of $\vec{\mathcal{E}}^\Delta$. By definition, $t \downarrow t' \mid \varphi$ holds iff for each ground constructor substitution[5] $\rho$ such that $\mathcal{E}_0 \vdash \varphi\rho$ there exist $u, v$ such that $t\rho \rightarrow^*_{\vec{\mathcal{E}}^\Delta} u =_B v \; {}^*_{\vec{\mathcal{E}}^\Delta}\!\leftarrow t'\rho$.

**The Shared Hierarchical Inference System.** We first introduce the inference rules applicable to *both* joinabilily and reducibility goals, i.e., goals either of the form (i) $t \downarrow t' \mid \varphi$, with $t, t'$ $\Sigma^\Delta$-terms; or of the form (ii) $red(f(u_1, \ldots, u_n)) \mid \varphi$, with $f \in \Delta$ and the $u_i$ $\Omega$-terms. The key feature shared by both kinds of goals is that their constraint $\varphi$ is a conjunction of $\Sigma_0$-equations. By assumption, $\vec{\mathcal{E}}_0$ is ground convergent, sufficiently complete with respect to $\Omega$, and the constructors $\Omega$ are free modulo $B_0$, with $B_0$ associative and/or commutative axioms.

In the shared inference system we assume constrained terms $p \mid \varphi$ of the form (i) or (ii). The shared inference rules are the following:

**Narrowing the Condition (NA)**

$$\frac{\{(p \mid \Gamma_i \wedge \varphi[r_i]_p)\alpha_{i,j}\}_{i \in I, \alpha_{i,j} \in Unif_B(f(\vec{v})=f(\vec{u}_i))}}{p \mid \varphi[f(\vec{v})]_p}$$

where the $\vec{v}$ are $\Omega$-terms, $f \in \Sigma_0 \setminus \Omega$ is defined by rules $\{[i] : f(\vec{u}_i) \rightarrow r_i \;\; if \;\; \Gamma_i\}_{i \in I}$ whose variables are always assumed disjoint of those in $p \mid \varphi$, and where $Unif_B(f(\vec{v}) = f(\vec{u}_i))$ denotes the set of $B$-unifiers of the equation $f(\vec{v}) = f(\vec{u}_i)$.

**Unification (UN)**

$$\frac{\{(p \mid \varphi)\theta\}_{\theta \in Unif_{U \cup E_0 \cup B_0}(\psi)}}{p \mid \varphi \wedge \psi}$$

where $\_ \wedge \_$ is assumed $AC$, and $\psi$ is a set of $\Sigma_0$-equations such that $\psi$ has a finite set of most general $U \cup E_0 \cup B_0$-unifiers. This is guaranteed to happen if either $\psi$ is a conjunction of $\Omega$-equations or, more generally, a conjunction of equations in a protected subtheory of $\vec{\mathcal{E}}_0$ that has the finite variant property [13].

---

[4] For a detailed definition of CCPs in an order-sorted setting see [11].

[5] The ground joinability of the CCP $\varphi \Rightarrow t = t'$ is normally stated as the joinability $t\alpha \downarrow t'\alpha$ for all ground substitution $\alpha$ such that $\mathcal{E}_0 \vdash \varphi\alpha$. However, since, by ground convergence and sufficient completeness of $\vec{\mathcal{E}}_0$ and the sort of all variables being in $S$, any such $\alpha$ can be normalized to a ground constructor substitution $\alpha!_{\vec{\mathcal{E}}_0}$, it can easily be shown that the CCP is ground joinable iff the property $t \downarrow t' \mid \varphi$ holds.

## Equality Simplification (ES)

$$\frac{p \mid \varphi!_{\vec{\mathcal{E}}_0^=}}{p \mid \varphi}$$

where $\varphi!_{\vec{\mathcal{E}}_0^=}$ denotes the canonical form of $\varphi$ in the ground convergent theory $\vec{\mathcal{E}}_0^=$ extending $\vec{\mathcal{E}}_0$ with *equality predicates* defined in [21], which allows formula simplifications such as, e.g., $(\top = \bot \land \psi)!_{\vec{\mathcal{E}}_0^=} = \bot$, and $(0 = s(u) \land \psi)!_{\vec{\mathcal{E}}_0^=} = \bot$.

## Case (CA)

$$\frac{\{(p \mid \varphi)\{x \mapsto v_i\}\}_{i \in I}}{p \mid \varphi}$$

with $x \in vars(p)$ a variable of sort $s$ and $\{v_i\}_{i \in I}$ a *generator set* for sort $s$ in $\Omega$, where all variables in $\{v_i\}_{i \in I}$ are assumed fresh.

## Split (SP)

$$\frac{\{p \mid \varphi \land \psi_i\}_{i \in I}}{p \mid \varphi}$$

where $T_{\Sigma_0/E_0 \cup B_0} \models \bigvee_{i \in I} \psi_i$ and $vars(\bigvee_{i \in I} \psi_i) \subseteq vars(p \mid \varphi)$.

## Generalization (GN)

$$\frac{p' \mid \psi}{p \mid \varphi}$$

where $\exists \theta \; p'\theta =_B p$, and $T_{\Sigma_0/E_0 \cup U \cup B_0} \models \varphi \Rightarrow (\psi\theta)$.

## Empty Goal ($\varnothing$)

$$\overline{p \mid \bot}$$

**Ground Joinability Inference System.** The goals to be proved are those associated to the conditional critical pairs in $\vec{\mathcal{E}}^\Delta$ that are not in $\vec{\mathcal{E}}_0$. The additional inference rules include: (i) constrained versions of the ground joinability rules in [12] (not needed for our running example); and (ii) the following two rules:

## Join (JN)

$$\overline{u \downarrow v \mid \varphi} \quad if \quad u =_B v$$

## Contextual Rewriting (CR)

$$\frac{w \downarrow w' \mid \varphi \qquad \overline{u} \to^*_{\vec{\mathcal{E}}^\Delta \cup \vec{\widehat{\varphi}}} \overline{w} \qquad \overline{v} \to^*_{\vec{\mathcal{E}}^\Delta \cup \vec{\widehat{\varphi}}} \overline{w'}}{u \downarrow v \mid \varphi}$$

where (i) $\overline{t}$ denotes the ground term obtained by replacing the variables in $t$ by corresponding *fresh constants*, (ii) $\widehat{\varphi}$ is the *rewrite condition* associated to $\varphi$ as explained in [11] (for example a condition $x \in U = \top \land x * y = 0$ yields the rewrite condition $x \in U \to \top \land x * y \to 0$, and (ii) $\vec{\widehat{\varphi}}$ denotes the set of ground rewrite rules $\overline{u_1} \to \overline{u_2}$ such that $u_1 \to u_2$ is a conjunct in $\vec{\widehat{\varphi}}$.

*Example 2.* For the theory inclusion $\vec{\mathcal{E}}_{.=.,\in.,\backslash .} \subset \vec{\mathcal{E}}$ in our running example, the ground joinability goals associated to the theory $\vec{\mathcal{E}}^{\Delta}$ are:

$$y, (x \cap (V \backslash y)) \downarrow x, (y \cap (V \backslash x)) \mid y \in V = \top \wedge x \in V = \top$$

$$y, ((x, W) \cap (V \backslash y)) \downarrow x, ((y, W) \cap (V \backslash x)) \mid y \in V = \top \wedge x \in V = \top$$

corresponding to the only two CCPs not already proved joinable for $\vec{\mathcal{E}}_0$ that cannot be proved joinable by Maude's Church-Rosser Checker tool. The proof of the first goal is shown below; we leave proving the second goal as an exercise. Note that, due to width constraints, the proof is broken up into named fragments. As is usual for proof trees, we place the root of the proof tree, fragment $P_0$, at the bottom and the remaining proof branch fragments ascending vertically.

---

$P_{2,3}$
$$\cfrac{\cfrac{}{x, (x \cap ((x, W) \backslash x)) \downarrow x, (x \cap ((x, W) \backslash x)) \mid \top} \text{ JN}}{y, (x \cap ((y, W) \backslash y)) \downarrow x, (y \cap ((y, W) \backslash x)) \mid x \doteq y = \top} \text{ NA}$$

$P_{2,2}$
$$\cfrac{\cfrac{\cfrac{}{y, x \downarrow x, y \mid \top} \text{ JN}}{y, (x \cap ((y, x) \backslash y)) \downarrow x, (y \cap ((y, x) \backslash x)) \mid \top} \text{ CR}}{y, (x \cap ((y, m) \backslash y)) \downarrow x, (y \cap ((y, m) \backslash x)) \mid x \doteq m = \top} \text{ NA}$$

$P_{2,1}$
$$\cfrac{\cfrac{\cfrac{}{y, x \downarrow x, y \mid \top} \text{ JN}}{y, (x \cap ((y, x, W') \backslash y)) \downarrow x, (y \cap ((y, x, W') \backslash x)) \mid \top} \text{ CR}}{y, (x \cap ((y, m, W') \backslash y)) \downarrow x, (y \cap ((y, m, W') \backslash x)) \mid x \doteq m = \top} \text{ NA}$$

$P_2$
$$\cfrac{\cfrac{P_{2,1} \qquad P_{2,2} \qquad P_{2,3}}{y, (x \cap ((y, W) \backslash y)) \downarrow x, (y \cap ((y, W) \backslash x)) \mid x \in (y, W) = \top} \text{ NA}}{y, (x \cap ((z', W) \backslash y)) \downarrow x, (y \cap ((z', W) \backslash x)) \mid y \doteq z' = \top \wedge x \in (z', W) = \top} \text{ NA}$$

$P_1$
$$\cfrac{\cfrac{\cfrac{\cfrac{}{y, (y \cap (y \backslash y)) \downarrow y, (y \cap (y \backslash y)) \mid \top} \text{ JN}}{y, (x \cap (y \backslash y)) \downarrow x, (y \cap (y \backslash x)) \mid x \doteq y = \top} \text{ NA}}{y, (x \cap (y \backslash y)) \downarrow x, (y \cap (y \backslash x)) \mid x \in y = \top} \text{ NA}}{y, (x \cap (z \backslash y)) \downarrow x, (y \cap (z \backslash x)) \mid y \doteq z = \top \wedge x \in z = \top} \text{ NA}$$

$P_0$
$$\cfrac{P_1 \qquad P_2}{y, (x \cap (V \backslash y)) \downarrow x, (y \cap (V \backslash x)) \mid y \in V = \top \wedge x \in V = \top} \text{ NA,}$$

---

**Ground Reducibility Inference System.** To show $\vec{\mathcal{E}}^{\Delta}$ sufficiently complete with respect to $\Omega$ we need to prove goals of the form $red(f(x_1, \ldots, x_n)) \mid \top$, with $f \in \Delta$, $f : s_1 \ldots s_n \to s$ a maximal typing for $f$ in $\vec{\mathcal{E}}$, and $x_i$ of sort $s_i$, $1 \leqslant i \leqslant n$. The inference rules that can be applied to prove such goals include all the shared rules plus the rule:

**Rewrite (RW)**

$$\overline{red(f(\vec{v})) \mid \psi}$$

where $f \in \Delta$ and there is a rule $(f(\vec{u}) \to r \;\; if \;\; \Gamma) \in \vec{E}_f$ and a substitution $\theta$ such that $f(\vec{v}) =_B f(\vec{u})\theta$ and $T_{\Sigma_0/E_0 \cup U \cup B_0} \models \psi \Rightarrow (\Gamma\theta)$.

*Example 3.* In our running example we leave as an exercise for the reader the proofs of ground convergence and sufficient completeness for the first theory inclusion $\vec{\mathcal{E}}_{.=.,\in} \subset \vec{\mathcal{E}}_{.=.,\in,.\backslash.}$ as well as the RPO-modulo-based proof of operational termination of $\vec{\mathcal{E}}$ (for which the MTA tool [20] can be used), and prove in detail both properties for the second theory inclusion $\vec{\mathcal{E}}_{.=.,\in,.\backslash.} \subset \vec{\mathcal{E}}$. The proof of sufficient completeness needs to prove the single goal $red(U \cap V \mid \top)$, with $U, V$ of sort *MSet*. Using the *generating set* $\{\varnothing, x, (y, W)\}$ for sort *MSet*, we get the proof tree (rendered in a table due to width constraints as was done previously):

| | |
|---|---|
| $P_3$ | $\dfrac{\overline{red((y,W) \cap V) \mid y \in V = \top}\;\text{RW} \qquad \overline{red((y,W) \cap V) \mid y \in V = \bot}\;\text{RW}}{red((y,W) \cap V) \mid \top}\;\text{SP}$ |
| $P_2$ | $\dfrac{\overline{red(x \cap V) \mid x \in V = \top}\;\text{RW} \qquad \overline{red(x \cap V) \mid x \in V = \bot}\;\text{RW}}{red(x \cap V) \mid \top}\;\text{SP}$ |
| $P_1$ | $\dfrac{}{red(\varnothing \cap V) \mid \top}\;\text{RW}$ |
| $P_0$ | $\dfrac{P_1 \qquad P_2 \qquad P_3}{red(U \cap V) \mid \top}\;\text{CA}$ |

Therefore, assuming that the remaining proof obligations left as exercises for the reader have already been discharged, we have proved for our running example that: (1) the theory $\vec{\mathcal{E}}^\Delta$ for multisets of natural numbers is ground convergent, and (2) all instances of the term $U \cap V$ by a ground constructor substitution $\rho$ are $\vec{\mathcal{E}}^\Delta$-reducible, i.e., $\vec{\mathcal{E}}^\Delta$ is sufficiently complete with respect to $\Omega$. Thanks to our methodology, we can conclude that (3) $\vec{\mathcal{E}}$ itself is also ground convergent and sufficiently complete with respect to $\Omega$, thus illustrating the entire hierarchical methodology. Fact (3) is a consequence of the following general theorem, whose proof can be found in Appendix A:

**Theorem 1.** *Under the already-stated assumptions on a theory inclusion $\vec{\mathcal{E}}_0 \subset \vec{\mathcal{E}}$, if $\vec{\mathcal{E}}^\Delta$ is ground convergent and sufficiently complete with respect to $\Omega$, then $\vec{\mathcal{E}}$ is also ground convergent and sufficiently complete with respect to $\Omega$.*

Of course, the *correctness* of the hierarchical proof methodology crucially depends on the soundness of its inference system, i.e., on the following theorem, whose proof can also be found in Appendix A:

**Theorem 2** *(Soundness Theorem). Under the stated assumptions for the theory inclusion $\vec{\mathcal{E}}_0 \subset \vec{\mathcal{E}}^\Delta$ and for the joinability and reducibility goals, if the inference system proves a joinability goal of the form $t \downarrow t' \mid \varphi$, then $t \downarrow t' \mid \varphi$ holds in $\vec{\mathcal{E}}^\Delta$. Likewise, if the inference system proves a reducibility goal of the form $red(f(u_1, \ldots, u_n)) \mid \varphi$, then $red(f(u_1, \ldots, u_n)) \mid \varphi$ holds in $\vec{\mathcal{E}}^\Delta$.*

# 4    Related Work and Conclusions

Research on sufficient completeness goes back to Guttag's thesis in the 1970's and includes, e.g., [5, 8, 22–25, 27, 29, 30, 38, 39, 42, 44].

Early papers on methods to prove ground confluence appeared in the 1980s,s, including [46] and [41]. Subsequent work includes, e.g., [1, 3, 4, 12, 14, 16, 17, 28]. Since confluence implies ground confluence, work on methods and tools to prove confluence, e.g., [2, 26, 39, 43, 45] is also relevant. However, there are many ground confluent specifications that are not confluent.

Both sufficient completeness (even for unconditional theories as soon as a symbol is associative or associative-commutative) and ground confluence (again, even for conditional theories) are *undecidable* properties (see, respectively, [29] and [28]). This is not surprising, since both are *inductive* properties.

On sufficient completeness, two papers most closely related to this work, because both deal with order-sorted theories, are [39] and [4]. The work in [39] provides some useful methods for proving sufficient completeness of order-sorted CafeOBJ specifications and shares with our work the feature of exploiting module hierarchies; however, the methods used in [39] do not seem to support rewriting modulo axioms. The work in [4] shares a number of important ideas with the present work, including: (i) it supports conditional order-sorted theories; and (ii) it emphasizes that proofs of sufficient completeness and of ground confluence help each other, and, like us, it provides an inference system to prove both properties. Differences from [4] include that it does not support rewriting modulo axioms and that—as in the SPIKE prover [6], whose implementation it extends—instantiation of variables by terms in a generating set are favored over unification and narrowing-based approaches like ours. In that sense, our work also bears some loose similarities to an extensive body of work on ground confluence proof methods originating in the "inductionless induction" approach to inductive theorem proving, including, [1, 3, 14, 16, 17], all of which use narrowing and unification in their inference rules. However, besides the fact that the inference systems in that body of work are quite different from ours, none of that work supports order-sorted theories or rewriting modulo axioms. The work on ground confluence that is most closely related to ours is the one in [12]. In fact, the present work should be seen as further progress along the lines initiated in [12]. Specifically: (i) as pointed out in Sect. 3, our inference rules for ground confluence include and extend those in [12]; and (ii) our hierarchical methods for proving ground confluence of conditional order-sorted theories extend and complement those presented in [12]. A key improvement in terms of greater applicability is that the theory inclusions $\vec{\mathcal{E}}_0 \subset \vec{\mathcal{E}}$ allowed in [12] had to obey the fairly restrictive assumption that some chosen sorts in $\vec{\mathcal{E}}_0$ could not have any extra terms in $\vec{\mathcal{E}}$. Our use of the theory inclusion $\vec{\mathcal{E}}_0 \subset \vec{\mathcal{E}}^\Delta$, for which that assumption holds by construction, completely obviates this restriction.

In conclusion, we have presented a new hierarchical methodology to prove conditional equational programs sufficiently complete and ground convergent. We have illustrated how the inference system works with the help of a running example. More inference rules can be added. For example, the **Unfeasibility** rule in [11] is an obvious addition. Also, a tool combining the inference systems in [12,33] and in this work would allow further experimentation and would be quite useful in many verification efforts, not just for Maude, but also for the less general cases of many-sorted or unsorted equational programs in any language.

**Acknowledgements.** We cordially thank the anonymous referees for their very helpful suggestions, that have helped us improve the manuscript.

# A    Proofs

### Proof of the Soundness Theorem 2

*Proof.* For each inference rule we must show that if the premises of the rule hold, then the conclusion follows. We do so for each inference rule. Recall that in all applications, i.e., to prove either a ground joinability or a ground reducibility property in $\vec{\mathcal{E}}^\Delta$, the meaning of $p \mid \varphi$ holding is that it does so for all its ground constructor substitutions $\rho$ such that $\varphi\rho$ holds in $\vec{\mathcal{E}}_0$.

**Shared Inference Rules.** Except for rule **GN**, all these rules correspond to *equivalences*. That is, the premises hold iff the conclusion does. Let us consider each inference rule.

- **NA.** For any ground constructor substitution $\rho$, at position $p$ in $\varphi$ the term $f(\vec{v}\rho)$ has constructor term arguments. Therefore, by sufficient completeness of $\vec{\mathcal{E}}_0$, there is a rewrite rule in $\vec{E}_{0_f}$, say rule $[i]$ whose lefthand side $f(\vec{u}_i)$ is $B$-matched by $f(\vec{v}\rho)$ with a ground constructor substitution $\gamma$, i.e., $f(\vec{v})\rho =_B f(u_i)\gamma$, and whose condition instance $\Gamma_i\gamma$ holds in $\vec{\mathcal{E}}_0$. Therefore, we can rewrite $f(\vec{v}\rho)$ to the instance $r_i\gamma$ of its righthand side. Therefore, there is a $B$-unifier $\alpha_{i,j}$ of the equation $f(\vec{v}) = f(u_i)$ and a ground constructor substitution $\delta$ such that $\rho \uplus \gamma = \alpha_{i,j}\delta$. Therefore $\varphi\rho$ holds in $\vec{\mathcal{E}}_0$ iff $(\Gamma_i \cup \varphi[r_i]_p)\alpha_{i,j}\delta$ does, and of course $u\rho =_B u\alpha_{i,j}\delta$. In brief, the equivalence summarizes symbolically (by narrowing) all the possible ways in which all ground constructor instances of condition $\varphi$ can be rewritten in one step at position $p$.

– **UN.** If $(\varphi \wedge \psi)\rho$ holds in $\vec{\mathcal{E}}_0$, then $\psi\rho$ does, i.e., $\rho$ is a $U \cup E_0 \cup B_0$-unifier of $\psi$. Therefore, there must be a $U \cup E_0 \cup B_0$-unifier $\theta$ of $\psi$ and a ground constructor substitution $\gamma$ such that $\rho =_{U \cup E_0 \cup B_0} \theta\gamma$. The equivalence follows naturally from this fact.

– **ES.** The main result about equality predicates in [21] is that for any Boolean formula $\varphi$ and ground constructor substitution $\rho$, $\varphi\rho$ holds in ground convergent $\vec{\mathcal{E}}_0$ iff $\varphi!_{\mathcal{E}_0}=\rho$ does. In particular, this equivalence holds when $\varphi$ is a conjunction of equalities.

– **CA.** The equivalence follows from the definition of a generating set for the sort $s$ of $x$, since for any ground constructor substitution $\rho$, $\rho(x)$ must be such that $\rho(x) =_B v_i\gamma$ for some $v_i$ in such a set and ground constructor substitution $\gamma$.

– **SP.** The equivalence between the premises and the conclusion follows from the semantic equivalence $T_{\Sigma_0/E_0 \cup B_0} \models \varphi \Leftrightarrow \bigvee_{i \in I} \psi_i \wedge \varphi$, plus the Boolean equivalence $(A \vee B) \Rightarrow C \equiv (A \Rightarrow C) \wedge (B \Rightarrow C)$.

– **GN.** This is the only shared rule tat is not an equivalence, i.e., where the premise implies the consequence but need not be equivalent to it. The property $p'\rho$ must hold (i.e., $p'\rho$'s ground reducibility, or $p'\rho$'s ground joinability, depending on $p$) whenever $\psi\rho$ does. In particular, if $\varphi\gamma$ holds, then $\psi\theta\gamma$ does, and therefore $p'\theta\gamma$ does. That is, $p'\theta \mid \varphi$ holds. But $p'\theta =_B p$. The result then follows from the fact that for either ground reducibility or ground joinability properties $q, q'$ such that $q =_B q'$, $q \mid \varphi$ holds iff $q' \mid \varphi$ does. This follows in either case from the assumption that the rules $\vec{U} \cup \vec{E}$ are strictly $B$-coherent.

– $\emptyset$. Since no ground substitution can satisfy $\bot$, $u \mid \bot$ holds trivially.

**Ground Joinability Inference System.** The proof of the constrained version of the ground confluence inference rules in [12] follows easily from that of the unconstrained inference rules in [12]. The soundness of rule **JN** holds trivially from the very notion of joinability. A proof of soundness for the **CR** inference rule can be found in [11].

**Ground Reducibility Inference System.** The only inference rule is **RW**. Suppose that $\psi\rho$ holds in $\vec{\mathcal{E}}_0$. Then, $\Gamma\theta\rho$ does; and by the rule's assumptions $f(\vec{v})\rho$ is reducible, as desired.

This finishes the proof of the Soundness Theorem.    □

## Proof of Theorem 1

*Proof.* First of all, note that, considering $T_\Sigma(X)$ and $T_{\Sigma^\Delta}(X)$ as *sets*, i.e., disregarding sorts, we have an inclusion $T_\Sigma(X) \subseteq T_{\Sigma^\Delta}(X)$. Also, for each $s \in S$ we have a set equality $T_{\Sigma_0,s}(X) = T_{\Sigma^\Delta,s}(X)$. In particular, $T_\Sigma \subseteq T_{\Sigma^\Delta}$, and $T_{\Sigma_0,s} = T_{\Sigma^\Delta,s}$ for each $s \in S$.

Second, $\vec{\mathcal{E}}$ and $\vec{\mathcal{E}}^\Delta$ have the exact same CCP's. To begin with, in both cases the rules not in $\vec{\mathcal{E}}_0$ are the same, namely $\vec{E}_\Delta$. Furthermore, in both cases, the only CCP's that do not come from $\vec{\mathcal{E}}_0$ can be of only two kinds: (i) between a unit rule in $\vec{U}$ and a rule in $\vec{E}_f$ for some $f \in \Delta$, where the unit rule's lefthand side unifies with a constructor subterm of the lefthand side of one of $f$'s constructor arguments; or (ii) between two, not necessarily different, rules in $\vec{E}_f$ for some $f \in \Delta$. In case (i), the unifier generating the CCP must be a constructor unifier so that the resulting CCP is the same in both $\vec{\mathcal{E}}$ and $\vec{\mathcal{E}}^\Delta$, and its condition is a $\Sigma_0$-condition. In case (ii), the CCP comes from two—not necessarily different, but variable-renamed if $i = j$ to ensure disjoint variables—rules $[i] : f(\vec{u_i}) \rightarrow r_i$ if $\Gamma_i$ and $[j] : f(\vec{u_j}) \rightarrow r_j$ if $\Gamma_j$ and its associated order-sorted unifier (in either $\vec{\mathcal{E}}$ or $\vec{\mathcal{E}}^\Delta$) solves the equation $f(\vec{u_i}) = f(\vec{u_j})$. We claim that the order-sorted unifiers of the equation $f(\vec{u_i}) = f(\vec{u_j})$ are the same in $\vec{\mathcal{E}}$ and in $\vec{\mathcal{E}}^\Delta$. Recall that, by assumption, $B_f$ is either empty or a commutativity axiom. If $B_f = \varnothing$, then $\alpha$ is a unifier of $f(\vec{u_i}) = f(\vec{u_j})$ iff it is a unifier of the system of equations $u_{i,1} = u_{j,1} \wedge \ldots \wedge u_{i,k} = u_{j,k}$, where $k$ is the number of arguments of $f$. If $f$ is commutative, the only difference is that in $\vec{\mathcal{E}}$ the axiom $f(x_1, x_2) = f(x_2, x_1)$ is such that $x_1, x_2$ have sort $s$ for $f : s\ s \rightarrow s_0$ the maximal typing of $f$, whereas in $\vec{\mathcal{E}}^\Delta$ $x_1, x_2$ have kind $[s]$. This, however, makes no difference, since, by the **Decomposition** inference rule for a commutative symbol of order-sorted unification (see [31] and [7] §15.1), $\alpha$ is a unifier of $f(u_{i,1}, u_{i,2}) = f(u_{j,1}, u_{j,2})$ iff it is a unifier of the disjunction of systems of equations $(u_{i,1} = u_{j,1} \wedge u_{i,2} = u_{j,2}) \vee (u_{i,1} = u_{j,2} \wedge u_{i,2} = u_{j,1})$. Therefore, the CCPs are the same and the unifiers are constructor unifiers, so that the CCP's condition is a $\Sigma_0$-condition.

Third, for ground terms we have proper inclusions of rewrite relations,

$$\rightarrow_{\vec{\mathcal{E}}_0} \subset \rightarrow_{\vec{\mathcal{E}}^\Delta} \subset \rightarrow_{\vec{\mathcal{E}}} \subset T_\Sigma^\Delta \times T_\Sigma^\Delta.$$

The first inclusion is proper because there are terms in $T_\Sigma \backslash T_{\Sigma_0}$ that can be rewritten with $\rightarrow_{\vec{\mathcal{E}}^\Delta}$. The second inclusion is proper because, by the definition of $\Sigma^\Delta$, a rule in the theory $\vec{\mathcal{E}}_\Delta$, say, $[i] : f(\vec{u_i}) \rightarrow r_i$ if $\Gamma_i$, can, only be enabled to rewrite a term $f(\vec{v})$ if the terms $\vec{v}$ are $\Sigma_0$-terms. That is, $\rightarrow_{\vec{\mathcal{E}}^\Delta}$ performs rewritings exactly like $\rightarrow_{\vec{\mathcal{E}}}$, but only in a "weakly innermost" manner ("weakly" because the $\Sigma_0$-terms $\vec{v}$ need not be constructors).

Fourth, for any $t \in T_\Sigma$, $t!_{\vec{\mathcal{E}}^\Delta}$ is a constructor term. Suppose not, i.e., there is a $t \in T_\Sigma$ such that $t!_{\vec{\mathcal{E}}^\Delta}$ is not a constructor term. But since we have an inclusion of rewrite relations $\rightarrow_{\vec{\mathcal{E}}_0} \subset \rightarrow_{\vec{\mathcal{E}}^\Delta}$ and $\vec{\mathcal{E}}_0$ is sufficiently complete, this means that $t!_{\vec{\mathcal{E}}^\Delta}$ must contain a subterm of minimal size of the form $f(\vec{v})$ with $f \in \Delta$ and the terms $\vec{v}$ constructor terms. But this is impossible, since all such terms have been proved $\vec{\mathcal{E}}^\Delta$-reducible.

Fifth, for any $t \in T_\Sigma$, if $t \rightarrow_{\vec{\mathcal{E}}}^* v$ and $v$ is in $\vec{\mathcal{E}}$-canonical form, then $v$ is a constructor term. This follows from the containments of rewrite relations $\rightarrow_{\vec{\mathcal{E}}_0} \subset \rightarrow_{\vec{\mathcal{E}}^\Delta} \subset \rightarrow_{\vec{\mathcal{E}}}$, the fourth property above, and the sufficient completeness of $\vec{\mathcal{E}}_0$.

Finally, we are now ready to prove that $\vec{\mathcal{E}}$ is ground convergent. Note that, by the fifth property above, $\vec{\mathcal{E}}$ is then also sufficiently complete with respect to $\Omega$. Since we have the containment of ground rewrite relations $\rightarrow_{\vec{\mathcal{E}}^\Delta} \subset \rightarrow_{\vec{\mathcal{E}}}$, the ground convergence of $\vec{\mathcal{E}}$ will follow from the fourth and fifth properties above if we can prove that for each $t \in T_\Sigma$ and each ground constructor term $v$ such that $t \rightarrow_{\vec{\mathcal{E}}}^* v$ we have $v =_B t!_{\vec{\mathcal{E}}^\Delta}$.

**Lemma 1.** *For each $t \in T_\Sigma$, if $t \to^*_{\vec{\mathcal{E}}} u$ and $u$ is a constructor term, then $u =_{B_\Omega} t!_{\vec{\mathcal{E}}^\Delta}$.*

*Proof.* Suppose not. Let us choose a term $t \in T_\Sigma$ such that: (i) $t \to^*_{\vec{\mathcal{E}}} u$, $u$ is a constructor term, and $u \neq_{B_\Omega} t!_{\vec{\mathcal{E}}^\Delta}$, and (ii) for $>$ the RPO order modulo proving $\vec{\mathcal{E}}$ operationally terminating, $t$ is a minimal element among the set of terms in $T_\Sigma$ such that (i) holds. This can only happen if $t$ is not a constructor term. Therefore, we have $t \to_{\vec{\mathcal{E}}} t' \to^*_{\vec{\mathcal{E}}} u$. Note that $t > t'$. Therefore, by the minimality assumption for $t$, we must have $u =_{B_\Omega} t'!_{\vec{\mathcal{E}}^\Delta}$. Let us now consider the one-step rewrite $t \to_{\vec{\mathcal{E}}} t'$. This means that there is a rule $f(\vec{u}) \to r$ if $\Gamma$ in $\vec{U} \cup \vec{E}$ with the $\vec{u}$ constructor terms (rules in $\vec{U}$, though unconditional, also have this form), a ground substitution $\alpha$ and a term position $p$ such that $t|_p =_B f(\vec{u})\alpha$, $\Gamma\alpha$ holds in $\vec{\mathcal{E}}$, and $t' = t[r\alpha]_p$. Since $>$ is a $B$-compatible RPO order and all rules are assumed $>$-operationally-terminating, for each equality $w = w'$ in $\Gamma$ we must have $t > w\alpha, w'\alpha$. Therefore, by the minimality hypothesis on $t$, we must have $(w\alpha)!_{\vec{\mathcal{E}}^\Delta} =_{B_\Omega} (w\alpha)!_{\vec{\mathcal{E}}} =_{B_\Omega} (w'\alpha)!_{\vec{\mathcal{E}}} =_{B_\Omega} (w'\alpha)!_{\vec{\mathcal{E}}^\Delta}$, so that $\Gamma\alpha$ also holds in $\vec{\mathcal{E}}^\Delta$ and, for the same reason, $\Gamma\rho$ holds in $\vec{\mathcal{E}}^\Delta$ for the constructor substitution $\rho = \alpha!_{\vec{\mathcal{E}}^\Delta}$ obtained by normalizing each $\alpha(x)$ with $x$ in the domain of $\alpha$. Therefore, we have a rewrite $t[f(\vec{u})\rho]_p \to_{\vec{\mathcal{E}}^\Delta} t[r\rho]_p$. Furthermore, $t =_B t[f(\vec{u})\alpha]_p$, and we have rewrite sequences $t[f(\vec{u})\alpha]_p \to^*_{\vec{\mathcal{E}}^\Delta} t[f(\vec{u})\rho]_p$, and $t[r\alpha]_p \to^*_{\vec{\mathcal{E}}^\Delta} t[r\rho]_p$, and since $t > t' = t[r\alpha]_p$, we must have $u =_{B_\Omega} t[r\rho]_p!_{\vec{\mathcal{E}}^\Delta}$. In summary, we have the sequence of rewrites in $\vec{\mathcal{E}}^\Delta$,

$$t[f(\vec{u})\alpha]_p \to^*_{\vec{\mathcal{E}}^\Delta} t[f(\vec{u})\rho]_p \to_{\vec{\mathcal{E}}^\Delta} t[r\rho]_p \to^*_{\vec{\mathcal{E}}^\Delta} t[r\rho]_p!_{\vec{\mathcal{E}}^\Delta}$$

with $u =_{B_\Omega} t[r\rho]_p!_{\vec{\mathcal{E}}^\Delta}$. But by $t =_B t[f(\vec{u})\alpha]_p$ and the convergence of $\vec{\mathcal{E}}^\Delta$ we also must have $t!_{\vec{\mathcal{E}}^\Delta} =_{B_\Omega} t[r\rho]_p!_{\vec{\mathcal{E}}^\Delta} =_{B_\Omega} u$, contradicting the assumption $u \neq_{B_\Omega} t!_{\vec{\mathcal{E}}^\Delta}$, as desired. □

This finishes the proof of Theorem 1. □

# References

1. Aoto, T., Toyama, Y.: Ground confluence prover based on rewriting induction. In: 1st International Conference on Formal Structures for Computation and Deduction, FSCD 2016. LIPIcs, vol. 52, pp. 33:1–33:12. Schloss Dagstuhl - Leibniz-Zentrum für Informatik (2016)
2. Aoto, T., Yoshida, J., Toyama, Y.: Proving confluence of term rewriting systems automatically. In: Treinen, R. (ed.) RTA 2009. LNCS, vol. 5595, pp. 93–102. Springer, Heidelberg (2009). https://doi.org/10.1007/978-3-642-02348-4_7
3. Becker, K.: Proving ground confluence and inductive validity in constructor based equational specifications. In: Gaudel, M.-C., Jouannaud, J.-P. (eds.) CAAP 1993. LNCS, vol. 668, pp. 46–60. Springer, Heidelberg (1993). https://doi.org/10.1007/3-540-56610-4_55
4. Bouhoula, A.: Simultaneous checking of completeness and ground confluence for algebraic specifications. ACM Trans. Comput. Log. **10**(3), 20:1–20:33 (2009)

5. Bouhoula, A., Jouannaud, J.P.: Automata-driven automated induction. Inf. Comput. **169**(1), 1–22 (2001)
6. Bouhoula, A., Rusinowitch, M.: Implicit induction in conditional theories. J. Autom. Reason. **14**(2), 189–235 (1995). https://doi.org/10.1007/BF00881856
7. Clavel, M., et al.: All About Maude - A High-Performance Logical Framework. LNCS, vol. 4350. Springer, Heidelberg (2007). https://doi.org/10.1007/978-3-540-71999-1
8. Comon, H.: Sufficient completeness, term rewriting systems and "anti-unification". In: Siekmann, J.H. (ed.) CADE 1986. LNCS, vol. 230, pp. 128–140. Springer, Heidelberg (1986). https://doi.org/10.1007/3-540-16780-3_85
9. Dershowitz, N., Jouannaud, J.P.: Rewrite systems. In: van Leeuwen, J. (ed.) Handbook of Theoretical Computer Science, vol. B, pp. 243–320. North-Holland (1990)
10. Durán, F., Lucas, S., Meseguer, J.: Termination modulo combinations of equational theories. In: Ghilardi, S., Sebastiani, R. (eds.) FroCoS 2009. LNCS (LNAI), vol. 5749, pp. 246–262. Springer, Heidelberg (2009). https://doi.org/10.1007/978-3-642-04222-5_15
11. Durán, F., Meseguer, J.: On the Church-Rosser and coherence properties of conditional order-sorted rewrite theories. J. Algebraic Log. Program. **81**, 816–850 (2012)
12. Durán, F., Meseguer, J., Rocha, C.: Ground confluence of order-sorted conditional specifications modulo axioms. J. Log. Algebraic Methods Program. **111**, 100513 (2020)
13. Escobar, S., Sasse, R., Meseguer, J.: Folding variant narrowing and optimal variant termination. J. Algebraic Log. Program. **81**, 898–928 (2012)
14. Fribourg, L.: A strong restriction of the inductive completion procedure. J. Symb. Comput. **8**(3), 253–276 (1989)
15. Futatsugi, K., Diaconescu, R.: CafeOBJ Report. World Scientific, Singapore (1998)
16. Ganzinger, H.: Ground term confluence in parametric conditional equational specifications. In: Brandenburg, F.J., Vidal-Naquet, G., Wirsing, M. (eds.) STACS 1987. LNCS, vol. 247, pp. 286–298. Springer, Heidelberg (1987). https://doi.org/10.1007/BFb0039613
17. Göbel, R.: Ground confluence. In: Lescanne, P. (ed.) RTA 1987. LNCS, vol. 256, pp. 156–167. Springer, Heidelberg (1987). https://doi.org/10.1007/3-540-17220-3_14
18. Goguen, J., Meseguer, J.: Order-sorted algebra I: equational deduction for multiple inheritance, overloading, exceptions and partial operations. Theoret. Comput. Sci. **105**, 217–273 (1992)
19. Goguen, J., Winkler, T., Meseguer, J., Futatsugi, K., Jouannaud, J.P.: Introducing OBJ. In: Software Engineering with OBJ: Algebraic Specification in Action, pp. 3–167. Kluwer (2000)
20. Gutiérrez, R., Meseguer, J., Skeirik, S.: The Maude termination assistant. In: Preproceedings of WRLA (2018)
21. Gutiérrez, R., Meseguer, J., Rocha, C.: Order-sorted equality enrichments modulo axioms. Sci. Comput. Program. **99**, 235–261 (2015)
22. Guttag, J.V., Horning, J.J.: The algebraic specification of abstract data types. Acta Inform. **10**, 27–52 (1978)
23. Hendrix, J., Clavel, M., Meseguer, J.: A sufficient completeness reasoning tool for partial specifications. In: Giesl, J. (ed.) RTA 2005. LNCS, vol. 3467, pp. 165–174. Springer, Heidelberg (2005). https://doi.org/10.1007/978-3-540-32033-3_13
24. Hendrix, J., Meseguer, J.: On the completeness of context-sensitive order-sorted specifications. In: Baader, F. (ed.) RTA 2007. LNCS, vol. 4533, pp. 229–245. Springer, Heidelberg (2007). https://doi.org/10.1007/978-3-540-73449-9_18

25. Hendrix, J., Meseguer, J., Ohsaki, H.: A sufficient completeness checker for linear order-sorted specifications modulo axioms. In: Furbach, U., Shankar, N. (eds.) IJCAR 2006. LNCS (LNAI), vol. 4130, pp. 151–155. Springer, Heidelberg (2006). https://doi.org/10.1007/11814771_14
26. Hirokawa, N., Klein, D.: Saigawa: a confluence tool. In: Proceedings of 1st International Workshop on Confluence (IWC 2012), p. 57 (2011). http://cl-informatik.uibk.ac.at/iwc/iwc2012.pdf
27. Jouannaud, J.P., Kounalis, E.: Automatic proofs by induction in theories without constructors. Inf. Comput. **82**(1), 1–33 (1989)
28. Kapur, D., Narendran, P., Otto, F.: On ground-confluence of term rewriting systems. Inf. Comput. **86**(1), 14–31 (1990)
29. Kapur, D., Narendran, P., Rosenkrantz, D.J., Zhang, H.: Sufficient-completeness, ground-reducibility and their complexity. Int. J. Biometeorol. **36**(4), 311–350 (1991). https://doi.org/10.1007/BF01212959
30. Kikuchi, K., Aoto, T.: Simple derivation systems for proving sufficient completeness of non-terminating term rewriting systems. In: 41st IARCS Annual Conference on Foundations of Software Technology and Theoretical Computer Science, FSTTCS 2021. LIPIcs, vol. 213, pp. 49:1–49:15. Schloss Dagstuhl - Leibniz-Zentrum für Informatik (2021)
31. Kirchner, C.: Order-sorted equational unification. Technical report 954. INRIA Lorraine & LORIA, Nancy, France, December 1988
32. Lucas, S., Meseguer, J.: Normal forms and normal theories in conditional rewriting. J. Log. Algebr. Meth. Program. **85**(1), 67–97 (2016)
33. Meseguer, J.: Checking sufficient completeness by inductive theorem proving. In: In: Bae, K. (ed.) WRLA 2022. LNCS, vol. 13252, pp. 171–190. Springer, Cham (2022)
34. Meseguer, J., Skeirik, S.: Inductive reasoning with equality predicates, contextual rewriting and variant-based simplification. In: Escobar, S., Martí-Oliet, N. (eds.) WRLA 2020. LNCS, vol. 12328, pp. 114–135. Springer, Cham (2020). https://doi.org/10.1007/978-3-030-63595-4_7
35. Meseguer, J.: Conditional rewriting logic as a unified model of concurrency. Theoret. Comput. Sci. **96**(1), 73–155 (1992)
36. Meseguer, J.: Membership algebra as a logical framework for equational specification. In: Presicce, F.P. (ed.) WADT 1997. LNCS, vol. 1376, pp. 18–61. Springer, Heidelberg (1998). https://doi.org/10.1007/3-540-64299-4_26
37. Meseguer, J.: Variant-based satisfiability in initial algebras. Sci. Comput. Program. **154**, 3–41 (2018)
38. Meseguer, J., Skeirik, S.: Equational formulas and pattern operations in initial order-sorted algebras. Formal Aspects Comput. **29**(3), 423–452 (2017). https://doi.org/10.1007/s00165-017-0415-5
39. Nakamura, M., Ogata, K., Futatsugi, K.: Incremental proofs of termination, confluence and sufficient completeness of OBJ specifications. In: Iida, S., Meseguer, J., Ogata, K. (eds.) Specification, Algebra, and Software. LNCS, vol. 8373, pp. 92–109. Springer, Heidelberg (2014). https://doi.org/10.1007/978-3-642-54624-2_5
40. Peterson, G.E., Stickel, M.E.: Complete sets of reductions for some equational theories. J. Assoc. Comput. Mach. **28**(2), 233–264 (1981)
41. Plaisted, D.A.: A logic for conditional term rewriting systems. In: Kaplan, S., Jouannaud, J.-P. (eds.) CTRS 1987. LNCS, vol. 308, pp. 212–227. Springer, Heidelberg (1988). https://doi.org/10.1007/3-540-19242-5_16

42. Shiraishi, T., Kikuchi, K., Aoto, T.: A proof method for local sufficient complete-
    ness of term rewriting systems. In: Cerone, A., Ölveczky, P.C. (eds.) ICTAC 2021.
    LNCS, vol. 12819, pp. 386–404. Springer, Cham (2021). https://doi.org/10.1007/
    978-3-030-85315-0_22
43. Sternagel, T., Middeldorp, A.: Conditional confluence (system description). In:
    Dowek, G. (ed.) RTA 2014. LNCS, vol. 8560, pp. 456–465. Springer, Cham (2014).
    https://doi.org/10.1007/978-3-319-08918-8_31
44. Thiel, J.J.: Stop losing sleep over incomplete data type specification. In: Kennedy,
    K. (ed.) Proceedings of Eleventh Symposium on Principles of Programming Lan-
    guages. Association for Computing Machinery (1984)
45. Zankl, H., Felgenhauer, B., Middeldorp, A.: CSI – a confluence tool. In: Bjørner,
    N., Sofronie-Stokkermans, V. (eds.) CADE 2011. LNCS (LNAI), vol. 6803, pp.
    499–505. Springer, Heidelberg (2011). https://doi.org/10.1007/978-3-642-22438-
    6_38
46. Zhang, H., Remy, J.-L.: Contextual rewriting. In: Jouannaud, J.-P. (ed.) RTA 1985.
    LNCS, vol. 202, pp. 46–62. Springer, Heidelberg (1985). https://doi.org/10.1007/
    3-540-15976-2_2

# Automating Safety Proofs About Cyber-Physical Systems Using Rewriting Modulo SMT

Vivek Nigam[2,3(✉)] and Carolyn Talcott[1]

[1] SRI International, Menlo Park, USA
clt@csl.sri.com
[2] Federal University of Paraíba, João Pessoa, Brazil
vivek.nigam@gmail.com
[3] Huawei Munich Research Center, Munich, Germany

**Abstract.** Cyber-Physical Systems, such as Autonomous Vehicles (AVs), are operating with high-levels of autonomy allowing them to carry out safety-critical missions with limited human supervision. To ensure that these systems do not cause harm, their safety has to be rigorously verified. Existing works focus mostly on using simulation-based methods which execute simulations on concrete instances of logical scenarios in which systems are expected to function. The level of assurance obtained by these methods is, therefore, limited by the number of simulations that can be carried out. A complementary approach is to produce, instead, proofs that vehicles are safe for all instances of logical scenarios. This paper investigates how Rewriting modulo SMT applied to Soft Agents, a rewriting framework for the specification and verification of Cyber-Physical system, can be used to generate such proofs in an automated fashion. In particular, rewrite rules specify the executable semantics of systems on logical scenarios instead of concrete scenarios. This is accomplished by generating at each execution step a set of (non-linear) constraints whose satisfiability are checked by using SMT-solvers. Intuitively, a model of such set of constraints corresponds to a concrete execution on an instance of the corresponding logical scenario. We demonstrate how to specify and verify scenarios in this framework using an example involving a vehicle platoon. Finally, we investigate the trade-offs between how much of the verification is delegated to search engines (namely Maude) and how much is delegated to SMT-solvers (e.g., Z3).

## 1 Introduction

Autonomous Vehicles (AVs) are expected to soon reach higher-levels of autonomy, being able to drive through complex environments with no or little human supervision. To achieve this, however, it is necessary to produce a rigorous safety assurance argument [12]. An assurance strategy based on collecting data by running AVs on the streets is not feasible [13] as it would require billions of miles of data for achieving confidence in the results. Symbolic methods based on formal models have been advocated [23] as a means for safety assurance.

© Springer Nature Switzerland AG 2022
K. Bae (Ed.): WRLA 2022, LNCS 13252, pp. 212–229, 2022.
https://doi.org/10.1007/978-3-031-12441-9_11

A safety assurance strategy begins by first identifying abstract scenarios, called logical scenarios [19], such as lane changing or platooning or pedestrian crossing, in which AVs have to avoid harm. These logical scenarios contain details about the situations in which a vehicle shall be able to safely operate,[1] such as which types and number of actors, e.g., vehicles, pedestrians, operating assumptions, e.g., range of speeds, and road topology, e.g., number of lanes. The system safety is then verified with respect to each scenario. The challenge, however, is that there are infinitely many instances for any given logical scenario.

To overcome this challenge, existing work can be divided into two different approaches. The first approach [5,9,16] is to use simulation-based methods that run a sufficiently large number of simulations using vehicle simulators [8]. A limitation of this approach is that a possibly large number of simulations need to be generated for each logical scenario, and even then critical situations may be missed. The second approach is to use algorithms [1,22] that are proved to generate safe trajectories under the assumption that the remaining agents behave correctly. These safe planners can then be integrated with advanced (high-performance, but not safe) controllers as fall-back options whenever safety assurance is low [7]. There are two limitations with this approach. The first limitation is that safety proofs have to be constructed manually. The second limitation is that these proofs consider only planning and not other aspects such as sensing, knowledge bases, and communication channels that are used in AV applications [5,16].

This paper's main goal is to address the limitations of these two types of approaches by proposing a rewriting framework, based on Soft Agents [24], that enables the automated construction of vehicle level safety proofs, i.e., produce proofs that AVs are safe *for all instances of a logical scenario*. Such safety proofs provide greater confidence on the safety of AVs, complementing other verification evidence such as simulation-based verification techniques.

Towards achieving this goal, we make the following contributions:

- **Soft Agents Framework with Rewriting Modulo SMT:** We propose an executable symbolic Soft Agents framework [24] where instead of considering concrete values for attributes such as agent's speed, position and acceleration, it represents these values as symbols whose possible values are specified by a set of (real non-linear) constraints. This is accomplished by extending the current Soft Agents framework with Rewriting Modulo SMT [20]. Soft Agent specifications can be executed by using Maude extensions with SMT [14]. In contrast to existing frameworks that can execute only instances of logical scenarios, symbolic soft agents can execute logical scenarios producing symbolic traces, each denoting a possibly infinite number of concrete executions of the logical scenario.
- **Vehicle Platooning Specification:** We demonstrate the Soft Agent framework by using a simple, but realistic vehicle platooning application. We illustrate how vehicle behavior and safety properties can be specified in Soft Agents, explaining how design choices may affect verification performance.

---

[1] Also called Operational Design Domain (ODD).

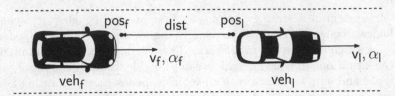

**Fig. 1.** Platooning Logical Scenario: The follower vehicle $veh_f$ and $veh_l$ are in a straight lane with respectively velocities and accelerations $v_f$, $\alpha_f$ and $v_l$, $\alpha_l$. $pos_f$ is the position of front of $veh_f$ and $pos_l$ is the position of the back of the $veh_l$. We consider vehicle positions to be only the x-component increasing with as one follows to the right direction of the road. The distance between the vehicles $dist = pos_l - pos_f$.

- **Verification Trade-off between Rewriting and Constraint Solving:** For the verification of systems, Soft Agents make uses of rewriting (through Maude [4]) and of SMT-solvers (through Z3 [6]). In particular, rewriting captures the evolution of the system by accumulating constraints. The constraint-solver, on the other hand, generates proofs that a property is satisfied or that a property is unsatisfiable. We investigate in this paper the trade-offs between how much of verification is delegated to rewriting and how much to the constraint-solver. On the one hand, the more fine grained is the rewriting, e.g., searching with more constrained system evolutions, the greater is the number of states the search engine has to traverse leading to a greater number of calls to the SMT-solver, but the simpler are the problems that the solver has to solve. On the other hand, the more coarse is the rewriting, e.g., searching with less constrained system evolutions, the fewer are the calls to the SMT-solver, but the larger are problems that the constraint solver has to solve. Our experiments indicate that these trade-offs need to be considered in order to verify more challenging properties.

*Plan.* We start in Sect. 2 by describing a motivating example: a logical scenario from a vehicle platooning case study, which is used as running example. Section 3 introduces symbolic rewriting, then recalls the soft agents framework and its generalization to symbolic rewriting with SMT solving. Section 4 presents key elements of the symbolic vehicle platooning logical scenario, including control decisions, safety properties, and search patterns for reachability analysis. Section 5 presents experiments evaluating trade-offs between size of search space and complexity of constraints to solve. We conclude by discussing related work in Sect. 6 and future work in Sect. 7.

## 2   Motivating Example

Our motivating example is a platooning scenario which is a typical Level 3 autonomy[2] use-case. This scenario takes place in a highway as illustrated by Fig. 1. The vehicle $veh_f$, called follower vehicle, follows autonomously, i.e., only with

---

[2] For the levels of autonomy, see the SAE classification described in [11].

human supervision, vehicle veh$_l$, called leader vehicle. The vehicles are driving in a highway lane and therefore are expected to have a speed within some given range of values normally obtained by considering legal speeds and the vehicle's capabilities, e.g., speeds between 60 km/h and 130 km/h. Moreover, the acceleration (and deceleration) capabilities of the vehicles are also bounded, typically between $-8\,\mathrm{m/s^2}$ and $2\,\mathrm{m/s^2}$.

The goal of the follower vehicle is to maintain a safe distance to the leader vehicle, but still be close enough to profit from the wind shadow of the leader vehicle yielding upto 17% of fuel savings [26]. Since the speed of the vehicles may vary, it is not appropriate to define a safe distance as an absolute quantity, but in terms of *time to react*. That is, the distance will depend on the relative speeds of the vehicles.

As an example, building on ideas from [7], we define the following three properties for the platooning logical scenario:

$$P_{\mathsf{safer}} := \mathsf{dist} \geq \mathsf{v_f} \times (1[s] + \mathsf{gap_{safer}}) - \mathsf{v_l} \times 1[s], \tag{1a}$$

$$P_{\mathsf{safe}} := \mathsf{v_f} \times (1[s] + \mathsf{gap_{safer}}) - \mathsf{v_l} \times 1[s]) > \mathsf{dist} \geq \mathsf{v_f} \times (1[s] + \mathsf{gap_{safe}}) - \mathsf{v_l} \times 1[s], \tag{1b}$$

$$P_{\mathsf{unsafe}} := \mathsf{dist} < \mathsf{v_f} \times (1[s] + \mathsf{gap_{safe}}) - \mathsf{v_l} \times 1[s] \tag{1c}$$

Intuitively, their satisfaction is conditional on the distance (dist) between the vehicles; their speeds (v$_l$ and v$_f$); and the parameters gap$_{\mathsf{safer}}$ and gap$_{\mathsf{safe}}$ which are time to react parameters, typically a few seconds. Moreover, gap$_{\mathsf{safer}}$ > gap$_{\mathsf{safe}}$, which means that the instance of a logical scenario satisfies $P_{\mathsf{safer}}$ (or simply safer) if the vehicles veh$_l$ and veh$_f$ have a greater distance between them. Finally, an instance of a logical scenario satisfies $P_{\mathsf{unsafe}}$ (or simply unsafe) if distance is to small to satisfy $P_{\mathsf{safe}}$ or $P_{\mathsf{safer}}$.

A description of the function of a vehicle, such as platooning, using formal notations and ranges of parameters is called a *logical scenario* [15]. The objective is to prove that an implementation of a controller for the platooning function is safe, that is either $P_{\mathsf{safer}}$ or $P_{\mathsf{safe}}$ is satisfied for all concrete instances of this logical scenario. This is challenging as there are infinitely many such instances.

# 3  Symbolic Soft Agents Framework

We begin with an overview of challenges in modeling cyber-physical systems (CPSs), then recall the main features of soft agent specifications, and then briefly discuss the generalization to symbolic form.

## 3.1  Overview

A soft agent (SA) model of a CPS makes explicit both discrete changes (cyber actions, control settings) and continuous change (in the physical environment). Following ideas developed in Real Time Maude [18], soft agent models have instantaneous rules that specify agents decision processes that generate actions such as communication or setting control parameters; and a timeStep rule that models the passage of some interval of time, updating the state according to a model of the time-dependent aspects of the state.

In contrast to the usual realtime specifications, soft agent CPS specifications involve variables, such as speed, distance, etc., that are dense and their evolutions over time are not discrete events. Moreover, system properties, such as safety properties, are expressed using these variables, e.g., keeping a given distance to the vehicle ahead rather than timing properties such as network delay or execution time. Verification of safety properties for CPS specifications involves reasoning about possibly infinitely many states and properties whose parameters may change continuously over time.

Two challenges for safety analysis of CPS specifications are (1) soundness of discrete time sampling execution; and (2) checking for reachability of unsafe states from a possibly infinite set of instances of a logical scenario. Challenge (1) includes choosing the timestep intervals small enough so that no unsafe situations are missed, while not being so fine grained that the state space becomes unmanageable. This is a design time concern, for example choosing the frequency with which sensors are read and control settings are updated. The latter challenge (2) involves the coverage and state space management with time properties.

Real Time Maude addresses (1) in [17], defining conditions on a timed rewrite theory that guarantee soundness and completeness of model checking based on maximal time elapsed discrete time sampling. Unfortunately, soft agent analysis problems generally do not meet these conditions. Narrowing is one approach to checking reachability from a possibly infinite initial set of system states. Maude supports narrowing modulo a rich collection of equational theories, but narrowing using conditional rules is not supported [4], and soft agent relies heavily on conditional rules.

New ideas are needed to address the verification challenges. We propose a form of symbolic rewriting that combines rewriting and constraint solving.

1. We represent *logical scenarios* as symbolic system states, representing a set of concrete states. A logical scenario consists of a pattern (a term with pattern variables called symbols) together with a set of constraints on values of the symbols.[3]
2. A symbolic rewrite rule introduces new symbols and additional constraints representing new values of the pattern variables. The resulting logical scenario represents the instances reachable from instances of the starting pattern using the rewrite rule.
3. Symbolic rule conditions use symbolic function evaluation to generate new symbols and their constraints.

The point of symbolic analysis is to check properties of concrete systems represented by concrete scenarios. Thus we want to connect symbolic executions to concrete executions. The concrete executions may be obtained from a concrete

---

[3] Mathematically, a logical scenario is a term with variables. To be able to rewrite logical scenarios in Maude, we replace variables by symbols, which formally are uninterpreted constants.

form of the rewrite rules, or simply using the symbolic rules with grounding constraints of the form, sym == ground term.

To describe the desired symbolic-concrete connection, we need a little notation. The basic idea is analogous to that presented in [20]. We assume a rewrite theory $\mathcal{T} = (\Sigma, B \cup E, R)$ with signature $\Sigma$, axioms $B$, equations $E$, and rules $R$. Assume further an equational subtheory $\mathcal{T}_0$ of $\mathcal{T}$ axiomatizing the theory in which the constraints are solved by the SMT solver. We use sS, $\mathsf{sS}_0$, $\mathsf{sS}_1$ .... to denote logical scenarios (symbolic states) and cS, $\mathsf{cS}_0$, $\mathsf{cS}_1$ ... to denote concrete states (ground states with no symbols). Let $\sigma$, $\sigma_0$, $\sigma_1$, ... denote substitutions mapping symbols to concrete terms (values). A logical scenario is structured as a pair (sP, sC) consisting of a pattern, sP, and a constraint, sC, on the symbols of sP. sC represents a quantifier free formula in the language of $\mathcal{T}_0$.

Application of a substitution, $\sigma$, to a logical scenario, sS = (sP, sC) (written ($\sigma$sP)), gives an instance of sS if the domain of $\sigma$ contains all the symbols of sS and $\sigma$ satisfies sC ($\mathcal{T}_0 \models$ sC$\sigma$). We say $\sigma_1$ extends $\sigma_0$, written $\sigma_1 \gg \sigma_0$ if the domain of $\sigma_1$ contains the domain of $\sigma_0$ and $\sigma_0(v) = \sigma_1(v)$ (wrt. $\mathcal{T}$) for $v$ in the domain of $\sigma_0$. Finally, we let $\longrightarrow_c$ denote the concrete rewrite relation induced by $\mathcal{T}$, and $\longrightarrow_s$ denote the symbolic rewrite relation induced by $\mathcal{T}$. Then the desired connection between the rewrite relations is give by the following *Soundness* and *Completeness* properties. These correspond to Theorems 1 and 2 of [20] and can be proved by analogous arguments.

*Soundness.* If $\mathsf{sS}_0 \longrightarrow_s \mathsf{sS}_1$ and $\sigma_0$ gives an instance of $\mathsf{sS}_0$, then there exists $\sigma_1 \gg \sigma_0$ such that $\mathsf{cS}_1$ is equivalent (in $\mathcal{T}$) to $\mathsf{sP}_1\sigma_1$ and $\sigma_0(\mathsf{sP}_0) \longrightarrow_c \mathsf{cS}_1$.

*Completeness.* If $\sigma_0$ gives an instance of $\mathsf{sS}_0$ and $\sigma_0(\mathsf{sP}_0) \longrightarrow_c \mathsf{cS}_1$ then there exists $\mathsf{sS}_1$, and $\sigma_1 \gg \sigma_0$ such that $\sigma_1$ gives an instance of $\mathsf{sS}_1$ with $\mathsf{cS}_1$ equivalent to $\sigma_1(\mathsf{sP}_1)$ and $\mathsf{sS}_0 \longrightarrow_s \mathsf{sS}_1$ where $\sigma_1$ gives an instance of $\mathsf{sS}_1$.

## 3.2    The Structure of Soft Agent Rewriting

In soft agents, a system state consists of a set of agent terms together with a unique environment term. Abstractly an agent term has the form A(id,attrs) where id is the agent identifier, and attrs is a set of named attributes including the agents local knowledge base (local KB), and a set of pending tasks and actions each labeled by the time until ready for execution. An environment term has the form E(ekb) where ekb is a knowledge base representing the physical state of the system and contextual information such as location of features or bounds on location.

There are two rewrite rules: doTask and timeStep. The doTask rule has the form

```
crl[doTask]: A(id,attrs) E(ekb) => A(id,attrs') E(ekb) if taskConds
```

where taskConds has clauses for reading sensors from the environment, evaluating possible actions, and updating the local KB, pending tasks, and actions. The timeStep rule has the form

```
crl[timeStep]: A(id1,attrs1) ... A(idk,attrsk) E(ekb)  =>
               A(id1,attrs1') ... A(idk,attrsk') E(ekb') if stepConds
```

where `stepConds` has a clause to execute ready actions (with time delay 0) and update time-dependent symbols to capture the passing of time. There are also clauses to update time parameters (clocks, delays...), transmit messages, and share knowledge amongst the agents. Executing actions affects parameters that control how the physical state evolves (change of acceleration, direction, on/off switches ...). Passing time lets the physical model run for the specified interval of time, updating the physical state (position, energy level, ...) according to laws parameterized by the control settings.

### 3.3  Symbolic Soft Agent Rewriting

To enable symbolic execution of soft agent specifications we abstract system states as terms of the form `SA[uu] SE[vv]` where `SA` is a pattern with symbols `uu` whose structure captures the state aspects that are not changed during execution, for example the number of agents, their ids, attribute names, and any persistent structure in attribute values. Similarly, `SE[vv]` is a pattern, with symbols `vv` capturing the persistent structure in the environment knowledge base. `uu` and `vv` are disjoint lists of symbols. For example, in a platooning scenario, symbols in `vv` would represent values including the position, acceleration, and velocity of each vehicle. Mathematically, we represent the symbolic constraint as a separate state component. In practice, we represent it as an element of the environment knowledge base.

Intuitively, the execution of a logical scenario constructs new constraints containing fresh symbols representing new values of the system's physical attributes. As for (concrete) soft agents, there are two rewrite rules for symbolic soft agents. At the framework level, the symbolic rules are obtained by replacing the clauses in the rule conditions of concrete rules by symbolic versions that refer to symbolic versions of the functions involved. It is the job of the specifier to define these symbolic functions and their symbolic evaluation equations. In the vehicle platooning case, symbolic functions were obtained by systematically transforming the original concrete versions. In the next section we give examples of key elements of the symbolic vehicle platooning system.

## 4  Vehicle Specifications

This section details how one can specify logical scenarios including safety properties by specifying the vehicle platooning example described in Sect. 2. While the specifications below are declarative, i.e., closely resemble textbook formulas, we do assume that the reader is familiar with the Maude syntax [4]. Our starting point is a concrete specification of the vehicle platooning described in [5]. It contains several features, such as vehicle controllers and communication protocol specifications, which have been ported to the symbolic machinery described below. The complete code can be found at https://github.com/SRI-CSL/VCPublic.git in the folder `symbolic-platooning`. To execute this code you will need the Maude integration with Z3 which can be found at [14].

## 4.1   Basic Symbolic Sorts

`RealSym` is the sort of real values. It contains concrete values, i.e., real numbers, or symbols of the form `vv(i)` or `vv(i,str)` where `i` is a `Nat` uniquely identifying a symbol and `str` is a string describing the intuitive meaning of the symbol, used for improved readability. The term `mkNuVar(i,id,str)` evaluates to a (fresh) symbol with identifiers `id,str`, where `id` is an agent identifier and `str` is a string with a short description of the fresh symbol.

*Example 1.* The following symbols represent the initial conditions for the follower `ag1`, namely, its position, speed, maximum acceleration, maximum deceleration, and initial acceleration.

```
eq v1posx = vv(2,"ag1-positionX") .  eq v1posy = vv(3,"ag1-positionY") .
eq v1vel = vv(5,"ag1-speed") .          eq maxacc1 = vv(9,"ag1-maxAcc") .
eq maxdec1 = vv(10,"ag1-maxDec") .   eq acc1 = vv(32,"ag1-acc") .
```

`SymTerm` is the sort of symbolic terms containing arithmetic expressions constructed inductively using basic arithmetic operators (e.g., addition, subtraction, division, multiplication) and elements of `RealSym`. They are used to specify constraints of sort `Boolean` involving symbols.

*Example 2.* The following constraint using the symbols in Example 1 specifies that `ag1`'s acceleration is bounded by the maximum acceleration and deceleration: `(acc1 <= maxacc1) and (acc1 >= maxdec1)`.

## 4.2   Knowledge Specifications

Cyber-physical systems reason using knowledge about their locations, speeds, direction, and accelerations and of the surrounding objects. Such knowledge is represented using a sort `Info`. Knowledge base elements are of the form `info @ t` where `t` is a logical time, i.e., the number of time steps since the beginning.

Vehicle locations are two-dimensional, speeds are real values, and directions are vectors specified using two locations and a magnitude:

```
op loc : SymTerm SymTerm -> Loc .
op speed : Id RealSym -> Info .
op dir : Id Loc Loc SymTerm -> Info .
```

*Example 3.* The agent `ag1`'s initial knowledge base, that is, at logical tick 0, contains the following terms, specifying its initial position, speed, acceleration and direction:

```
(at(ag1,loc(v1posx,v1posy)) @ 0) (speed(ag1,v1vel) @ 0)
(accel(ag1,acc1) @ 0)
(dir(v(1),loc(v1ix,v1iy),loc(v1tx,v1ty),v1mag) @ 0)
```

Based on the above notation, we can specify symbolically typical definitions, such as the distance between two locations:

```
op ldist : Nat Loc Loc -> NatSymTermBoolean .
eq ldist(i,loc(x0,y0),loc(x1,y1))
 = {s(i),vv(i,"dist"), (vv(i,"dist") >= 0/1) and
    vv(i,"dist") * vv(i,"dist") === ((y1 - y0) * (y1 - y0) +
                                      (x1 - x0) * (x1 - x0)) } .
```

This definition creates a fresh symbol, vv(i,"dist") together with the constraint specifying the Euclidean distance. Notice that we need to specify that the distance is a non-negative value. Similar specifications can be made for other distance measures, such as, Manhattan distance.

The following operator specifies how an agent's location, loc(x,y), is updated to loc(nuVarX,nuVarY) given an (average) speed, spd, and a direction.

```
op upVLoc : Nat Id Loc SymTerm Info -> NatLocBoolean .
ceq upVLoc(i,id,loc(x,y),spd,dir(id,loc(x0,y0),loc(x1,y1),mag))
 = {i + 2,loc(nuVarX,nuVarY),cond}
 if nuVarX := mkNuVar(i,id,"-positionX")
 /\ nuVarY := mkNuVar(i + 1,id,"-positionY")
 /\ cond1 := (x0 === x1) and (not (y0 === y1)) and
             (nuVarX === x) and (nuVarY === y + spd)
 /\ cond2 := (not (x0 === x1)) and (y0 === y1) and
             (nuVarX === x + spd) and (nuVarY === y)
 /\ cond3 := (not (x0 === x1)) and (not (y0 === y1)) and
             (nuVarX === (x + spd * (x1 - x0) / mag)) and
             (nuVarY === (y + spd * (y1 - y0) / mag))
 /\ cond := cond1 or cond2 or cond3 .
```

We made some design choices in this definition. The first design choice is to split it into three different cases. The first case (cond1) is when the agent is moving vertically, the second case (cond2) horizontally, and the third case (cond3) when it is moving in the quadrant. In this way we help the constraint solver to avoid to solve the harder non-linear constraint involved in the third case whenever the agent is moving only along the x-axis and only along the y-axis. The second design choice was to include the magnitude in the definition of dir which may seem redundant as it can be specified from the two associated locations. However, by doing so, we avoid the need to generate fresh symbols and new constraints whenever the magnitude is needed as in the third case of upVLoc.

Finally, we also capture symbolically the fact that the physical system is continuous while the cyber part of the system works in logical ticks. The size of the tick is specified by the term tickSize(dt), where dt is symbol denoting the size of the tick. Typically it is fixed during the whole execution by using a constraints, e.g., dt === 1/10, specifying a tick duration of $100ms$. We assume here for simplicity that all agents use the same tick duration. However, agents with different tick duration can also be specified. When the soft agent machinery updates the agent's positions using upVLoc it scales accordingly the speed to the tick size by multiplying the speed with dt.

## 4.3   Soft-Constraint Controller

Agents decide which action to take based on their local knowledge base, which
is updated by reading their sensors, and taking into account different concerns,
such as safety and efficiency. For vehicle platooning, as described in detail in [5],
there are two main concerns, *safety*, i.e., maintaining a safe distance between
vehicles, and *fuel-efficiency*, i.e., maintaining a distance between vehicles that is
not too great.

The controller is specified in a similar way to the knowledge functions
described above by using existing symbols, creating new symbols, and using
contraints to determine its possible values.

The following equation specifies the controller evaluation to rank the possible
actions that the vehicle can take from a safety perspective. In particular, it takes
as input i, for creating fresh symbols, vmin,vmax, respectively, the minimum
and maximum speeds that the vehicle is allowed to use, vminD,vmaxD, the minimum and maximum desired speeds according to the safety parameters (gap$_{safe}$,
gap$_{safer}$), and the constraints cond on the existing symbols. It then returns a
range of speeds that are safe specified by the interval between the fresh symbols vv(i)and vv(i + 1). However, the concrete values for vv(i),vv(i + 1)
depend on the relation between the possible speeds (vmin,vmax) and the desired
speeds vminD,vmaxD as detailed by the constraints cond11,cond21,...,cond61.

```
ceq symValSpeedRed(i,str,vmin,vmax,vminD,vmaxD,cond) =
{i + 2, [vv(i),vv(i + 1),nuCond and cond]}
if cond1 := vmin >= vmaxD
/\ cond11 := vv(i) === vmin and
             vv(i + 1) === ((vmin + vmax) / 2/1) and cond1
/\ cond2 := vmax <= vminD
/\ cond21 := vv(i) === ((vmin + vmax) / 2/1)
        and vv(i + 1)  === vmax and cond2
...
/\ cond6 := vmin >= vminD and vmax < vmaxD
/\ cond61 := vv(i) === vmin and vv(i + 1)  === max and cond6
/\ nuCond := (cond11 or cond21 or cond31 or cond41 or cond51 or cond61) .
```

In the definition above, the effort of determining which condition applies is delegated to the constraint solver. As we will investigate in Sect. 5, this will lead to
great performance penalties.

An alternative way to expressing the same controller is to return six possibilities as specified by the following equation, rather than the single disjunction
nuCond:

```
ceq symValSpeedRed-Split(i,str,vmin,vmax,vminD,vmaxD,cond) =
 {i + 2, [vv(i),vv(i + 1),cond11 and cond]}
 {i + 2, [vv(i),vv(i + 1),cond21 and cond]}
 ...
 {i + 2, [vv(i),vv(i + 1),cond61 and cond]}
 if cond1 := vmin >= vmaxD
 ...
/\ cond61 := vv(i) === vmin and vv(i + 1)  === vmax and cond6.
```

With this new definition the choice of which condition is applicable is left to the search engine, i.e., Maude.

A similar choice occurs when specifying how the time advancement affects agent's speeds. Several cases occur due to the fact that logical scenarios assume that vehicle's speeds are bounded. For example, depending on the tick duration, current speed and maximum acceleration, an agent's speed may reach the maximum speed or not before completing a logical tick. For analyzing the impact of delegating such enumeration of cases to the SMT-solver or to the search engine, we implemented two versions of time advancement: `timestep` that returns one output with a constraint with a disjunct for each case, as in `symValSpeedRed`; and `timestep-split` that returns several outputs, one for each possible case as `symValSpeedRed-split`.

## 4.4   System Configurations

As described in Sect. 3, a system configuration of sort `ASystem` is a collection of agent configurations and an environment configuration.

An agent configuration has the form [id : class | attrs ], where `id` is the agent's unique identifier, `class` is its class, e.g., vehicle, and `attrs` are its attributes which include its local knowledge base written `lkb : kb`, where `lkb` is a label and `kb` is the local knowledge base contents.

An environment configuration has the form [eId | ekb] where `ekb` is the environment knowledge base which specifies state of the world. The environment knowledge base contains the knowledge item `constraints(i,cond)` where `i` is the current index of fresh variables, and `cond` is the constraints (accumulated) on the existing symbols.

*Example 4.* The initial configuration of a platooning scenario described in Sect. 2 is as follows:

```
asysI = { [eid | (kb constraint(i,condI))]
          [v(0) : veh | lkb : kb0 ] [v(1) : veh | lkb : kb1 ] }
```

where `kb` is the environment knowledge base specifying among other things, the vehicles's actual locations and speeds, while `kb0` and `kb1` are the vehicle `v(0)` and `v(1)`'s local knowledge bases. The constraint `condI` contains the constraints on these values as per the logical scenario. It contains for example constraints on the acceleration of vehicles (see Example 2) and the following constraints:

```
(v1vel >= vellb1) and (v1vel <= velub1) and (v0posy > v1posy)
```

which specify that the follower vehicle's speed is bound within the bounds `vellb1` and `velub1`. Moreover, the following vehicle `v(1)` is behind the leader `v(0)`.

Notice that such a symbolic system configuration may correspond to infinitely many concrete system configuration, i.e., concrete instances of the specified platooning scenario.

## 4.5   Safety Properties

We are interested in generating proofs regarding the safety of logical scenarios, such as the one specified in Example 4. The specification of safety property is formalized using the operator:

```
op mkSPCond : SP ASystem -> SPSpec .
```

This function takes a property (an identifier in SP) and a system configuration, and returns a safety property of sort SPSpec of the form:

```
op {_,_,_,_} : Nat SymTerms Boolean Boolean -> SPSpec .
```

The first element is the new symbol index, the second is the new (auxiliary) symbols created for specifying the property, which are then constrained by the third element. The last element specifies the safety property based on the auxiliary symbols and the previously existing symbols in the given system configuration.

For example, the first safety property in Eq 1b is specified as follows:

```
ceq mkSPCond(saferSP, { conf env }) = {k + 1,dis,cond00,nucond}
if [id0 | kb] := env
/\ (atloc(v(0),10) @ t0) (atloc(v(1),11) @ t1)
   (speed(v(0),v0) @ t2) (speed(v(1),v1) @ t3)
   (gapSafety(v(1),gapSafer,gapSafe)) (constraint(n,cond)) kb1 := kb
/\ {k,dis,cond00} := ldist(n,11,10)
/\ nucond := (dis >= ((1/1 + gapSafer) * v1) - v0) .
```

Notice the use of the function ldist that creates the auxiliary fresh symbol dis.

Using mkSPCond, we specify an operator (definition elided)

```
op enforceSP : SP ASystem -> ASystem.
```

For example, enforce(saferSP,asysI) returns a configuration in which the conditions (cond00 and nucond from mkSPCond) are added to the set of constraints. This means that the resulting configuration will only have instances asysI that satisfy the saferSP. The term isSatModel(enforce(saferSP, asysI)) calls the SMT-Solver and returns an assignment for asysI symbols:

```
ag0-positionX |-> (0/1).Real,   ag0-positionY |-> (1/1).Real
ag1-positionX |-> (0/1).Real,   ag1-positionY |-> (0/1).Real,
ag0-speed |-> (7/1).Real,       ag1-speed |-> (2/1).Real,
ag1-safer |-> (3/1).Real
```

This state satisfies the saferSP property for a $gap_{safer}$ of value 3.

## 4.6   Verifying Logical Scenarios

We can now use Rewriting Modulo SMT [20] to verify and effectively generate safety proofs of the specifications above in an automated fashion. Consider the following search:

```
search enforceSP(safeSP,setStopTime(asysI,2)) =>*
  asys such that checkSP(unsafeSP,asys) .
No solution. states: 63  rewrites: 394686 in 20134ms
```

It attempts to find any instance of system configuration `asys` that satisfies `unsafeSP` (see Eq. 1c) starting from any instance of `asysI` that satisfies property `safeSP`. Moreover, the term `setStopTime(asysI,2)` specifies that the search is bound to two logical ticks, i.e., search stops after two tick rules. The search engine combined with the SMT-solver can generate proofs that no instance of reachable states are unsafe. However, as shown Sect. 5, the complexity of the problem greatly increases when considering larger logical tick bounds.

## 5   Trade-Offs Between Rewriting and Constraint Solving

The verification of logical scenario involves rewriting and constraint solving. Rewriting enumerates possible system states while the constraint solver attempts to check the satisfiability of constraints. As demonstrated in Sect. 4.3, how much of verification is delegated to rewriting and how much to the constraint solver can be adjusted by leaving the non-determinism in the constraints, e.g., by placing disjunctions in the constraints, or to the rewriting, e.g., returning instead for each disjunct an output, a rewriting choice.

Delegating verification to the rewriting engine means that the search tree is larger leading to more calls to the SMT-solver, but each call involves simpler constraints to solve, i.e., with less disjunctions and therefore less cases to consider. Delegating verification to the constraint solver, on the other hand, means a smaller search space traversed by the rewriting engine leading to less calls to the constraint solver, but with more complex constraints.

To demonstrate this, we considered three cases according to the specifications described in Sect. 4.3:

- **More SMT Less Search:** This case uses `symValSpeedRed` for the controller and `timestep` for the time step evolution. This means that all cases are specified as disjunctions in the constraint that will need to be solved by the solver.
- **Less SMT More Search:** This case uses `symValSpeedRed-split` for the controller and `timestep-split` for the time step evolution. This means that all cases are specified as different outputs that need to be traversed by the rewriting engine.
- **Balanced:** This case uses `symValSpeedRed` for the controller and the specification `timestep-split` for the time step evolution. This means that some cases are specified as constraints and others as outputs.

To evaluate the different cases, we executed the command:

```
search enforceSP(safeSP,setStopTime(asysI,Bound)) =>! asys
      such that isSat(asys) .
```

which enumerates all the reachable symbolic configurations that are satisfiable exactly in `Bound` time ticks, i.e., number of applications of the `timeStep` rule.

**Table 1.** Experiments with the Platooning Logical Scenario Verification. DNF denotes that the experiment was aborted after 5 h. The experiment results are expressed as *states/time*, where *states* is the total number of states in the search tree and *time* is the time needed to traverse all states. The experiments were carried out in a 2.2 GHz 6-Core Intel Core i7 machine with 16 GB memory.

| Time Bound | Pruning | More SMT Less Search | Balanced | Less SMT More Search |
|---|---|---|---|---|
| 2 | No | 19/20.4s | 71/2.5s | 1427/29.7s |
|   | Tick | 19/32.4s | 63/8.3s | 497/47.4s |
|   | All | 19/56.0s | 63/11.6s | 296/52.7s |
| 3 | No | DNF | DNF | 42827/3054s |
|   | Tick | DNF | DNF | 2484/3412s |
|   | All | DNF | DNF | 1976/5238s |

A second dimension that we investigated was on the way we can prune the search tree. We considered the following cases:

- **All Pruning:** At each rewrite rule for `doTask`, which evaluates an agent's actions, and `tick`, which applies the agent's actions, we placed a check whether the resulting configuration is satisfiable. This means that the search tree has only satisfiable configurations with the price of calling the SMT-Solver at each step.
- **No Pruning:** As opposed to the `All Pruning` case, rewrites `doTask` nor `tick` did not check the satisfiability of the resulting configuration. The check was made only at the configuration resulting from applying the number to ticks specified by the bound. This means that the search tree is not pruned, and therefore, more states are traversed.
- **Tick Pruning:** The third case does a check on the configuration resulting from `timeStep` rewrites, but not on `doTask`. In this way, we still prune the search tree without calling the SMT-solver at each rewriting step.

Table 1 summarizes our experiments with these scenarios using bounds of two and three cs. The best case was not pruning the tree and delegating verification to the search tree when considering greater time bounds. The balanced case had better results when considering lower time bounds.

Interestingly, pruning the tree, while had a great effect on number of states, it did not improve the time required to traverse the tree. We believe that this can be further improved if the search engine uses the SMT-solver in a more clever way, in particular, using its incremental solving features. This would allow the solver to re-use work done in previous calls.

# 6   Related Work

Existing work for the verification of autonomous cyber-physical systems can be divided into three different approaches.

The first approach [9] is to use simulation-based methods that run a sufficiently large number of simulations using simulators [8]. A main advantage of this approach is that it can be used to verify the actual artifacts, e.g., machine learning artifacts, used in applications and rely on vehicle simulators to generate very complicated and high-fidelity scenarios. However, as already mentioned, as each simulation is run using a concrete instance of a logical scenario, a limitation of this approach is that possibly a large number of simulations need to be generated for each logical scenario. Our work complements this work by enabling the specification and verification of vehicle behavior using symbolic methods covering all instances of a logical senario, and enables early verification of designs before expensive artifacts are built.

The second approach is to use safe controllers [1,22] that are guaranteed to generate safe trajectories under the assumption that the remaining agents behave correctly. A limitation of this type of work is that it focuses only on individual functions, typically control algorithms without taking into account other functions needed for AVs, e.g., sensing, knowledge bases, and communication channels. As shown in [7], safe controllers can be integrated with advanced (high-performance, but not safe) controllers as fall-back options whenever safety assurance is low. In particular, a formal framework for Run Time Assurance (RTA) is presented, and conditions are given that, if satisfied by a safe controller and associated monitor, guarantee that integration with an untrusted control maintains safe operation. The paper leaves open methods to verify that a controller satisfies its RTA requirements. Our work has been greatly inspired by [7] and the result is complimentary. Symbolic rewriting combined with SMT solving provides automated methods to verify correctness of time sampling mechanisms and safety requirements.

The third approach [16,18,25], similar to the non-symbolic Soft Agents, are formal frameworks that enable the specification and verification of other functions, besides trajectory planning [5,10]. However, as with the first approach, the evidence that can be produced by these frameworks is based on running simulations or model checking concrete scenario instances. Therefore, it also suffers the limitation that a large number of simulations need to be carried out, or a large sample of senario instances must be model checked.

The Soft Agent execution strategy is based on the Real Time Maude maximal time elapse (MTE) execution strategy for real time theories [18]. In [17] two conditions for soundness and completeness of model checking Real Time Maude specifications based on the MTE execution strategy are given. The first condition, time robustness, is a property of the rewrite theory. It requires that timesteps of any duration are allowed, and a timestep can be subdivided without changing the end result. The second condition requires that atomic propositions are stable with respect to time: at most one change during a time step. These conditions hold for a wide range of Real Time Maude specifications, timing of protocols, network performance, or discrete events used for defining system behavior of, e.g., manufacturing plants. SA specifications are concerned with physical properties of a system such as bounds on distance, change of position, use of resources to express both safety and goal satisfaction properties. SA specifications are time robust, but the properties of interest are generally not stable with respect to time. Thus, we can not directly use the Real Time Maude results. Work is in progress to define an analog to stability for system properties that evolve over time.

A formal mathematical foundation for symbolic rewriting modulo SMT is presented in [20]. Our work is essentially a mapping of these ideas to be executable in Maude with an integrated SMT solver. The soft agents doTask rule is not technically topmost, but could easily be modified to be topmost without changing any behavior in our examples. Also, the theory $\mathcal{T}$ has non-axiom equations that are not in $\mathcal{T}_0$ These equations define functions is a straight forward way, so they do not cause a problem for our symbolic rewriting but may challenge narrowing. Our logical scenarios are ground terms from Maude's perspective and correspond to terms whose only variables have builtin sorts (in $\mathcal{T}_0$). On the other hand, search starts with terms that possibly have non builtin variables in [20]. Generating new symbols to update values plays a similar role to the *fresh* substitution used in the symbolic rewrite relation of [20]. Important future work is to better understand criteria for allowing equations over non-builtin sorts, to make symbolic rewriting modulo SMT more generally applicable.

A notion of guarded term is introduced in [2] as a method to reduce the search state space in symbolic rewriting modulo SMT. A guarded term is a pair consisting of a term and a constraint, or the disjunction of a set of guarded terms. The paper develops the formal theory of rewriting with guarded terms and presents experiments based on the CASH protocol showing state space reduction for various forms of guard. Although the paper motivates guards by a need to also reduce complexity of constraints sent to the SMT solver, no results on constraint size are reported. The results in the present paper seem to suggest that not only the size of state space matters for automation, but also the size of constraints that are sent to the SMT-Solver. It will be interesting to see if guards can be used to control the tradeoffs between search space size and constraint size explored in the present paper.

## 7    Conclusions

This paper proposes an extension of Soft Agents frameworks with Rewriting Modulo SMT to enable the automated generation of safety proofs of CPS. We demonstrate its expressiveness with a vehicle platoon scenario which is a common feature of autonomous vehicles. We carry out a collection of experiments demonstrating that delagating verification to rewriting has a positive impact in verification performance.

We are planning to use this framework in several directions that complement related work. We are currently automating the verification conditions for RTA [7]. We also believe that our framework is applicable to problems other than vehicle safety, for example it could be used to enable symbolic security verification by extending our previous work [5].

Inspired by the presentation at WRLA 2022 on the Python bindings for Maude [21], we adapted our implementation to use the Python bindings instead of MaudeSE [14]. This enables full access to SMT-solver interface, including to new SMT-solvers such as CVC5 [3]. In the future, we plan to implement Python libraries based on these Python bindings for Maude to improve usability of the Soft Agents framework and quick integration to other tools/methods.

**Acknowledgments.** Talcott was partially supported by the U. S. Office of Naval Research under award numbers N00014-15–1-2202 and N00014-20–1-2644, and NRL grant N0017317-1-G002.

# References

1. Althoff, M., Dolan, J.M.: Online verification of automated road vehicles using reachability analysis. IEEE Trans. Robot. **30**(4), 903–918 (2014)
2. Bae, K., Rocha, C.: Symbolic state space reduction with guarded terms for rewriting modulo SMT. Sci. Comput. Program. **178**, 20–42 (2019)
3. Barbosa, H., et al.: cvc5: a versatile and industrial-strength SMT solver. In: Fisman, D., Rosu, G. (eds.) Tools and Algorithms for the Construction and Analysis of Systems. TACAS 2022. Lecture Notes in Computer Science, vol. 13243. Springer, Cham (2022). https://doi.org/10.1007/978-3-030-99524-9_24
4. Clavel, M.: All About Maude - A High-Performance Logical Framework. LNCS, vol. 4350. Springer, Heidelberg (2007). https://doi.org/10.1007/978-3-540-71999-1
5. Dantas, Y.G., Nigam, V., Talcott, C.L.: A formal security assessment framework for cooperative adaptive cruise control. In: IEEE Vehicular Networking Conference, VNC 2020, New York, NY, USA, pp. 16–18 December 2020, pp. 1–8. IEEE (2020)
6. de Moura, L., Bjørner, N.: Z3: an efficient SMT solver. In: Ramakrishnan, C.R., Rehof, J. (eds.) TACAS 2008. LNCS, vol. 4963, pp. 337–340. Springer, Heidelberg (2008). https://doi.org/10.1007/978-3-540-78800-3_24
7. Desai, A., Ghosh, S., Seshia, S.A., Shankar, N., Tiwari, A.: SOTER: a runtime assurance framework for programming safe robotics systems. In: 49th Annual IEEE/IFIP International Conference on Dependable Systems and Networks, DSN 2019, Portland, OR, USA, 24–27 June 2019, pp. 138–150. IEEE (2019)
8. Dosovitskiy, A., Ros, G., Codevilla, F., López, A.M., Koltun, V.: CARLA: an open urban driving simulator. In: 1st Annual Conference on Robot Learning, CoRL 2017, Mountain View, California, USA, 13–15 November 2017, Proceedings, vol. 78 of Proceedings of Machine Learning Research, pp. 1–16. PMLR (2017)
9. Fremont, D.J., Dreossi, T., Ghosh, S., Yue, X., Sangiovanni-Vincentelli, A.L., Seshia, S.A.: Scenic: a language for scenario specification and scene generation. In: McKinley, K.S., Fisher, K. (eds.) Proceedings of the 40th ACM SIGPLAN Conference on Programming Language Design and Implementation, PLDI 2019, Phoenix, AZ, USA, 22–26 June 2019, pp. 63–78. ACM (2019)
10. Mason, I.A., Nigam, V., Talcott, C., Brito, A.: A framework for analyzing adaptive autonomous aerial vehicles. In: Cerone, A., Roveri, M. (eds.) SEFM 2017. LNCS, vol. 10729, pp. 406–422. Springer, Cham (2018). https://doi.org/10.1007/978-3-319-74781-1_28
11. SAE J3016. https://www.sae.org/news/2019/01/sae-updates-j3016-automated-driving-graphic (2021)

12. Jha, S., Rushby, J., Shankar, N.: Model-centered assurance for autonomous systems. In: Casimiro, A., Ortmeier, F., Bitsch, F., Ferreira, P. (eds.) SAFECOMP 2020. LNCS, vol. 12234, pp. 228–243. Springer, Cham (2020). https://doi.org/10.1007/978-3-030-54549-9_15

13. Kalra, N., Paddock, S.M.: Driving to safety. https://www.rand.org/content/dam/rand/pubs/research_reports/RR1400/RR1478/RAND_RR1478.pdf (2021)

14. MaudeSE. https://github.com/maude-se/maude-se.github.io (2021)

15. Menzel, T., Bagschik, G., Maurer, M.: Scenarios for development, test and validation of automated vehicles. In: 2018 IEEE Intelligent Vehicles Symposium, IV 2018, Changshu, Suzhou, China, 26–30 June 2018, pp. 1821–1827. IEEE (2018)

16. Moradi, F., Asadollah, S.A., Sedaghatbaf, A., Causevic, A., Sirjani, M., Talcott, C.L.: An actor-based approach for security analysis of cyber-physical systems. In: ter Beek, M.H., Nickovic, D. (eds.) FMICS 2020. LNCS, vol. 12327, pp. 130–147. Springer, Cham (2020). https://doi.org/10.1007/978-3-030-58298-2_5

17. Ölveczky, P.C., Meseguer, J.: Abstraction and completeness for real-time maude. In: Denker, G., Talcott, C.L. (eds.) Proceedings of the 6th International Workshop on Rewriting Logic and its Applications, WRLA 2006, Vienna, Austria, 1–2 April 2006, vol. 174 of Electronic Notes in Theoretical Computer Science, pp. 5–27. Elsevier (2006)

18. Ölveczky, P.C., Meseguer, J.: The real-time maude tool. In: Ramakrishnan, C.R., Rehof, J. (eds.) TACAS 2008. LNCS, vol. 4963, pp. 332–336. Springer, Heidelberg (2008). https://doi.org/10.1007/978-3-540-78800-3_23

19. Riedmaier, S., Ponn, T., Ludwig, D., Schick, B., Diermeyer, F.: Survey on scenario-based safety assessment of automated vehicles. IEEE Access 8, 87456–87477 (2020)

20. Rocha, C., Meseguer, J., Muñoz, C.: Rewriting modulo SMT and open system analysis. J. Logical Algebraic Methods Program. 86(1), 269–297 (2017)

21. Rubio, R.: Maude as a library: an efficient all-purpose programming interface. In: Rewriting Logic and its Applications (WRLA) (2022)

22. Shalev-Shwartz, S., Shammah, S., Shashua, A.: On a formal model of safe and scalable self-driving cars. CoRR, abs/1708.06374 (2017)

23. Sifakis, J.: Autonomous systems - an architectural characterization. CoRR, abs/1811.10277 (2018)

24. Talcott, C., Nigam, V., Arbab, F., Kappé, T.: Formal specification and analysis of robust adaptive distributed cyber-physical systems. In: Bernardo, M., De Nicola, R., Hillston, J. (eds.) SFM 2016. LNCS, vol. 9700, pp. 1–35. Springer, Cham (2016). https://doi.org/10.1007/978-3-319-34096-8_1

25. Talcott, C., Arbab, F., Yadav, M.: Soft agents: exploring soft constraints to model robust adaptive distributed cyber-physical agent systems. In: De Nicola, R., Hennicker, R. (eds.) Software, Services, and Systems. LNCS, vol. 8950, pp. 273–290. Springer, Cham (2015). https://doi.org/10.1007/978-3-319-15545-6_18

26. van de Hoef, S., Johansson, K.H., Dimarogonas, D.V.: Fuel-efficient en route formation of truck platoons. IEEE Trans. Intell. Transp. Syst. 19(1), 102–112 (2018)

# Executable Semantics and Type Checking for Session-Based Concurrency in Maude

Carlos Alberto Ramírez Restrepo[1] and Jorge A. Pérez[2]([✉])

[1] Pontificia Universidad Javeriana Cali, Cali, Colombia
carlosalbertoramirez@javerianacali.edu.co
[2] University of Groningen, Groningen, The Netherlands
j.a.perez@rug.nl

**Abstract.** Session types are a well-established approach to communication correctness in message-passing programs. We present an executable specification of the operational semantics of a session-typed $\pi$-calculus, implemented in the Maude system. We also develop an executable specification of its associated algorithmic type checking, and describe how both specifications can be integrated. We further explore how our executable specification enables us to detect well-typed but deadlocked processes by leveraging reachability and model checking tools in Maude. Our developments define a promising new approach to the (semi)automated analysis of communication correctness in message-passing concurrency.

## 1 Introduction

This paper presents an executable rewriting semantics for a $\pi$-calculus equipped with *session types*. Widely known as the paradigmatic calculus of interaction, the $\pi$-calculus [5,9] offers a rigorous platform for reasoning about message-passing concurrency. Session types are arguably the most prominent representative of *behavioral type systems* [3], which can statically ensure that processes respect their ascribed *interaction protocols* and never exhibit errors and mismatches.

The integration of (variants of) the $\pi$-calculus with different formulations of session types has received much attention from foundational and applied perspectives. As a result, our understanding about (abstract) communicating processes and their typing disciplines steadily reaches maturity. Despite this progress, rigorous connections with more concrete representation models fall short. In particular, the study of session-typed $\pi$-calculi within frameworks and systems like Maude [2] seems to remain unexplored. This gap is an opportunity to investigate the formal systems underlying session-typed $\pi$-calculi (reduction semantics and type systems) from a fresh yet rigorous perspective, taking advantage of the concrete representation given by executable semantics in Maude.

Looking at session-typed $\pi$-calculi from the perspective of Maude is insightful, for several reasons. First, Maude enables the systematic validation of such formal systems and their results, improving over pen-and-paper developments. Second, as there is not a canonical session-typed $\pi$-calculus, but actually many

K. Bae (Ed.): WRLA 2022, LNCS 13252, pp. 230–250, 2022.
https://doi.org/10.1007/978-3-031-12441-9_12

different formulations (with varying features and properties), an implementation in Maude could provide a concrete platform for uniformly representing them all. Third, resorting to Maude as a host representation framework for session-typed $\pi$-calculi could help in addressing known limitations of static type checking for deadlock detection, leveraging tools already available in Maude.

This paper presents our work on pursuing these three directions. We adopt the session-typed $\pi$-calculus developed by Vasconcelos [13] as the basis for our implementation in Maude. For this typed language, dubbed $s\pi$, we first implement its (untyped) reduction semantics as a rewriting semantics, essentially extending prior work on representing the $\pi$-calculus in Maude (see below). Then, we implement its associated algorithmic type system, also given in [13]. Well-typedness in [13] ensures *fidelity* (i.e., well-typed processes respect at runtime their ascribed protocols) but does not rule out deadlocks and other kinds of insidious circular dependencies. To address this, we leverage reachability and model checking in Maude. Our Maude developments are publicly available online.[1]

To our knowledge, we are the first to represent session-typed $\pi$-calculi using Maude. Prior works have used rewriting logic to investigate the operational semantics for variants of the $\pi$-calculus. Viry [14,15] defines the reduction semantics of a synchronous $\pi$-calculus as a rewrite theory, which is implemented in ELAN. The work of Thati et al. [12] considers an untyped, asynchronous $\pi$-calculus, whose labeled transition semantics is implemented as a rewrite theory, which is used to formalize an associated may-testing preorder. The work of Pitsiladis and Stefaneas [7] concerns a typed process calculus but in a different context, in which types are used to enforce privacy properties. Indeed, such work gives a Maude implementation of the labeled transition semantics of a privacy-oriented variant of the $\pi$-calculus and a Maude implementation of its associated type system, which is implemented as a membership equational theory.

The rest of this paper is organized as follows. Next, Sect. 2 summarizes the syntax and semantics of $s\pi$. Section 3 describes the definition of our rewriting semantics for $s\pi$ in Maude, whereas Sect. 4 presents the rewriting implementation of the algorithmic type checking. Section 5 presents our developments on deadlock detection. Section 6 closes with some concluding remarks. An extended version, available online, contains additional material [8].

## 2    The Typed Process Model

The typed process calculus $s\pi$, formalized by Vasconcelos [13], is a variant of the synchronous $\pi$-calculus with constructs for session-based concurrency. Here we summarize its syntax and semantics.

The calculus $s\pi$ relies on a base set of *variables*, ranged over by $x, y, \ldots$. Variables denote *channels* (or *names*). Processes interact to exchange values, which can be variables or booleans. Variables can be seen as consisting of (dual) *endpoints* on which interaction takes place. Rather than non-deterministic choices

---

[1] See https://gitlab.com/calrare1/session-types.

$$
\begin{array}{ll}
P \mid Q \equiv Q \mid P & P \mid \mathbf{0} \equiv P \\
P \mid (Q \mid R) \equiv (P \mid Q) \mid R & (\boldsymbol{\nu} xy)\, \mathbf{0} \equiv \mathbf{0} \\
(\boldsymbol{\nu} xy)(\boldsymbol{\nu} wz)P \equiv (\boldsymbol{\nu} wz)(\boldsymbol{\nu} xy)P & (\boldsymbol{\nu} xy)P \mid Q \equiv (\boldsymbol{\nu} xy)(P \mid Q) \quad \text{If } x, y \notin \mathtt{fv}(Q) \\
\text{if true then } P_1 \text{ else } P_2 \equiv P_1 & \text{if false then } P_1 \text{ else } P_2 \equiv P_2
\end{array}
$$

**Fig. 1.** Structural congruence Rules for s$\pi$

among prefixed processes, there are two complementary operators: one for offering a finite set of alternatives (called *branching*) and one for choosing one of such alternatives (*selection*). More formally, the syntax of *values*, *qualifiers*, and *processes* is presented below:

$$
v ::= x \quad \mid \quad \mathsf{true} \quad \mid \quad \mathsf{false} \qquad\qquad q ::= \mathsf{un} \quad \mid \quad \mathsf{lin}
$$
$$
P ::= \mathbf{0} \quad \mid \quad \overline{x}v.P \quad \mid \quad q\, x(y).P \quad \mid \quad P_1 \mid P_2 \quad \mid \quad (\boldsymbol{\nu} xy)P \quad \mid
$$
$$
\text{if } v \text{ then } P_1 \text{ else } P_2 \quad \mid \quad x \triangleright \{l_i : P_i\}_{i \in I} \quad \mid \quad x \triangleleft l.P
$$

The inactive process is denoted as $\mathbf{0}$. The output process $\overline{x}v.P$ sends the value $v$ along $x$ and continues as $P$. The process $q\, x(y).P$ denotes an input action on $x$, which prefixes $P$. The qualifier $q$ is used for inputs, which can be linear (to be executed exactly once) or shared. The process un $x(y).P$ denotes a persistent input action, which corresponds to (input-guarded) replication in the $\pi$-calculus. The parallel composition $P_1 \mid P_2$ denotes the concurrent execution of $P_1$ and $P_2$. The process $(\boldsymbol{\nu} xy)P$ declares the scope of *co-variables* $x$ and $y$ to be $P$. These co-variables are intended to be the complementary ends of a communication channel. Given a boolean $v$, the process if $v$ then $P_1$ else $P_2$ continues as $P_1$ if $v$ is true; otherwise it continues as $P_2$. The branching process $x \triangleright \{l_i : P_i\}_{i \in I}$ offers multiple alternative branches $P_1, P_2, \ldots$ (each with a *label* $l_1, l_2, \ldots$), along $x$; it is meant to interact with a selection process $x \triangleleft l.P$, which uses $x$ to indicate the choice of the alternative labeled $l$ and then continues as $P$.

As usual, $q\, x(y).P$ binds variable $y$ in $P$ and $(\boldsymbol{\nu} xy)P$ binds co-variables $x, y$ in $P$. The sets of free and bound variables of a process $P$, denoted $\mathtt{fv}(P)$ and $\mathtt{bv}(P)$, are defined accordingly. Process $P[v/y]$ denotes the capture-avoiding substitution of variable $y$ by value $v$ in process $P$.

The operational semantics for s$\pi$ is given as a *reduction semantics*, which, as customary, relies on a structural congruence relation, the smallest congruence relation on processes that satisfy the axioms in Fig. 1. Structural congruence includes the usual axioms for inaction and parallel composition as well as adapted axioms for scope restriction, scope extrusion, and conditionals. Armed with structural congruence, the rules of the reduction semantics are presented in Fig. 2. Rules [R-LinCom] and [R-UnCom] induce different patterns for process communication, depending on the qualifier of their corresponding input action. Indeed, processes $\overline{x}v.P$ and $q\, y(z).Q$ can synchronize if $x$ and $y$ are co-variables. This is only possible if both processes are underneath a scope restriction $(\boldsymbol{\nu} xy)$. When this occurs, processes $\overline{x}v.P$ and $q\, y(z).Q$ continue respectively as $P$ and

$$(\boldsymbol{\nu}xy)(\overline{x}v.P \mid \operatorname{lin} y(z).Q \mid R) \longrightarrow (\boldsymbol{\nu}xy)(P \mid Q[v/z] \mid R) \qquad \text{[R-LinCom]}$$

$$(\boldsymbol{\nu}xy)(\overline{x}v.P \mid \operatorname{uny}(z).Q \mid R) \longrightarrow (\boldsymbol{\nu}xy)(P \mid Q[v/z] \mid \operatorname{un} y(z).Q \mid R) \qquad \text{[R-UnCom]}$$

$$\frac{j \in I}{(\boldsymbol{\nu}xy)(x \triangleleft l_j.P \mid y \triangleright \{l_i : Q_i\}_{i \in I} \mid R) \longrightarrow (\boldsymbol{\nu}xy)(P \mid Q_j \mid R)} \qquad \text{[R-Case]}$$

$$\frac{P \longrightarrow P'}{P \mid Q \longrightarrow P' \mid Q} \qquad \frac{P \longrightarrow P'}{(\boldsymbol{\nu}xy)P \longrightarrow (\boldsymbol{\nu}xy)P'} \qquad \text{[R-Par] [R-Res]}$$

$$\frac{P \equiv P' \quad P' \longrightarrow Q' \quad Q \equiv Q'}{P \longrightarrow Q} \qquad \text{[R-Struct]}$$

Fig. 2. Reduction semantics for s$\pi$

$Q[v/z]$. When $q = \operatorname{un}$ then process $q\,y(z).Q$ remains (Rule [R-UnCom]); otherwise, process $q\,y(z).Q$ disappears (Rule [R-LinCom]). Rule [R-Case] stands for the case synchronization: processes $x \triangleleft l_j.P$ and $y \triangleright \{l_i : Q_i\}_{i \in I}$ can synchronize if they are underneath a scope restriction $(\boldsymbol{\nu}xy)$. Process $x \triangleleft l_j.P$ reduces to process $P$ and process $y \triangleright \{l_i : Q_i\}_{i \in I}$ reduces to process $Q_j$. Rules for parallel composition, scope restriction, and structurally congruent processes (Rules [R-Par], [R-Res], [R-Struct]) are as usual.

As an example, consider the processes:

$$P_1 = \operatorname{un} y_1(t).\overline{t}\mathsf{false}.0 \qquad P_2 = \operatorname{lin} y_1(w).\overline{w}\mathsf{true}.0 \qquad P_3 = \overline{x_1}x_2.y_2(z).\overline{a}z.0$$
$$P = (\boldsymbol{\nu}x_1y_1)(\boldsymbol{\nu}x_2y_2)(P_1 \mid P_2 \mid P_3)$$

Starting from $P$, there are two possible sequences of reductions depending on the processes involved in the initial synchronization in the co-variables $x_1$, $y_1$. If the synchronization involves $P_1$ and $P_3$ then we have:

$$P \longrightarrow \longrightarrow (\boldsymbol{\nu}x_1y_1)(\boldsymbol{\nu}x_2y_2)(P_1 \mid P_2 \mid \overline{a}\mathsf{false}.0)$$

On the other hand, if $P_2$ and $P_3$ synchronize then we have:

$$P \longrightarrow \longrightarrow (\boldsymbol{\nu}x_1y_1)(\boldsymbol{\nu}x_2y_2)(P_1 \mid \overline{a}\mathsf{true}.0)$$

The *standard form* of a process, defined in [13], will be crucial for the executable specification of the reduction semantics. We say $P$ is in standard form if it matches the pattern expression $(\boldsymbol{\nu}x_1y_1)(\boldsymbol{\nu}x_2y_2)\ldots(\boldsymbol{\nu}x_ny_n)(P_1 \mid P_2 \mid \ldots \mid P_k)$, where each $P_i$ is a process of the form $\overline{x}v.Q$, $qx(y).Q$, $x \triangleleft l.Q$ or $\triangleright\{l_i : Q_i\}_{i \in I}$. Every process is structurally congruent to a process in standard form.

## 3  Rewriting Semantics for s$\pi$

*Syntax.* Our rewriting semantics for s$\pi$ adapts the one in [12], which is defined for an untyped $\pi$-calculus without sessions. There is a direct correspondence between the syntactic categories (values, variables, qualifiers, and terms) and Maude sorts (Value, Chan, Qualifier, and Trm, respectively). We also have some auxiliary sorts such as Guard, Choice, and Choiceset.

```
sorts Value Chan Qualifier Trm Guard Choice Choiceset .
subsort Choice < Choiceset .
subsort Chan < Value .

op _{_} : Qid Nat -> Chan [prec 1] .
ops lin un : -> Qualifier [ctor] .
ops True False : -> Value [ctor] .
op __(_) : Qualifier Chan Qid -> Guard [ctor prec 5] .
op _<_> : Chan Value -> Guard [ctor prec 6] .
op nil : -> Trm [ctor] .
op new[__]_ : Qid Qid Trm -> Trm [ctor prec 10] .
op _|_ : Trm Trm -> Trm [ctor assoc comm prec 12 id: nil] .
op if_then_else_fi : Value Trm Trm -> Trm [ctor prec 8] .
op _ << _._ : Chan Qid Trm -> Trm [ctor prec 15] .
op _ >> {_} : Chan Choiceset -> Trm [ctor prec 17] .
op _._ : Guard Trm -> Trm [ctor prec 7] .
op _:_ : Qid Trm -> Choice .
op empty : -> Choiceset [ctor] .
op __ : Choiceset Choiceset -> Choiceset [ctor assoc comm id: empty] .
```

Following the syntax in Sect. 2, values can be variables or booleans. We represent booleans as the constructors True and False whereas we distinguish variables (sort Chan) as values through the subsort relation. The only constructor for variables _{_} takes a Qid and a natural number. Each production rule for processes is represented using a constructor, as expected. Notice that the constructor for input guards __(_) is preceded by a qualifier. Process 0 is denoted as nil and a single guarded term is represented by the constructor _._. The constructor for scope restriction new[__]_ uses two instances of Qid, since it declares a pair of co-variables. The constructor for conditionals is parametric on an instance of Value. We add constructors for selection and branching process terms; their definition is as expected. In particular, the constructor for branching processes relies on instances of Choiceset, which consists of sets of pairs of Qid and process terms. We use instances of Qid to represent labels.

*Substitutions.* As we have seen, the semantics of s$\pi$ relies on substitutions of variables with values. To deal with substitutions in Maude, we follow Thati et al.'s approach [12] and use Stehr's CINNI calculus [11], an explicit substitution calculus, which provides a mechanism to implement $\alpha$-conversion at the language level. The idea behind CINNI is to syntactically associate each use of a variable $x$ to an index, which acts as a counter of the number of binders for $x$ that are found before it is used. In CINNI, there are three types of substitution operations: A simple substitution of a variable a for a variable x takes place if the index of x is 0; the index is decreased by 1 otherwise. A shift substitution over a

| Type | Meaning |
|---|---|
| Simple substitution | $[a := x]\, a\{0\} \mapsto x \qquad [a := x]\, a\{n+1\} \mapsto a\{n\}$ <br> $[a := x]\, b\{m\} \mapsto b\{m\}$ |
| Shift substitution | $\uparrow_a a\{n\} \mapsto a\{n+1\} \qquad \uparrow_a b\{m\} \mapsto b\{m\}$ |
| Lift substitution | $\Uparrow_a (S)\, a\{0\} \mapsto a\{0\} \qquad \Uparrow_a (S)\, a\{n+1\} \mapsto \uparrow_a (S\, a\{n\})$ <br> $\Uparrow_a (S)\, b\{m\} \mapsto \uparrow_a (S\, b\{m\})$ |

increases by 1 the index and a substitution S can be lifted to skip one index. Any substitution over a variable a has no effect on other variables.

We now present the definition of explicit subtitutions for s$\pi$ using an approach similar to the one in [11]. We firts present the definition of the variable substitutions. We use the sort `Subst` and the substitution application is performed by the operator `__`, which takes a substitution and a variable. We define the three substitutions above as presented there, by means of some equations.

```
sort Subst .
op [_:=_] : Qid Value -> Subst .
op [shiftup_] : Qid -> Subst .
op [lift__] : Qid Subst -> Subst .
op __ : Subst Chan -> Chan .

eq [ a := v ] a{0} = v .
eq [ a := v ] a{s(n)} = a{n} .
ceq [ a := v ] b{n} = b{n}  if a =/= b .
eq [ shiftup a ] a{n} = a{s(n)} .
ceq [ shiftup a ] b{n} = b{n}  if a =/= b .
eq [ lift a S ] a{0} = a{0} .
eq [ lift a S ] a{s(n)} = [ shiftup a ] S a{n} .
ceq [ lift a S ] b{n} = [ shiftup a ] S b{n} if a =/= b .
```

Equipped with these elements, we adapt to the s$\pi$ syntax the equations associated to the explicit substitutions for the process terms as follows:

```
op  __ : Subst Trm -> Trm [prec 3] .
op  subst-aux : Subst Choiceset -> Choiceset .
eq  S nil = nil .
eq  S (new [x y] P) = new [x y] ([lift x S] [lift y S] P) .
eq  S (q a(y) . P ) = q (S a)(y) . ([lift y S] P) .
eq  S (a < b > . P) = (S a) < (S b) > . (S P) .
ceq S (a < v > . P) = (S a) < v > . (S P) if v == True or v == False .
ceq S (if v then P else Q fi) = if v then (S P) else (S Q) fi
                                if v == True or v == False .
eq  S (a >> {CH}) = (S a) >> { subst-aux(S, CH) } .
eq  S (a << x . P) = (S a) << x . (S P) .
eq  S (P | Q) = (S P) | (S Q) .
eq  subst-aux(S, empty) = empty .
eq  subst-aux(S, (x : P) CH) = (x : (S P)) subst-aux(S, CH) .
```

In each equation, we deal with a specific production rule for process terms. In each process, the substitution S is applied in each variable and each subprocess as expected. Particularly, a lift substitution is performed over x, y and S to skip the index 0 and perform the substitution in the remaining indices for the scope restriction operator. In this way, the substitution S has the expected effect.

*Structural Congruence.* To represent the rules in Fig. 1, we exploit the Maude equational attributes `assoc`, `comm`, and `id` to declare the associative, commutative, and identity axioms for parallel composition, with process `nil` acting as its identity. This suffices to cover the rules on the two first lines of Fig. 1. The remaining rules are explicitly declared as equations below:

```
eq new[x y] nil = nil .
ceq P | new[x y] Q = new [x y] (Q | [shiftup x] [shiftup y] P)
      if P =/= nil /\ Q =/= nil /\ CS := freevars(P) /\
          x{0} in CS and y{0} in CS .
eq if True then P else Q fi = P .
eq if False then P else Q fi = Q .
ceq P | new[x y] Q = new [x y] (Q | [shiftup x] P)
      if P =/= nil /\ Q =/= nil /\ CS := freevars(P) /\
          x{0} in CS and not y{0} in CS .
ceq P | new[x y] Q = new [x y] (Q | [shiftup y] P)
      if P =/= nil /\ Q =/= nil /\ CS := freevars(P) /\
          not x{0} in CS and y{0} in CS .
ceq P | new[x y] Q = new [x y] (Q | P)
      if P =/= nil /\ Q =/= nil /\ CS := freevars(P) /\
          not x{0} in CS /\ not y{0} in CS .
```

Scope extrusion is represented through four equations corresponding to the four cases in the presence of x, y in the free variables of process P. Function `freevars` stands for the Maude implementation for function `fv` over processes.

*Operational Semantics.* Combined, the Maude rewriting rules, the equational attributes, and the explicit equations associated to variables of sort `Trm` can appropriately express the reduction semantics of s$\pi$ and manipulate terms in a compositional fashion. A process is reduced to a simpler equivalent form by virtue of the equational theory; a process is rewritten as long as it satisfies the structure required for a rule wherever the process is located. As a consequence, subprocesses are also rewritten and we do not need to explicitly represent the contextual rules ([R-PAR] and [R-RES]).

A process is converted into standard form using the explicit congruence rules. This way, the scope of every unguarded occurrence of the `new` operator is extended to the top level.

Process interaction in s$\pi$ can only occur through co-variables and therefore processes that are involved must be underneath a scope restriction over such co-variables. Nonetheless, since in the standard form the order of the unguarded ocurrences of the new operator is irrelevant, it would be necessary to explicitly look for the processes that are enabled to interact, which would affect the efficiency of the rewriting specification. To counter this, we include an auxiliary operator, dubbed new*, which declares a list of pairs of new co-variables, rather than just a single pair. This is equivalent to using nested new operators, i.e., the term new* [x1 y1 x2 y2 ... xn yn] P is equivalent to the term

new [x1 y1] new [x2 y2] ... new [xn yn] P.

We declare the constructor for the sort QidSet with the equational attribute comm to impose that the order among the pairs of new co-variables is not distinguished. In this way, whatever they are the process to interact, these will be underneath a scope restriction new* and the interaction will be enabled.

```
sorts QidPair QidSet . subsort QidPair < QidSet .
op __ : Qid Qid -> QidPair [ctor] .
op mt : -> QidSet [ctor] .
op __ : QidSet QidSet -> QidSet [ctor comm assoc id: mt] .
op new* [_] _ : QidSet Trm -> Trm [ctor] .
```

Given a process $P$, let us write $[\![P]\!]$ to denote its representation in Maude. A reduction rule $P \longrightarrow Q$ can be associated to a rewriting rule $l : [\![P]\!] \Rightarrow [\![Q]\!]$. The reduction rules can be stated as follows:

```
crl [FLAT] : P => P' if P' := flatten(P) /\ P =/= P' .
rl [LINCOM] : new* [(x y) nl] x{N} < v > . P | lin y{N}(z) . Q | R =>
                new* [(x y) nl] P | [z := v] Q | R .
rl [UNCOM] : new* [(x y) nl] x{N} < v > . P | un y{N}(z) . Q | R =>
                new* [(x y) nl] P | [z := v] Q | un y{N}(z) . Q | R .
rl [CASE] : new* [(x y) nl] (x{N} << w . P) |
                (y{N} >> { (w : Q) CH }) | R => new* [(x y) nl] P | Q | R .
```

Rule FLAT normalizes the whole process. In this sense, additional to the implicit rewriting performed by the equations associated to the congruence rules, the nested new declarations are stated as a flat declaration new*. We use an auxiliary operation flatten, which is defined as follows:

```
op flatten : Trm -> Trm .
eq flatten(new [x y] P) = flatten(new* [x y] P) .
eq flatten(new* [nl] new [x y] P) = flatten(new* [nl x y] P) .
eq flatten(new* [nl] new* [nl'] P) = flatten(new* [nl nl'] P) .
eq flatten(P) = P [owise] .
```

Rules LINCOM, UNCOM and CASE correspond to the specification of the reduction rules related to synchronization in the calculus semantics (see Fig. 2). In these rules, nl stands for the additional co-variables being declared. As expected, Rules LINCOM, and UNCOM perform a substitution of the variable z for the value v.

We include also some equations which capture natural equivalences for processes involving the auxiliary operator new*.

```
eq new* [nl] nil = nil .
eq new* [x y nl] y{N} < v > . P | q x{N}(z) . Q | R =
   new* [y x nl] y{N} < v > . P | q x{N}(z) . Q | R .
eq new* [x y nl] (y{N} << w . P) | (x{N} >> { CH }) | R =
   new* [y x nl] (y{N} << w . P) | (x{N} >> { CH }) | R .
```

Given a pair x y of co-variables, we assume that the first action of x is an output or a selection and the first action y is an input or a branching. The last two equations swap x and y when this is not the case, to enable the execution of the rewriting rules.

Our rewriting specification enables us to directly execute a possible sequence of reductions over a process using the Maude command 'rew'. In this way, we can obtain a stable (final) reachable process, which cannot reduce further. Moreover, we can use the reachability command 'search' to: (i) perform all possible sequence of reductions of a process and obtain every possible stable process and (ii) check whether a process that fits some pattern is reachable or if a specific process is reachable. In Sect. 4, we leverage commands 'search' and 'modelCheck' to detect deadlocked s$\pi$ processes.

*Specification Correctness.* The transition system associated to our rewrite theory in Maude can be shown to coincide with the reduction semantics in Sect. 2. This operational correspondence result is detailed in [8].

## 4    Algorithmic Type Checking for s$\pi$

### 4.1    Type Syntax

We present a Maude implementation of the algorithmic type checking given in [13]. The type system considers *typing contexts*, denoted $\Gamma$, which associate each variable to a specific type, denoted $T$. Typing contexts and types are defined inductively as follows:

$$\Gamma ::= \emptyset \ | \ \Gamma, x : T \qquad\qquad q ::= \text{lin} \ | \ \text{un}$$
$$p ::= \ ?T.T \ | \ !T.T \ | \ \&\{l_i : T_i\}_{i \in I} \ | \ \oplus\{l_i : T_i\}_{i \in I}$$
$$T ::= \text{bool} \ | \ \text{end} \ | \ q\,p \ | \ a \ | \ \mu a.T$$

where $q$ stands for qualifiers and $p$ stands for pretypes. Moreover, $x$ denotes a variable, each $l_i$ denotes a label and $a$ denotes a general variable. For simplicity, we assume a single basic type for values (bool). Each variable is associated to a (session) type, which represents its intended protocol. In the above grammar, these types correspond to qualified pretypes. The pretype $?T_1.T_2$ (resp. $!T_1.T_2$) is assigned to a variable that first receives (resp. sends) a value of type $T_1$ and then proceeds to type $T_2$. The pretype $\&\{l_i : T_i\}_{i \in I}$ (resp. $\oplus\{l_i : T_i\}_{i \in I}$) is assigned to a variable that can offer (resp. select) $l_i$ options and continues with type $T_i$ depending on the label selected. The type end (empty sequence) denotes the type of a variable where no interaction can occur. Recursive types can express infinite sequences of actions; in the type $\mu a.T$, $a$ corresponds to a type variable that must occur guarded in $T$.

We encode session types in Maude by associating the non-terminals context, qualifiers, pretypes, and types to sorts Context, Qualifier, Pretype, and Type.

```
sorts Pretype Type Context ChoiceT ChoiceTset .
subsort ChoiceT < ChoiceTset .
op ?_._ : Type Type -> Pretype .   op !_._ : Type Type -> Pretype .
op +{_} : ChoiceTset -> Pretype .  op &{_} : ChoiceTset -> Pretype .
ops bool end : -> Type .            op __ : Qualifier Pretype -> Type .
op u [_] _ : Qid Type -> Type .    op var : Qid -> Type .
ops nil invalid-context : -> Context .
op _:_ : Value Type -> Context .
op _,_ : Context Context -> Context [ctor assoc comm id: nil] .
op _:_ : Qid Type -> ChoiceT . op empty : -> ChoiceTset .
op __ : ChoiceTset ChoiceTset -> ChoiceTset [assoc comm id: empty] .
```

Each production rule is given as a specific constructor. In particular, constructors +{_} and &{_} represent the pretypes $\oplus\{l_i : T_i\}_{i \in I}$ and $\&\{l_i : T_i\}_{i \in I}$, respectively. The pairs of labels $l_i$ and subtypes $T_i$ are defined as instances of the sort ChoiceTset. The recursive type $\mu a.T$ is given as the constructor u [_] _ and the type variables are given as the constructor var. Typing contexts are defined as expected. An empty context is denoted as nil whereas a single context is associated to the constructor _:_. General contexts are provided by the constructor _,_, which is annotated with the equational attributes assoc, comm and id since the order is irrelevant in typing contexts and the construction is associative. Finally, we added a constant invalid-context to be used in the type checking to denote a typing error.

## 4.2   Algorithmic Type Checking

We follow the algorithmic type checking proposed in [13]. This type system enables to type check the sπ processes from Sect. 2, with a minor caveat: algorithmic type checking uses processes in which the restriction operator has a corresponding type annotation, i.e., it uses $(\nu xy : T)P$ instead of $(\nu xy)P$. Consequently, we add a constructor for the sort Trm in the Maude specification:

```
op new[__:_]_ : Qid Qid Type Trm -> Trm [ctor prec 28] .
```

Following [13], we implement the type checking algorithm by relying on some auxiliary functions for type duality (i.e., compatibility), type equality, and context update and difference, among others. They are implemented by means of functions and equations in Maude. The details of the Maude implementation for type duality (function dual), context update (function +), and the context difference (function \) can be found in [8].

$$\Gamma \vdash \text{true} : \text{bool}; \Gamma \quad [\text{A-TRUE}] \qquad \dfrac{}{\Gamma_1, x : \text{lin } p, \Gamma_2 \vdash x : \text{lin } p; (\Gamma_1, \Gamma_2)} \quad [\text{A-LINVAR}]$$

$$\Gamma \vdash \text{false} : \text{bool}; \Gamma \quad [\text{A-FALSE}] \qquad \dfrac{\text{un}(T)}{\Gamma_1, x : T, \Gamma_2 \vdash x : T; (\Gamma_1, x : T, \Gamma_2)} \quad [\text{A-UNVAR}]$$

**Fig. 3.** Typing rules for values, $\Gamma \vdash v : T; \Gamma$

Algorithmic type checking is expressed by sequents of the form $\Gamma_1 \vdash v : T; \Gamma_2$ (for values) and $\Gamma_1 \vdash P : \Gamma_2; L$ (for processes). These sequents have an input-output reading: sequent $\Gamma_1 \vdash v : T; \Gamma_2$ denotes an algorithm that takes $\Gamma_1$ and $v$ as input and returns $T$ and $\Gamma_2$ as output; similarly, sequent $\Gamma_1 \vdash P : \Gamma_2; L$ denotes an algorithm that takes $\Gamma_1$ and $P$ as input and produces $\Gamma_2$ and $L$ as output. While $\Gamma_2$ is a residual context, the set $L$ collects linear variables occurring in subject position. Intuitively, $L$ tracks the linear variables that are used in $P$ to prevent that they are used again in another process. Both algorithms are given by means of typing rules, which we specify in Maude as an equational theory.

Figure 3 shows the typing rules for values, which correspond to the rules in [13]. The rules for boolean values [A-TRUE] and [A-FALSE] produce as results the type bool and the input context $\Gamma$ remains unaltered. There are two rules for a variable $x$: if $x$ has a linear type lin $p$ then the entry $x : \text{lin } p$ is removed from the returned context (Rule [A-LINVAR]); otherwise, if $x$ is unrestricted then the entry $x : T$ is kept in the returned context (Rule [A-UNVAR]). The algorithm for type checking of values is then implemented as a function type-value, which is defined as follows:

```
op type-value : Context Value -> TupleTypeContext .
eq type-value(C, True) = [C bool] .                         ---[A-TRUE]
eq type-value(C, False) = [C bool] .                        ---[A-FALSE]
ceq type-value(((a : T), C), a) = [((a : T), C) unfold(T)]  ---[A-UNVAR]
    if unrestricted(T) .
eq type-value(((a : lin p), C), a) = [C (lin p)] .          ---[A-LINVAR]
eq type-value(((a : u [x] T), C), a) =
    type-value(((a : unfold(u [x] T)), C), a) .             ---[A-LINVAR]
eq type-value(C, v) = ill-typed [owise] .
```

Function type-value produces an instance of the sort TupleTypeContext. This sort groups a context and a type or a set of variables and it has only one constructor [_ _]. The equations related to the typing of boolean values arise as expected, according to the corresponding typing rule. In those cases, a tuple that contains the unmodified context and the type bool is produced. For unrestricted variables, given that some types are infinite then, before the update, the unrestricted types are *unfolded* (cf. the unfold operation). Unfolding is the mechanism defined in [13] to deal with infinite types: If a type $T$ is a recursive type $\mu a.U$ then the substitution $U[\mu a.U/a]$ is performed. Otherwise, the type $T$

$$\Gamma \vdash \mathbf{0} : \Gamma; \emptyset \qquad \dfrac{\Gamma_1 \vdash P : \Gamma_2; L_1 \qquad \Gamma_2 \div L_1 \vdash Q : \Gamma_3; L_2}{\Gamma_1 \vdash P \mid Q : \Gamma_3; L_2} \qquad \text{[A-Inact] [A-Par]}$$

$$\dfrac{\Gamma_1, x : T, y : \overline{T} \vdash P : \Gamma_2; L}{\Gamma_1 \vdash (\boldsymbol{\nu} xy : T)\, P : \Gamma_2 \div \{x, y\}; L \backslash \{x, y\}} \qquad \text{[A-Res]}$$

$$\dfrac{\Gamma_1 \vdash v : q\ \mathsf{bool}; \Gamma_2 \qquad \Gamma_2 \vdash P : \Gamma_3; L \qquad \Gamma_2 \vdash Q : \Gamma_3; L}{\Gamma_1 \vdash \mathsf{if}\ v\ \mathsf{then}\ P\ \mathsf{else}\ Q : \Gamma_3; L} \qquad \text{[A-If]}$$

$$\dfrac{\Gamma_1 \vdash x : q!T.U; \Gamma_2 \qquad \Gamma 2 \vdash v : T; \Gamma_3 \qquad \Gamma_3 + x : U \vdash P : \Gamma_4; L}{\Gamma_1 \vdash \overline{x} v.P : \Gamma_4; L \cup (\text{if } q = \mathsf{lin\ then}\ \{x\}\ \mathsf{else}\ \emptyset)} \qquad \text{[A-Out]}$$

$$\dfrac{\Gamma_1 \vdash x : q_2?T.U; \Gamma_2 \quad (\Gamma_2, y : T) + x : U \vdash P : \Gamma_3; L \quad q_1 = \mathsf{un} \Rightarrow L \backslash \{y\} = \emptyset}{\Gamma_1 \vdash q_1 x(y).P : \Gamma_3 \div \{y\}; L \backslash \{y\} \cup (\text{if } q_2 = \mathsf{lin\ then}\ \{x\}\ \mathsf{else}\ \emptyset)} \quad \text{[A-In]}$$

$$\dfrac{\Gamma_1 \vdash x : q \& \{l_i : T_i\}_{i \in I}; \Gamma_2 \quad \Gamma_2 + x : T_i \vdash P_i : \Gamma_3; L_i \quad \forall_{i \in I, j \in I}\ L_i \backslash \{x\} = L_j \backslash \{x\}}{\Gamma_1 \vdash x \triangleright \{l_i : P_i\}_{i \in I} : \Gamma_3; L \cup (\text{if } q = \mathsf{lin\ then}\ \{x\}\ \mathsf{else}\ \emptyset)}$$
$$\text{[A-Branch]}$$

$$\dfrac{\Gamma_1 \vdash x : q \oplus \{l_i : T_i\}_{i \in I}; \Gamma_2 \qquad \Gamma_2 + x : T_j \vdash P : \Gamma_3; L \qquad j \in I}{\Gamma_1 \vdash x \triangleleft l_j.P : \Gamma_3; L \cup (\text{if } q = \mathsf{lin\ then}\ \{x\}\ \mathsf{else}\ \emptyset)} \quad \text{[A-Sel]}$$

**Fig. 4.** Typing Rules for Processes, $\Gamma \vdash P : \Gamma; L$

remains unaltered. For linear variables, we also unfold the type when necessary and the linear type is returned and removed from the context.

Figure 4 shows some of the typing rules for $\mathsf{s}\pi$ processes; they largely correspond to the rules in [13].

Rule [A-Inact] proceeds as expected. Process $\mathbf{0}$ is well-typed and the typing context $\Gamma$ remains unaltered and the set of linear variables is empty. Rule [A-Par] handles parallel composition: to check a process $P \mid Q$ over a context $\Gamma_1$, the type of $P$ is checked and the resulting context $\Gamma_2$ is used to type-check process $Q$, making sure that the linear variables used for $P$ are first removed by using the context difference function $(\Gamma_2 \div L_1)$. This ensures that free linear variables are used only once. The output of the algorithm for $Q$ (context $\Gamma_3$ and set $L_2$) then corresponds to the ouput of the entire process $P \mid Q$. Rule [A-Res] type-checks a process $(\boldsymbol{\nu} xy : T)P$ in a context $\Gamma_1$: it first checks the type of sub-process $P$ in the context $\Gamma_1$ extended with the association of variables $x$, $y$ to the type $T$ and its dual type, denoted $\overline{T}$. It is expected that if type $T$ $(\overline{T})$ is linear then it should not be in the resulting context $\Gamma_2$; otherwise, if type $T$ $(\overline{T})$ is unrestricted then it will appear in $\Gamma_2$. We require that variables $x$, $y$ are deleted from the residual context $(\Gamma_2 \div \{x, y\})$ and from the set $L$ of linear variables.

Rule [A-If] verifies that type of value $v$ is $\mathsf{bool}$ in the context $\Gamma_1$, and requires that the typecheck of $P$ and $Q$ in the context $\Gamma_2$ generate the same residual context $\Gamma_3$ and the same set $L$, since both processes should use the same linear variables. Rule [A-Out] handles output processes: it uses the incoming context

$\Gamma_1$ to check the type of $x$, which should be of the form $q!T.U$. Then, it checks that the type of $v$ in the residual context $\Gamma_2$ is $T$. The type of the continuation $P$ is checked in a new context $\Gamma_3$ extended with the association of $x$ and the continuation type $U$. The rule enforces that types $q!T.U$ and $U$ must be equivalent when $x$ is unrestricted (i.e., $q = \mathsf{un}$). The rule returns a context $\Gamma_4$ and a set of variables $L$ joined with $x$, if linear. Rule [A-IN] presents some minor modifications with respect to the one in [13]. We require that in the case of replication there are no (free) subjects on linear variables in process $P$ except possibly the input variable $y$. Other than this, this rule is similar to Rule [A-OUT].

Rule [A-SEL] looks the type of $x$ in the incoming context $\Gamma_1$. This type must be of the form $q \oplus \{l_i : T_i\}_{i \in I}$. Subsequently, the continuation $P$ is type-checked under the resulting context $\Gamma_2$ updated with a new assumption for $x$, which is associated to a type $T_j$. In this way, when $q = \mathsf{un}$ we must have $\oplus\{l_i : T_i\}_{i \in I} = T_j$. This rule produces as result the context $\Gamma_3$ and the set of linear variables $L$ is augmented with $x$ if linear. Context $\Gamma_3$ and set $L$ also corresponds to the output of the type checking of process $P$. Finally, we have Rule [A-BRANCH], which has some minor modifications with respect to the rule in [13]. More precisely, this rule has been changed to require that the sets of (free) subjects on linear variables $L_i$ only differ in the input variable $y$. The additional details of this rule is quite similar to Rule [A-SEL].

As an example of type checking, if $T = \mathsf{lin}\ !\mathsf{bool.lin}\ ?\mathsf{bool.end}$ then we can establish the following sequent:

$$a : \mathsf{bool} \vdash (\boldsymbol{\nu} x_1 y_1 : T)(\mathsf{lin}\ y_1(v).\overline{y_1}v.\mathbf{0} \mid \overline{x_1}a.\mathsf{lin}\ x_1(z).\mathbf{0}) : (a : \mathsf{bool}); \{x_1, y_1\}$$

The algorithm for type-checking processes is implemented as a function `type-term` that receives an instance of the sort `Context` and an instance of the sort `Trm`. Moreover, it produces an instance of the sort `TupleTypeContext` that groups the resulting typing context and the set $L$ of linear variables that were collected during type-checking. Each rule is implemented by an equation:

```
op type-term : Context Trm -> TupleTypeContext .
eq type-term(C, nil) = [C mt] .                        --- [A-INACT]
ceq type-term(C, P | Q) = [C2 L2]                       --- [A-PAR]
    if [C1 L1] := type-term(C, P) /\
       [C2 L2] := type-term(C1 / L1, Q) .
ceq type-term(C, new [x y : T] P) =                     --- [A-RES]
       [(C1 / (x{0} y{0})) remove(remove(L1, x{0}), y{0})]
   if [C1 L1] := type-term((C, (x{0} : T), (y{0} : dual(T))), P) .
ceq type-term(C, if v then P else Q fi) = [C2 L1]        --- [A-IF]
   if [C1 bool] := type-value(C, v) /\
      [C2 L1] := type-term(C1, P) /\ [C2 L1] := type-term(C1, Q) .
ceq type-term(C, a < v > . P) =                         --- [A-OUT]
        [C3 (if q == lin then (L1 a) else L1 fi)]
    if [C1 (q ! T . U)] := type-value(C, a) /\
       [C2 T'] := type-value(C1, v) /\ /\ equal(T, T')
       [C3 L1] := type-term((C2 + a : U), P) .
ceq type-term(C, un a(y) . P) = [(C2 / y{0}) mt]        --- [A-IN]
```

```
    if [C1 (un ? T . U)] := type-value(C, a) /\
       [C2 L] := type-term((C1, (y{0} : T)) + a : U, P) /\
       remove(L, y{0}) == mt .
ceq type-term(C, lin a(y) . P) =                           --- [A-IN]
       [((C2 / y{0}) (remove(L, y{0})
        (if q == lin then a else mt fi))]
    if [C1 (q ? T . U)] := type-value(C, a) /\
       [C2 L] := type-term((C1, (y{0} : T)) + a : U, P) .
ceq type-term(C, a>>{CH}) = check-branch(C1, a, CH, CHT,q) ---[A-BRANCH]
    if [C1 (q & { CHT })] := type-value(C, a) .
ceq type-term(C, a << x . P) =                             ---[A-SEL]
       [C2 (if q == lin then (L1 a) else L1 fi)]
    if [C1 (q + { (x : T) CHT })] := type-value(C, a) /\
       [C2 L1] := type-term((C1 + a : T), P) .
eq type-term(C, P) = ill-typed [owise] .
```

When type checking is successful, function type-term produces an outgoing type context and a set of variables. Those elements are grouped using the constructor [_,_], which is associated to the sort TupleTypeContext. We use a Maude comment to annotate each equation with the corespondent typing rule. The correspondence is quite intuitive; we highlight some important details. An empty set of variables is represented with the constant mt. We remark that the operator / stands for the context difference operation that removes some variables of a type context, whereas operator 'remove' drops a variable of a variable set. In the equation for Rule [A-OUT], we do not use the same variable T in the type associated to variable a and the type associated to value v as it would be expected, since the types are possibly infinite and there are many possible representations for the same infinite type. Instead, we use another variable T' and we check that T and T' are equivalent, using function equal.

We divide Rule [A-IN] in two different equations for linear and unrestricted inputs. In the linear case, it is possible that the type of the subject a is linear or unrestricted; when the variable is linear it must be included in the returned set of linear variables. In the unrestricted case, the type of subject a is required to be unrestricted inasmuch as the attempt to use a linear variable in an unrestricted fashion must be rejected. Moreover, we require that the only free linear variable used in process P is y{0} (condition remove(L, y{0}) == mt).

## 4.3 Type Soundness

Vasconcelos [13] established that the type system for $s\pi$ is *sound*: a closed, well-typed process is guaranteed to have a well-defined behavior according to the ascribed protocols and the reduction semantics of the calculus. Also, the algorithmic type checking, as implemented in this section, is proven correct. With these elements in mind, we can integrate both the rewriting specification of the operational semantics and the implementation of the algorithmic type checking. This way, we only execute well-typed processes. For this purpose, we use two auxiliary functions well-typed and erase. The former checks whether a process does not have typing errors:

```
op well-typed : Trm -> Bool .
eq well-typed(P) = (type-term(nil, P) =/= ill-typed) .
```

Function `well-typed` applies the algorithm for type checking `type-term` over a process P and returns `true` when type-checking is successful, i.e. when the result is not `ill-typed`. Function `erase` proceeds inductively on the structure of a process; when it reaches an annotated subprocess 'new [x y : T] P', it removes the annotation to produce 'new [x y] P'—see [8] for details.

Correspondingly, we extend our specification of the reduction semantics to enable the execution of annotated processes, i.e., processes that use the operator $(\nu xy : T)P$ instead of the operator $(\nu xy)P$:

```
rl [TYPED] : new [x y : T] P => if well-typed(new [x y : T] P)
                 then erase(new [x y] P) else ill-typed-process fi .
```

We check whether process `new [x y : T] P` is well-typed; if so, we rewrite it as an equivalent process in which each occurrence of `new [x y : T]` is replaced by `new [x y]` through the function `erase`. Otherwise, process `new [x y : T] P` is rewritten as `ill-typed-process`, a constant that denotes that the process has a typing error and cannot be executed.

## 5   Lock and Deadlock Detection in Maude

Although the type system for $s\pi$ given in [13] enables us to statically detect processes whose variables are used according to their ascribed protocols (expressed as session types), there are processes that are well-typed but that exhibit unwanted behaviors, in particular deadlocks. For example, consider the process

$$P = \overline{x_3}\mathsf{true}.\overline{x_1}\mathsf{true}.\overline{y_2}\mathsf{false}.0 \mid \mathsf{lin}\ y_3(z).\mathsf{lin}\ x_2(w).\mathsf{lin}\ y_1(t).0$$

Process $P$ is well-typed in a context $x_1$ : lin !bool.end, $y_1$ : lin ?bool.end, $x_2$ : lin ?bool.end, $y_2$ : lin !bool.end, $x_3$ : lin !bool.end, $y_3$ : lin ?bool.end. Then, process $(\nu x_1 y_1 x_2 y_2 x_3 y_3)P$ can reduce but becomes deadlocked after such a synchronization, due to a circular dependency on variables $x_1, y_1, x_2, y_2$.

### 5.1   Definitions

Here we characterize deadlocks in $s\pi$ and we show how to use the rewrite specification of the operational semantics and the Maude tools for detecting processes with deadlocks. We follow the formulation of deadlock and lock freedom given by Padovani [6], which uses the notion of *pending communication*. We start by defining the reduction contexts $\mathcal{C}$:

$$\mathcal{C} ::= [\ ] \mid (\mathcal{C} \mid P) \mid (\nu xy)\mathcal{C}$$

The notion of pending communication in a process $P$ with respect to variables $x$, $y$ is defined with the following auxiliary predicates:

$$\mathsf{in}(x, P) \overset{\text{def}}{\Longleftrightarrow} P \equiv \mathcal{C}[\mathsf{lin}\ x(y).Q] \ \wedge\ x \notin \mathsf{bv}(\mathcal{C})$$

$$\mathsf{in}^*(x, P) \overset{\text{def}}{\Longleftrightarrow} P \equiv \mathcal{C}[\mathsf{un}\ x(y).Q] \ \wedge\ x \notin \mathsf{bv}(\mathcal{C})$$

$$\mathsf{out}(x, P) \overset{\text{def}}{\Longleftrightarrow} P \equiv \mathcal{C}[\overline{x}v.Q] \ \wedge\ x \notin \mathsf{bv}(\mathcal{C})$$

$$\mathsf{sync}(x, y, P) \overset{\text{def}}{\Longleftrightarrow} (\mathsf{in}(x, P) \ \vee\ \mathsf{in}^*(x, P)) \ \wedge\ \mathsf{out}(y, P)$$

$$\mathsf{wait}(x, y, P) \overset{\text{def}}{\Longleftrightarrow} (\mathsf{in}(x, P) \ \vee\ \mathsf{out}(y, P)) \ \wedge\ \neg\mathsf{sync}(x, y, P)$$

where we assume the extension of function $\mathsf{bv}(.)$ to reduction contexts. Intuitively, the first three predicates express the existence of a pending communication on a variable $x$. More in details:

- Predicate $\mathsf{in}(x, P)$ holds if $x$ is free in $P$ and there is a subprocess of $P$ that is able to make a linear input on $x$. Predicate $\mathsf{in}^*(x, P)$ is its analog for unrestricted inputs.
- Predicate $\mathsf{out}(x, P)$ holds if $x$ is free in $P$ and a subprocess of $P$ is waiting to send a value $v$.
- Predicate $\mathsf{sync}(x, y, P)$ denotes a pending input on $x$ for which a synchronization on $y$ is immediately possible.
- Predicate $\mathsf{wait}(x, y, P)$ denotes a pending input/output for which a synchronization on $x$, $y$ is not immediately possible.

Let us write $\longrightarrow^*$ to denote the reflexive, transitive closure of $\longrightarrow$. Also, write $P \nrightarrow$ if there is no $Q$ such that $P \longrightarrow Q$. With these elements, we may now characterize the deadlock and lock freedom properties. We say process $P$ is

- *deadlock free* if for every $Q$ such that $P \longrightarrow^* (\nu x_1 y_1)(\nu x_2 y_2)\ldots(\nu x_n y_n)Q \nrightarrow$ it holds that $\neg\mathsf{wait}(x_i, y_i, Q)$ for every $x_i$.
- *lock free* if for every $Q$ such that $P \longrightarrow^* (\nu x_1 y_1)(\nu x_2 y_2)\ldots(\nu x_n y_n)Q$ and $\mathsf{wait}(x_i, y_i, Q)$ there exists $R$ such that $Q \longrightarrow^* R$ and $\mathsf{sync}(x_i, y_i, R)$ hold.

This way, a process is deadlock free if there are not stable states with pending inputs or outputs; a process is lock free if it is able to eventually perform a synchronization in any pending input or output.

We can use Maude to verify deadlock freedom and lock freedom for typed processes. Indeed, we can use the reachability tool `search` and the LTL model checker `modelCheck`. We first represent the previous predicates over process terms as functions in Maude over instances of the sorts `Trm` and `Chan`:

```
ops in out in* : Chan Trm -> Bool .
ops sync wait : Chan Chan Trm -> Bool .
op wait-aux : QidSet Trm -> Bool .
eq in(a, lin a(x) . Q | R) = true .
eq in(a, P) = false [owise] .
eq in*(a, un a(x) . Q | R) = true .
eq in*(a, P) = false [owise] .
```

```
eq out(a, a < v > . Q | R) = true .
eq out(a, P) = false [owise] .
eq sync(a, b, P) = (in(a, P) or in*(a, P)) and out(b, P) .
eq wait(a, b, P) = (in(a, P) or out(b, P)) and not sync(a, b, P) .
eq wait-aux(mt, P) = false .
eq wait-aux((x y) nl, P) = wait(x{0}, y{0}, P) or
   wait(y{0}, x{0}, P) or wait-aux(nl, P) .
```

Above, we use function `wait-aux` to determine if a group of pairs of co-variables contains a pair for which there is a pending communication.

The deadlock freedom property imposes that there should be no stable states in which there are pending communications. Consequently, we can use the Maude command `search` as follows to determine whether a process is deadlock free:

```
search init =>!
         new* [nl:QidSet] P:Trm such that wait-aux(nl:QidSet, P:Trm) .
```

where `init` denotes for the process to be checked. We recall that the `search` command with the arrow `=>!` looks for final (stable) states. In this way, `init` is deadlock free if the search returns no solution.

For the lock freedom property, we can not use the reachability tool since this property requires the checking some intermediate states. Consequently, we represent the lock freedom property as an LTL formula and use the built-in LTL model checker in Maude. Below, we define the Maude predicates `psync` and `pwait` that we will use in the LTL model checker:

```
ops pwait psync : Chan Chan -> Prop [ctor] .
eq new* [(x y) nl] P |= pwait(x{0}, y{0}) =
                        wait(x{0}, y{0}, P) or wait(y{0}, x{0}, P) .
eq new* [(x y) nl] P |= psync(x{0}, y{0}) =
                        sync(x{0}, y{0}, P) or sync(y{0}, x{0}, P) .
```

In the predicates `psync` and `pwait`, we use normalized processes, i.e., processes where the nested scope restrictions are flattened in an equivalent process that uses the operator `new*`. This assumption simplifies the definitions. Both `psync` and `pwait` predicates use the functions `in`, `in*`, `out`, `sync`, and `wait` as expected according to the definition.

The Kripke structure that is generated for Maude will use such normalized process term as states. The Maude predicates `pwait` and `psync` hold with respect to a pair of dual variables if there is a pending communication and there is a synchronization in the process associated to a state. The lock freedom property imposes for each variable that if in any state there is a pending communication then eventually there will be a synchronization. Formalizing the lock freedom property requires to check each possible subject. For that reason, the LTL formula associated to this property depends on the variables being used in the process. We define a function `build-lock-formula` that takes the used variables and builds the corresponding LTL formula as follows:

```
op build-lock-formula : QidSet -> Formula .
eq build-lock-formula(mt) = True .
eq build-lock-formula((x y) nl) =
    [] (<> pwait(x{0}, y{0}) -> <> psync(x{0}, y{0})) /\
       build-lock-formula(nl) .
```

This way, the resulting LTL formula corresponds to the conjunction of subformulas associated to each dual variable. The model checker can be used as follows:

```
red modelCheck(init, build-lock-formula(vars)) .
```

where `init` stands for the process term and `vars` stands for a set of pairs of co-variables. If the `init` is lock-free then the invocation of `modelCheck` will produce `true`. Otherwise, the invocation will show a counterexample with a sequence of rules that produces a state where the formula is not fulfilled.

## 5.2   Examples

We give a couple of examples of well-typed processes in $s\pi$, with different lock- and deadlock-freedom properties. (See [8] for additional examples.)

$$P_1 = (\nu x_1 y_1)(\nu x_2 y_2)(\nu x_3 y_3)(\overline{x_3}\text{true}.\overline{x_1}\text{true}.\overline{y_1}\text{false}.0 \mid \text{lin } y_3(z).\text{lin}y_2(x).\text{lin}x_2(w).0)$$
$$P_2 = (\nu x_1 y_1)(\nu x_2 y_2)(\nu ab)(\overline{x_1}b.0 \mid \overline{a}\text{true}.0 \mid \text{un } y_1(z).\overline{x_2}z.0 \mid \text{un } y_2(w).\overline{x_1}w.0)$$

Process $P_1$ is a simple process that reduces to a deadlock immediately after a synchronization on the co-variables $x_3$, $y_3$. Process $P_2$ represents an infinite process where the variable $b$ is repeatedly shared through communications on $x_1, y_1, x_2, y_2$. The process is a not lock-free: $b$ is never used to synchronize with its co-variable $a$. Figure 5 gives the Maude terms associated to these processes.

```
ops P1 P2 P3 P4 P5 : -> Trm .
eq P1 = new* [('y1' 'x1')('y2' 'x2')('y3' 'x3')]
       ('x3'{0} < True > . 'x1'{0} < True > . 'y1'{0} < False > . nil |
       lin 'y3'{0}('z') . lin 'y2'{0}('x') . lin 'x2'{0}('w') . nil) .
eq P2 = new* [('x1' 'y1')('x2' 'y2')('a' 'b')]
             ('x1'{0} < 'b'{0} > . nil | 'a'{0} < True > . nil |
             un 'y1'{0}('z') . 'x2'{0} < 'z'{0} > . nil |
             un 'y2'{0}('w') . 'x1'{0} < 'w'{0} > . nil ) .
```

**Fig. 5.** Processes in Maude

We analyze P1 using Maude by executing:

```
search P1 =>! new* [nl:QidSet] P:Trm
                    such that wait-aux(nl:QidSet, P:Trm) .
red modelCheck(P1,
        build-lock-formula(('y1' 'x1')('y2' 'x2')('y3' 'x3'))) .
```

We obtain the following results, which confirm that P1 is not deadlock free and not lock free:

```
search in TEST : P1 =>! new*[nl:QidSet]P:Trm
                    such that wait-aux(nl:QidSet, P:Trm) = true .
Solution 1 (state 1)
nl:QidSet --> ('x3' 'y3') ('y1' 'x1') 'y2' 'x2'
P:Trm --> 'x1'{0} < True > . 'y1'{0} < False > . nil |
          lin 'y2'{0}('x') . lin 'x2'{0}('w') . nil

No more solutions.

result ModelCheckResult: counterexample(
   {new*[('x3' 'y3') ('y1' 'x1') 'y2' 'x2']
       'x3'{0} < True > . 'x1'{0} < True > . 'y1'{0} < False > . nil |
       lin 'y3'{0}('z') . lin 'y2'{0}('x') . lin 'x2'{0}('w') . nil,
    'LINCOM},
   {new*[('x3' 'y3') ('y1' 'x1') 'y2' 'x2']
       'x1'{0} < True > . 'y1'{0} < False > . nil |
       lin 'y2'{0}('x') . lin 'x2'{0}('w') . nil,
    deadlock})
```

Consider now a similar execution for process P2:

```
search P2 =>! new* [nl:QidSet] P:Trm
                    such that wait-aux(nl:QidSet, P:Trm) .
red modelCheck(P2, build-lock-formula(('x1' 'y1')('x2' 'y2')('a' 'b'))) .
```

We obtain the following results, which confirm that P2 is an infinite process that is deadlock free but not lock free:

```
search in TEST : P2 =>! new*[nl:QidSet]P:Trm
                    such that wait-aux(nl:QidSet, P:Trm) = true .

No solution.

result ModelCheckResult: counterexample(nil,
   {new*[('a' 'b') ('x1' 'y1') 'x2' 'y2']
       'a'{0} < True > . nil | 'x1'{0} < 'b'{0} > . nil |
       un 'y1'{0}('z') . 'x2'{0} < 'z'{0} > . nil |
       un 'y2'{0}('w') . 'x1'{0} < 'w'{0} > . nil, 'UNCOM}
```

```
{new*[('a' 'b') ('x1' 'y1') 'x2' 'y2']
    'a'{O} < True > . nil | 'x2'{O} < 'b'{O} > . nil |
    un 'y1'{O}('z') . 'x2'{O} < 'z'{O} > . nil |
    un 'y2'{O}('w') . 'x1'{O} < 'w'{O} > . nil, 'UNCOM})
```

## 6   Closing Remarks

In this paper, we have reported on an executable specification in Maude of
the operational semantics and the associated algorithmic type-checking of s$\pi$,
a session-typed $\pi$-calculus proposed by Vasconcelos in [13]. We integrated both
specifications closely following his formulation. To our knowledge, ours is the
first Maude implementation of a session-typed process language. Because typing
in [13] does not exclude deadlocks, we leverage built-in tools in Maude and executable specifications to detect well-typed dead-locked processes. In our view,
these developments establish a promising starting point to the automated analysis of message-passing concurrency specifications.

As future work, we intend to adapt our equational theories to leverage the
confluence checker tool available in Maude. We also plan to extend our executable
specifications to perform behavioral analysis of the processes that implement
*multiparty session types*, in the spirit of [10]. Likewise, we aim to explore the
automated analysis of communication correctness of an extension of s$\pi$ with
*higher-order* process communication, in which values can be abstractions (functions from names to processes) [4]. Finally, we plan to consider connections
between our executable implementations and type systems derived from the
Curry-Howard correspondence between session types and linear logic [1,16].

**Acknowledgements.** This work has been partially supported by the Dutch Research
Council (NWO) under project No. 016.Vidi.189.046 (Unifying Correctness for Communicating Software).

We are grateful to Camilo Rocha for his useful suggestions on this work, to the
anonymous reviewers for their careful reading, and to the WRLA'22 attendees for
constructive remarks.

## References

1. Caires, L., Pfenning, F., Toninho, B.: Linear logic propositions as session types.
   Math. Struct. Comput. Sci. **26**(3), 367–423 (2016)
2. Clavel, M., et al.: All About Maude - A High-Performance Logical Framework.
   LNCS, vol. 4350. Springer, Heidelberg (2007). https://doi.org/10.1007/978-3-540-
   71999-1
3. Hüttel, H., et al.: Foundations of session types and behavioural contracts. ACM
   Comput. Surv. **49**(1), 3:1–3:36 (2016). https://doi.org/10.1145/2873052
4. Kouzapas, D., Pérez, J.A., Yoshida, N.: On the relative expressiveness of higher-
   order session processes. Inf. Comput. **268** (2019). https://doi.org/10.1016/j.ic.
   2019.06.002
5. Milner, R., Parrow, J., Walker, D.: A calculus of mobile processes. I. Inf. Comput.
   **100**(1), 1–40 (1992)

6. Padovani, L.: Deadlock and lock freedom in the linear $\pi$-calculus. In: Proceedings of the Joint Meeting of the Twenty-Third EACSL Annual Conference on Computer Science Logic (CSL) and the Twenty-Ninth Annual ACM/IEEE Symposium on Logic in Computer Science (LICS), CSL-LICS 2014. Association for Computing Machinery, New York (2014). https://doi.org/10.1145/2603088.2603116

7. Pitsiladis, G.V., Stefaneas, P.: Implementation of privacy calculus and its type checking in Maude. In: Margaria, T., Steffen, B. (eds.) ISoLA 2018. LNCS, vol. 11245, pp. 477–493. Springer, Cham (2018). https://doi.org/10.1007/978-3-030-03421-4_30

8. Ramírez Restrepo, C.A., Pérez, J.A.: Executable rewriting semantics for session-based concurrency (Extended Version). Technical report, Pontificia Universidad Javeriana Cali/University of Groningen (2022). https://japerezp.github.io/files/wrla22-long.pdf

9. Sangiorgi, D., Walker, D.: The Pi-Calculus: A Theory of Mobile Processes. Cambridge University Press (2003). http://books.google.com/books?id=QkBL_7VtiPgC

10. Scalas, A., Yoshida, N.: Less is more: multiparty session types revisited. Proc. ACM Program. Lang. 3(POPL), 30:1–30:29 (2019). https://doi.org/10.1145/3290343

11. Stehr, M.: CINNI - a generic calculus of explicit substitutions and its application to $\lambda$-, $\varsigma$- and $\pi$-calculi. In: Futatsugi, K. (ed.) The 3rd International Workshop on Rewriting Logic and its Applications, WRLA 2000, Kanzawa, Japan, 18–20 September 2000. Electronic Notes in Theoretical Computer Science, vol. 36, pp. 70–92. Elsevier (2000). https://doi.org/10.1016/S1571-0661(05)80125-2

12. Thati, P., Sen, K., Martí-Oliet, N.: An executable specification of asynchronous pi-calculus semantics and may testing in Maude 2.0. Electron. Notes Theor. Comput. Sci. 71, 261–281 (2002). https://doi.org/10.1016/S1571-0661(05)82539-3

13. Vasconcelos, V.T.: Fundamentals of session types. Inf. Comput. 217, 52–70 (2012). https://doi.org/10.1016/j.ic.2012.05.002

14. Viry, P.: Input/output for ELAN. Electron. Notes Theor. Comput. Sci. 4, 51–64 (1996). https://doi.org/10.1016/S1571-0661(04)00033-7. First International Workshop on Rewriting Logic and its Applications, WRLA 1996, Asilomar, USA, 1996

15. Viry, P.: A rewriting implementation of pi-calculus. Technical report, University of Pisa (1996). http://eprints.adm.unipi.it/id/eprint/1952

16. Wadler, P.: Propositions as sessions. J. Funct. Program. 24(2–3), 384–418 (2014)

# Tool Papers

# Parallel Maude-NPA for Cryptographic Protocol Analysis

Canh Minh Do[1]([✉])[iD], Adrián Riesco[2][iD], Santiago Escobar[3][iD],
and Kazuhiro Ogata[1][iD]

[1] Japan Advanced Institute of Science and Technology (JAIST), Nomi, Japan
{canhdominh,ogata}@jaist.ac.jp
[2] Universidad Complutense de Madrid, Madrid, Spain
ariesco@fdi.ucm.es
[3] VRAIN, Universitat Politècnica de València,Valencia, Spain
sescobar@upv.es

**Abstract.** Maude-NPA is a symbolic model checker for analyzing cryptographic protocols in the Dolev-Yao strand space model modulo an equational theory defining the cryptographic operations, which starts from an attack state to find counterexamples by performing a backward narrowing reachability analysis. Although Maude-NPA is a powerful analyzer, its running performance can be improved by taking advantage of parallel and/or distributed computing when dealing with non-trivial protocols in which the state space is huge. This paper describes a parallel version of Maude-NPA and a tool that supports it. We report on some experiments of various kinds of protocols that demonstrate that the tool can increase the running performance of Maude-NPA by 30% on average for most non-trivial case studies in which the number of states located at each layer is considerably large.

**Keywords:** Maude · Cryptographic Protocol Analysis · Parallel Maude-NPA

## 1 Introduction

With the emergence of the Internet and network-based services, many cryptographic protocols, also called security protocols, have been developed over

This work was supported by JST SICORP Grant Number JPMJSC20C2, Japan, by grant S2018/TCS-4339 (BLOQUES-CM) funded by Comunidad de Madrid co-funded by EIE Funds of the European Union, by grant PID2019-108528RB-C22 (ProCode-UCM) funded by MICIN. S. Escobar has been partially supported by the grant RTI2018-094403-B-C32 funded by MCIN/AEI/10.13039/501100011033 and ERDF A way of making Europe, by the grant PROMETEO/2019/098 funded by Generalitat Valenciana, and by the grant PCI2020-120708-2 funded by MICIN/AEI/10.13039/501100011033 and by the European Union NextGenerationEU/PRTR.

© Springer Nature Switzerland AG 2022
K. Bae (Ed.): WRLA 2022, LNCS 13252, pp. 253–273, 2022.
https://doi.org/10.1007/978-3-031-12441-9_13

decades to provide information security, such as confidentiality and authentication, in an insecure network. The design of cryptographic protocols, such as authentication protocols, is difficult, error-prone, and hard to detect flaws [13]. Many protocols contain flaws even a long time after they were published. For example, Lowe found an attack on the Needham-Schroeder public key authentication protocol after seventeen years [27,33]. Therefore, it is important to have automated tools to verify the properties of cryptographic protocols. There are several tools dedicated to security protocol analysis, such as Athena [34], ProVerif [6], OFMC [5], Avispa [1], Scyther [10], TAMARIN [28], Maude-NPA [16], and Verifpal [22]. Most analyzers based on model checking suffer from the notorious state space explosion problem, which prevents some model checking experiments from being carried out. Another challenge is to increase the running performance of model checking. One promising approach to this challenge is to parallelize model checking, which can make best use of multicore architectures.

Maude-NPA is a powerful symbolic model checker for analyzing cryptographic protocols modulo an equational theory that uses the Dolev-Yao strand space model [12,18], which is capable of intercepting, modifying, and injecting messages to impersonate other protocol principals by intruders. Maude-NPA uses a backward narrowing reachability analysis, which starts from a final insecure state, an attack state, to determine whether or not it is reachable from an initial state, which has no further backward steps. If that is the case, the initial state is a counterexample. The advantage of Maude-NPA is that it supports an unbounded session model and different equational theories; as a counterpart, these theories often lead to a bigger state space that requires more time to conduct model checking. Although some techniques were devised to reduce the state space, such as grammar-based techniques, giving priority to input messages in strands, early detection of inconsistent states (never reaching an initial state), a relation of transition subsumption (to discard transitions and states already being processed in another part of the search space), and the super lazy intruder (to delay the generation of substitution instances as much as possible) [14,15], the state space explosion problem is inevitable in some cases. Therefore, improving the running performance of Maude-NPA to some extent is worth doing. We are aware that Maude-NPA basically uses a breadth-first search (BFS) to explore the state space. Given a set of states in layer $l$, for each state in the set, we can perform independently the backward narrowing to obtain its successor states in layer $l + 1$, which opens an opportunity for parallelization so as to improve the running performance of Maude-NPA (time challenge). Note that an attack state is located at layer 0 and all states that reach the attack state by one-step state transition are located at layer 1.

In the present paper, we describe a parallel version of Maude-NPA and a tool that supports it, where successor states are generated in parallel at each layer. Basically, we transform the breadth-first search in Maude-NPA into a parallel breadth-first search without altering the number or form of the states in the state space. If the number of states located at each layer is considerably large, our tool can effectively improve the running performance of Maude-NPA. The tool has been built in Maude [8] as an implementation language, which is one direct successor language of OBJ3 [19], an algebraic specification language, and based on rewriting logic [31] as its theoretical foundation. Maude-NPA is written in Maude, which supports adequate parallel facilities, making it possible to develop parallel tools, such as Parallel $L + 1$-DCA2L2MC [11], which is a new technique to mitigate the state explosion for leads-to model checking. Therefore, the use of Maude for the parallel development of Maude-NPA without extra conversion is superior to other programming languages. The architecture of the tool is a master-worker model where one master and multiple workers are involved. The tool uses a shared cache maintained by the master and a local cache maintained by each worker to avoid making unnecessary duplications of jobs. The tool uses a set of jobs and a queue of worker identifiers to distribute (or assign) jobs to workers in a well-balanced way. The present paper also reports on some case studies on various kinds of protocols that demonstrate that the tool can increase the running performance of Maude-NPA by 30% on average for most non-trivial case studies. The support tool is available at the webpage.[1]

The rest of the paper is organized as follows. Section 2 mentions some preliminaries in which narrowing is described. Section 3 describes the overview of Maude-NPA. Section 4 describes a parallel version of Maude-NPA and a tool that supports it. Section 5 reports on some experimental results. Section 6 mentions some existing work. Finally, Sect. 7 concludes the paper together with some future directions.

## 2  Preliminaries

We follow the classical notation and terminology from [21] for term rewriting and from [29,30] for rewriting logic and order-sorted notions. We assume an *order-sorted signature* $\Sigma$ with a finite poset of sorts $(\mathbf{S}, \leq)$ and a finite number of function symbols. We furthermore assume that: (i) each connected component in the poset ordering has a top sort, and for each $\mathbf{s} \in \mathbf{S}$ we denote by $[\mathbf{s}]$ the top sort in the component of $\mathbf{s}$; and (ii) for each operator declaration $f : \mathbf{s}_1 \times \ldots \times \mathbf{s}_n \to \mathbf{s}$ in $\Sigma$, there is also a declaration $f : [\mathbf{s}_1] \times \ldots \times [\mathbf{s}_n] \to [\mathbf{s}]$. We assume an $\mathbf{S}$-sorted family $\mathcal{X} = \{\mathcal{X}_\mathbf{s}\}_{\mathbf{s} \in \mathbf{S}}$ of disjoint variable sets with each $\mathcal{X}_\mathbf{s}$ countably infinite. $\mathcal{T}_\Sigma(\mathcal{X})_\mathbf{s}$ is the set of terms of sort $\mathbf{s}$, and $\mathcal{T}_{\Sigma,\mathbf{s}}$ is the set of ground terms of sort $\mathbf{s}$. We write $\mathcal{T}_\Sigma(\mathcal{X})$ and $\mathcal{T}_\Sigma$ for the corresponding term algebras. The set of positions of a term $t$ is written $Pos(t)$, and the set of non-variable positions $Pos_\Sigma(t)$. The root of a term is $\Lambda$. The subterm of $t$ at position $p$ is $t|_p$ and $t[u]_p$ is the term obtained from $t$ by replacing $t|_p$ with $u$. A *substitution* $\sigma$ is a sorted mapping from a finite subset of $\mathcal{X}$, written $Dom(\sigma)$, to $\mathcal{T}_\Sigma(\mathcal{X})$. The set of variables introduced by $\sigma$ is $Ran(\sigma)$. The identity function substitution is $id$. Substitutions are homomorphically extended to $\mathcal{T}_\Sigma(\mathcal{X})$. The restriction of $\sigma$ to a set of variables $V$ is $\sigma|_V$.

A $\Sigma$-*equation* is an unoriented pair $t = t'$, where $t, t' \in \mathcal{T}_\Sigma(\mathcal{X})_\mathbf{s}$ for some sort $\mathbf{s} \in \mathbf{S}$. Given $\Sigma$ and a set $E$ of $\Sigma$-equations such that $\mathcal{T}_{\Sigma,\mathbf{s}} \neq \emptyset$ for every sort $\mathbf{s}$, order-sorted equational logic induces a congruence relation $=_E$ on terms

---

[1] https://github.com/canhminhdo/parallel-maude-npa.

$t, t' \in \mathcal{T}_\Sigma(\mathcal{X})$ (see [30]). We assume that $\mathcal{T}_{\Sigma,\mathbf{s}} \neq \emptyset$ for every sort $\mathbf{s}$. The $E$-subsumption order on terms $\mathcal{T}_\Sigma(\mathcal{X})_\mathbf{s}$, written $t \preccurlyeq_E t'$ (meaning that $t'$ is more general than $t$), holds if $\exists \sigma : t =_E \sigma(t')$. The $E$-renaming equivalence on term $\mathcal{T}_\Sigma(\mathcal{X})_\mathbf{s}$, written $t \approx_E t'$, holds if $t \preccurlyeq_E t'$ and $t' \preccurlyeq_E t$. An $E$-unifier for two terms $t, t' \in \mathcal{T}_\Sigma(X)$ is a substitution $\sigma$ such that $\sigma(t) =_E \sigma(t')$. A complete set of $E$-unifiers of two terms $t$, $t'$ is written $CSU_E(t = t')$. We say $CSU_E(t = t')$ is finitary if it contains a finite number of $E$-unifiers. This notion can be extended to multiple pairs of two terms, written $CSU_E(t_1 = t'_1 \wedge \ldots \wedge t_n = t'_n)$.

A rewrite rule is an oriented pair $l \rightarrow r$, where $l \notin \mathcal{X}$ and $l, r \in \mathcal{T}_\Sigma(\mathcal{X})_\mathbf{s}$ for some sorts $\mathbf{s} \in \mathbf{S}$. An (unconditional) order-sorted rewrite theory is a triple $\mathcal{R} = (\Sigma, E, R)$ with $\Sigma$ an order-sorted signature, $E$ a set of $\Sigma$-equations, and $R$ a set of rewrite rules. A topmost rewrite theory is a rewrite theory such that for each $l \rightarrow r \in R, l, r \in \mathcal{T}_\Sigma(\mathcal{X})_{\mathbf{State}}$ for a top sort $\mathbf{State}$, $r \notin \mathcal{X}$, and no operator in $\Sigma$ has $\mathbf{State}$ as an argument sort. The rewriting relation $\rightarrow_R$ on $\mathcal{T}_\Sigma(\mathcal{X})$ is $t \xrightarrow{p}_R t'$ (or $\rightarrow_R$) if $p \in Pos_\Sigma(t), l \rightarrow r \in R, t|_p = \sigma(l)$, and $t' = t[\sigma(r)]_p$ for some $\sigma$. The relation $\rightarrow_{R/E}$ on $\mathcal{T}_\Sigma(\mathcal{X})$ is $=_E; \rightarrow_R; =_E$. Note that $\rightarrow_{R/E}$ on $\mathcal{T}_\Sigma(\mathcal{X})$ induces a relation $\rightarrow_{R/E}$ on $\mathcal{T}_{\Sigma/E}(\mathcal{X})$ by $[t]_E \rightarrow_{R/E} [t']_E$ if and only if $t \rightarrow_{R/E} t'$. $[t]_E$ is the equivalence class of term $t$ with respect to $=_E$. When $\mathcal{R} = (\Sigma, E, R)$ is a topmost rewrite theory we can safely restrict ourselves [32] to the rewriting relation $\rightarrow_{R,E}$ on $\mathcal{T}_\Sigma(\mathcal{X})$, where $t \xrightarrow{\Lambda}_{R,E} t'$ (or $\rightarrow_{R,E}$) if $l \rightarrow r \in R, t =_E \sigma(l)$, and $t' = \sigma(r)$. Note that $\rightarrow_{R,E}$ on $\mathcal{T}_\Sigma(\mathcal{X})$ induces a relation $\rightarrow_{R,E}$ on $\mathcal{T}_{\Sigma/E}(\mathcal{X})$ by $[t]_E \rightarrow_{R,E} [t']_E$ if and only if $\exists w \in \mathcal{T}_\Sigma(\mathcal{X})$ such that $t \rightarrow_{R,E} w$ and $w =_E t'$.

The narrowing relation $\rightsquigarrow_R$ on $\mathcal{T}_\Sigma(\mathcal{X})$ is $t \xrightarrow{p,\sigma}_R t'$ (or $\xrightarrow{\sigma}_R, \rightsquigarrow_R$) if $p \in Pos_\Sigma(t)$, $l \rightarrow r \in R, \sigma \in CSU_\emptyset(t|_p = l)$, and $t' = \sigma(t[r]_p)$. Assuming that $E$ has a finitary and complete unification algorithm, the narrowing relation $\rightsquigarrow_{R,E}$ on $\mathcal{T}_\Sigma(\mathcal{X})$ is $t \xrightarrow{p,\sigma}_{R,E} t'$ (or $\xrightarrow{\sigma}_{R,E}, \rightsquigarrow_{R,E}$) if $p \in Pos_\Sigma(t), l \rightarrow r \in R, \sigma \in CSU_E(t|_p = l)$, and $t' = \sigma(t[r]_p)$. Note that $\rightsquigarrow_{R,E}$ on $\mathcal{T}_\Sigma(\mathcal{X})$ induces a relation $\rightsquigarrow_{R,E}$ on $\mathcal{T}_{\Sigma/E}(\mathcal{X})$ by $[t]_E \xrightarrow{\sigma}_{R,E} [t']_E$ if and only if $\exists w \in \mathcal{T}_\Sigma(\mathcal{X}) : t \xrightarrow{\sigma}_{R,E} w$ and $w =_E t'$.

## 3   Maude-NPA

Maude-NPA [16] is a model checker for analyzing cryptographic protocols modulo equations, which is written in Maude with about 18,000 lines of code. This section gives an overview of Maude-NPA focusing on those pieces of code that will be used for parallelization while omitting the rest.

A protocol specification in Maude-NPA is done by overwriting the three predefined modules: PROTOCOL-EXAMPLE-SYMBOLS, PROTOCOL-EXAMPLE-ALGEBRAIC, and PROTOCOL-SPECIFICATION that specify the syntax of the protocol, which consist of sorts, subsorts, and operators, the algebraic properties of the operators, which consist of equational rules (equations) and equational axioms (axioms), and the actual behaviors of the protocol using the Dolev-Yao strand space model [12,18], which consists of the intruder strands, regular strands, and attack states. Maude-NPA starts from an attack state, a final insecure state, to perform a backward reachability analysis which determines whether or not it is reachable from an initial state, which has no further backward steps. If that is the

case, the initial state is a counterexample. The backward search is performed
by a backward narrowing with a symbolic execution since the attack state is
a term with logical variables. Each backward narrowing step can be regarded
as a state transition, such as sending or receiving a message by principals, or
manipulating a message by intruders. Given a symbolic state, a backward narrowing step is performed to return a previous symbolic state in the protocol. By
that we can obtain all successor states (in the backward sense) from the state. In
Maude-NPA, each state found during the backward analysis is represented by six
sections separated by the symbol | | in the following order: (1) state id, (2) set
of current protocol and intruder strands, (3) intruder knowledge, (4) sequence
of messages, (5) ghost list, and (6) never pattern. For instance, the following is a
state found during the backward analysis of the Needham-Schroeder public key
protocol:

```
< 1 . 9 > (
:: nil ::
[ nil |
   -(n(b, #0:Fresh)),
   +(pk(b, n(b, #0:Fresh))), nil] &
:: #0:Fresh ::
[ nil,
   -(pk(b, #1:NNSet ; a)),
   +(pk(a, #1:NNSet ; b * n(b, #0:Fresh))) |
   -(pk(b, n(b, #0:Fresh))), nil]
)
||
pk(b, n(b, #0:Fresh)) !inI,
n(b, #0:Fresh) inI,
irr(pk(b, n(b, #0:Fresh)))
||
-(n(b, #0:Fresh)),
+(pk(b, n(b, #0:Fresh))),
-(pk(b, n(b, #0:Fresh)))
||
nil
||
nil
```

The state id is a unique id assigned to each state during the backward analysis.
The set of current strands represents the messages that were sent or received in
the past (those messages before the symbol |) and the messages that will be sent
or received in the future (those messages after the symbol |) in each strand. The
strand set also indicates how to advance each strand in the execution process by
partial substitutions for the messages in each strand. The intruder knowledge
represents what messages the intruder knows (symbol _inI) or does not know
yet (symbol _!inI) at each state. The sequence of messages denotes the actual
sequence of messages communicated to reach the state. The ghost list is extra
information for optimization in the super lazy intruder technique to reduce the
state space. The never pattern is used for authentication attacks.

We can divide the whole process of Maude-NPA into two main stages. In the first stage, given a protocol specification $\mathcal{P}$ and an equational theory $E_\mathcal{P}$, Maude-NPA needs to do as follows:

- Extracting the attack state $St$ from the protocol given an attack state id file.
- Building rewrite rules $R_\mathcal{P}$ against the behavior of the protocol specified in form of intruder and regular strands along with some pre-defined rewrite rules in the Maude-NPA specification.
- Generating grammars that represent infinite sets of states unreachable for the intruder to reduce the state space.

In the second stage, Maude-NPA performs the backward narrowing reachability analysis from the attack state $St$ using the relation $\leadsto_{R_\mathcal{P}^{-1},E_\mathcal{P}}$ where $R_\mathcal{P}^{-1}$ is the set of rewrite rules derived from $R_\mathcal{P}$ by inverting its rewrite rules. Maude-NPA basically uses the breadth-first search to explore the state space. There are three main steps needed for the exploration of each layer as follows:

- The first step is to generate all successor states for the next layer given a set of states in the current layer. This step also consists of almost all techniques to reduce the state space except for the transition subsumption technique, which is used in the second step subsequently.
- The second step is to simplify the successor states by the transition subsumption technique for removing states that are subsumed by either other states in the successor states or visited states (history states).
- Ultimately, the third step will filter out states from the previous step by using history states to avoid state duplications and rule out initial states as counterexamples. The cycle continues until a depth bound is reached or no more states exist for the next layer.

The first step in the second stage actually performs the backward narrowing just by one step to obtain all successor states from a given set of states in a layer. The successor states then go through a series of optimization steps, such as giving priority to input messages in strands, early detection of inconsistent states, the super lazy intruder, and filtering states by the grammars. We are aware that this step can be executed independently for each given state from the set of states, which opens an opportunity for parallelization. Given a set of states in layer $l$, for each state in the set, we can perform the backward narrowing step independently to obtain its successor states in layer $l+1$. In the next section, we will describe a parallel version of Maude-NPA in which the successor states are generated in parallel for each layer. Note that such reduction techniques are also included in this step and the parallel version does not alter the number or form of the states in the state space. If the number of states located at each layer is considerably large, our tool can effectively improve the running performance of Maude-NPA.

In addition, the second step in the second stage plays an important role to reduce the state space in Maude-NPA, which may transform an infinite-state system into a finite one [17], and this is also time-consuming. Basically, it performs two series of transition subsumption tasks, also called the implication step throughout this paper. Firstly, for each state in the successor states obtained in the first step of the stage it will be checked whether or not the state is subsumed (implied) by another state in the successor states. If that is the case, the state is ignored. We only keep states which cannot be subsumed by other states after this process. Secondly, each state will be checked whether or not the state is subsumed (implied) by a state in history states again. If that is the case, the state is ignored. Otherwise, the state is stored. We plan to parallelize the whole process of this step in the near future as one piece of our future work.

## 4    Parallel Maude-NPA and Its Tool Support

The support tool is implemented in Maude to conveniently extend the implementation of what is developed in Maude-NPA. We use object-based programming that can model an object-based system, where objects can communicate to each other via message passing. In addition, Maude also supports communicating with external objects by using sockets so that objects inside an object-based system can interact with different objects inside another object-based system. We adopt such functionalities to make a parallel version of Maude-NPA based on a master-worker model, which is described in this section.

As mentioned above, we parallelize the backward narrowing step in Maude-NPA. We use a master-worker model to make a parallel version of Maude-NPA. In our tool, a master maintains a shared cache that is a set of states (history states), while each worker also maintains a local cache that is a set of states, which contains all states explored by the worker. Use of the shared cache prevents jobs that have been processed from being assigned to workers, while use of the local caches prevents jobs that have been processed from being made by workers. The very initial job is made by the master, while all the other jobs are made by workers and basically sent to the master. Jobs are assigned to workers by the master unless the jobs have been tackled.

There are two kinds of messages exchanged by the master and workers: job and getJob. A job message is in the form of a state. A job message is sent to a worker by the master, distributing (or assigning) a job to the worker. Meanwhile, a job message is sent to the master by a worker, delivering a job made by the worker to the master. A getJob message is sent to the master by a worker, asking the master to assign a job to the worker. In Maude, we can send data over sockets provided that the data must be a string. The getJob message is just literally a string "getJob," while the job message is in the form of the state described in Sect. 3 in which the state will be transformed into a string before sending at the sender side and be restored to the original state from the string at the receiver side. The following functions written in Maude are used to do the transformation and restoration.

```
op qidListToString : QidList -> String .
op qidListToString : QidList String -> String .
eq qidListToString(QIL) = qidListToString(QIL, "") .
eq qidListToString(nil, S) = S .
eq qidListToString(Q QIL, S) = qidListToString(QIL, S + string(Q) + " ") .

op stringToQidList : String -> QidList .
op stringToQidList : String QidList -> QidList .
eq stringToQidList(S) = stringToQidList(S, nil) .
eq stringToQidList("", QIL) = QIL .
eq stringToQidList(S, QIL) = QIL qid(S) [owise] .
ceq stringToQidList(S, QIL) = stringToQidList(S'', QIL qid(S') )
if N := find(S, " ", 0)
/\ S' := substr(S, 0, N)
/\ S'' := substr(S, N + 1, length(S)) .

op state2string : IdSystemSet -> String .
eq state2string(State) = qidListToString(
   metaPrettyPrint(SM, upTerm(State), none)) .

op string2state : String -> IdSystemSet .
eq string2state(S) = downTerm(getTerm(
   metaParse(SM, stringToQidList(S), 'IdSystemSet)), errIdSystemSet) .
```

where State and SM are Maude variables of Module and IdSystemSet sorts, respectively. The function string2state is used to transform a state into a string by doing the following order: (1) convert the state to its meta representation by using the function upTerm, (2) convert the meta representation of the state to a list of quoted identifiers that presents the string of tokens, and (3) convert the list of quoted identifiers to a string by the function qidListToString. The function string2state is used to restore the state from the string by doing the following order: (1) convert the string to a list of quoted identifiers by the function stringToQidList, (2) parse the list of quoted identifiers in the module SM by the function metaParse, (3) get the term, the meta representation of the state, from the output of the function metaParse by the function getTerm, and (4) convert the meta representation of the state to the original state by the function downTerm. Note that some essential functions, such as upTerm, downTerm, getTerm, metaPrettyPrint, metaParse, are built-in in Maude, while the two functions qidListToString and stringToQidList are defined above.

The master is in charge of collecting all successor states (jobs) from workers, then performing the implication step to remove implied states, checking state duplications with history states, ruling out initial states as counterexamples, and distributing (or assigning) unprocessed jobs to workers. Besides, the master can stop the tool whenever counterexamples are found or there are no unprocessed jobs left or a depth bound is reached. Meanwhile, each worker is responsible for processing a job, a state, assigned to it by the master. A worker generates all successor states reachable from the state by the backward narrowing and checks them with its local cache to avoid explored states. The worker may then

---

**Algorithm 1.** Job Scheduling by a Master

---

input  : $\mathcal{P}$ – a protocol specification
           $Id$ – an attack state id in the protocol specification
           $F$ – a filter
           $BStep$ – the maximum number of backward steps
           $N$ – a number of workers
output: empty or counterexamples

```
1   (workers, jobs, next, history) ← (empty, empty, empty, empty);
2   (M, GS, IS) ← initialize(P, Id, BStep, F);
3   jobs ← {IS}; history ← {IS};
4   while True do
5       for k ← 1 to N do
6           if DATA ← recv(worker_k) then
7               if DATA = getJob then
8                   enqueue(workers, worker_k);
9               else
10                  (IS) ← DATA;
11                  next ← insert(next, IS);
12      while not isEmpty(workers) and not isEmpty(jobs) do
13          worker ← dequeue(workers);
14          IS ← dequeue(jobs);
15          send(worker, IS);
16      if isEmpty(jobs) and size(workers) = N then
17          if not isEmpty(next) then
18              IST ← simplifyByImplication(F, history, next);
19              (INIT, IST') ← filterWithHistoryAndInit(M, history, IST);
20              (jobs, next, BStep) ← (IST', empty, BStep − 1);
21              history ← insert(history, IST');
22              if not isEmpty(INIT) then
23                  closeConnection();
24                  return INIT;
25          if BStep = 0 or isEmpty(jobs) then
26              closeConnection();
27              return empty;
```

---

construct new jobs and send them to the master as job messages. At last, when a worker has completed a job, the worker requests a new job by sending a getJob message to the master. The master uses a set of states and a queue of worker identifiers to distribute jobs to workers so that job distribution can be well-balanced, which means that all workers are processing jobs all the time except the beginning, the implication step, and ending of the backward narrowing.

Algorithm 1 shows the pseudo-code for job scheduling conducted by the master. workers is a queue data structure that contains worker identifiers, which are requesting jobs. jobs, history, and next are set data structures. jobs contains jobs (states) that are distributed to workers, while next contains all possible next jobs (successor states) of the next layer. history contains all states explored at the moment. Initially, workers is set to the empty queue, while jobs, next, and jobs are set to the empty set at line 1. In the first stage, Maude-NPA needs to build rewrite rules $R_\mathcal{P}$ in form of a module, an attack state, and grammars from a protocol specification $\mathcal{P}$. The module is used to perform the backward reachability analysis, the attack state is used as the beginning state, and the grammars are used to remove unreachable states for intruders. This stage is proceeded by initialize function at line 2 with $\mathcal{P}$, $Id$, $BStep$, and $F$ parameters that are a protocol specification, the id of an attack state in the specification, the maximum number of backward steps, and a filter which is $+parallel$ as default denoting the parallelization mode, respectively. The result of the function is deconstructed and stored in M, GS, and IS, which stand for the module, the grammar, and the attack state, respectively. jobs and history are updated to contain only the attack state at line 3.

For each $worker_k$, whenever the master receives DATA from $worker_k$, where DATA is one of the two kinds of messages described above, it checks whether DATA is getJob, meaning that the worker is requesting a job. If so, $worker_k$ is enqueued to workers so that a job can be assigned to $worker_k$ subsequently. When DATA is a job that has been made and sent from $worker_k$, the master deconstructs DATA into a state IS at line 10 and then inserts it to the set of successor states next at line 11. Note that if the state already exists in the set, it is ignored. Otherwise, it is added to the set. The code fragment at lines 12–15 checks whether workers and jobs are not empty. If that is the case, the master dequeues workers and jobs to obtain a job and a worker identifier and assigns the job to the worker by sending a job message to the worker. The code fragment at lines 16–27 checks whether there are neither unprocessed jobs left nor jobs being processed by workers. If that is the case, the master continuously checks if next is not empty. If that is the case, we need to process all successor states next before moving to explore the next layer. Firstly, the successor states in next are simplified with the implication step by simplifyByImplication function at line 18 in which states implied by other states in either next or history are ignored. The output is a new set of states IST that is filtered by using history states history and rule out initial states as counterexamples by filterWithHistoryAndInit function at line 19. Ultimately, the final successor states IST' and initial states INIT are obtained. We assign jobs to IST', reset next to empty, and decrease BStep by one. Note that if BStep is unbounded, it is unchanged regardless of the subtraction. history is also updated by inserting IST' at line 21. If INIT contains some initial states, we close all connections and return INIT as counterexamples at lines 22–24. After preparing jobs for the next layer, if either BStep is 0 or jobs is empty, the tool terminates and returns empty meaning that there is no counterexample up to the depth BStep given at the beginning. Note that the three functions initialize, simplifyByImplication, and filterWithHistoryAndInit are based on existing functions provided in Maude-NPA.

Algorithm 2 shows the pseudo-code for job handling conducted by workers. Each worker maintains a set of states history to avoid sending explored states

---

**Algorithm 2.** Job Handling by Workers

---

   **input** : $\mathcal{P}$ – a protocol specification
               $Id$ – an attack state id in the protocol specification
               $F$ – a filter
               $BStep$ – the maximum number of backward steps
               $N$ – a number of workers

1  $(M, GS, IS) \leftarrow initialize(\mathcal{P}, Id, BStep, F)$;
2  $history \leftarrow \{IS\}$;
3  $send(server, getJob)$;
4  **while** $isOpen()$ **do**
5     **if** $DATA \leftarrow recv(server)$ **then**
6        $IS \leftarrow DATA$;
7        $IST \leftarrow nextBackNarrowForParallel(M, GS, F, IS)$;
8        $IST' \leftarrow filterWithHistory(M, history, IST)$;
9        $history \leftarrow insert(history, IST')$;
10      **forall the** $IS' \in IST'$ **do**
11        $JOB \leftarrow IS'$;
12        $send(server, JOB)$;
13      $send(server, getJob)$;

---

by the worker to the master. Initially, we need to call `initialize` function at line 1 as the same mentioned above for the master. `history` is initially set to contain only the attack state `IS` at line 2. Each worker starts the job handling by sending a getJob message to the master to request a job at line 3. While the connection is open, whenever a worker receives `DATA` from the master, which must be a job, the worker deconstructs it into the state `IS` at line 6. Given `M`, `GS`, `F`, and `IS`, `nextBackNarrowForParallel` function performs the backward narrowing step to obtain successor states reachable from `IS` at line 7, which is the main task that the worker needs to do. `IST` is then filtered with the local cache `history` by using `filterStateWithHistory` function, which returns a new set of states `IST'` at line 8. `history` is then updated by inserting `IST'` at line 9. For each state in `IST'`, we produce a new job and send it to the master at lines 10–12. We intend to send each job one by one to the master because it achieves the best running performance in our experiments. Once all jobs are sent to the master, the worker sends a getJob message to request a job. Note that the workers terminate if and only if the master closes all connections. The two functions `initialize` and `filterWithHistory` are based on existing functions provided in Maude-NPA.

## 5  Experiments

We have used a MacPro computer that carries a 2.5 GHz microprocessor with 28 cores and 1.5 TB memory to conduct experiments. We use Maude-NPA and the parallel version of Maude-NPA in our case studies. The tool and the case studies for the experiments are publicly available at the webpage in the Footnote 1. Besides, the original source code of the cases studies and more protocols are listed at the webpage.[2]

We have conducted experiments on various kinds of protocols to confirm the usefulness of our parallel version of Maude-NPA such as Symmetric Key Protocols, Homomorphism Protocols, Exclusive OR Protocols, API Protocols, PKCS Protocols, Choice Protocols, and Distance-Bounding Protocols. The experimental data are shown in Tables 1–2. The first, second, and third columns denote the name of the protocols, the attack state id used in protocol specifications, and the depth bound, respectively. The fourth and fifth columns denote the verification time excluding the time taken to generate the grammars for protocols when conducting model checking with Maude-NPA and Parallel Maude-NPA, respectively. In a row, the bold value is either in the fourth column or the fifth column denoting the corresponding winning tool. The sixth column denotes the percentage of improvement when using Parallel Maude-NPA. If the value is a positive number, namely $X$, it means that the parallel Maude-NPA is $X\%$ faster than Maude-NPA. Conversely, if the value is a negative number, namely $-X$, it means that Maude-NPA is $X\%$ faster than the parallel Maude-NPA. The last column denotes the average number of states at each layer for each worker to handle, respectively. Furthermore, we inspect the number of states located at each layer for each protocol shown in Appendix A. Model checking experiments terminate as soon as counterexamples are found or the depth bound is reached.

The tool uses sockets to communicate between the master and the workers so that we can flexibly choose to use a shared-memory machine or a distributed environment. For all experiments, we use a master and eight workers with a shared-memory machine, the MacPro computer. The experimental data says that for simple case studies (24 experiments) in which the verification time is less than 40 s, Maude-NPA is obviously faster than the parallel Maude-NPA because the number of states located at each layer is very small and the verification time is so short that the cost of communication between the master and workers becomes burdensome. However, the parallel Maude-NPA still can finish in a reasonably short amount of time. For non-trivial case studies (35 experiments) in which the number of states located at each depth is larger, the parallel Maude-NPA has a very good performance that is 30% faster than Maude-NPA on average, demonstrating its potential. For the Diffie Hellman protocol, the percentage of improvement can be up to 49%. The average number of states at each layer for a worker is measured to let us know how busy the worker is, which reflects the number of states located at each layer. The more busy workers are and the deeper the depth bound is, the more benefit we may gain from the use of parallelization.

We select three protocols whose verification time is the largest among all protocols to conduct extra experiments with different numbers of workers used in our tool. The experimental data are shown in Table 3. The sixth column shows the number of workers used in the experiments. We can see that the average number of states at each layer for a worker is subject to the number of

---

[2] http://personales.upv.es/sanesro/Maude-NPA_Protocols/index.html.

**Table 1.** Results of Maude-NPA and Parallel Maude-NPA

| Protocol | Attack State | Depth | Maude-NPA (seconds) | Parallel Maude-NPA (seconds) | P(%) | States/ Layer/ Worker |
|---|---|---|---|---|---|---|
| **1. Symmetric Key Protocols** | | | | | | |
| Amended Needham Schroeder | 0 | 7 | 4588.61 | **2821.933** | 39 | 11.11 |
| Carlsen Secret Key Initiator | 0 | 5 | 224.175 | **148.732** | 34 | 3.73 |
| Denning Sacco | 0 | 11 | 35.243 | **33.986** | 4 | 0.43 |
| | 0 | 11 | 284.211 | **158.882** | 44 | 1.56 |
| Diffie Hellman | 1 | 12 | 286.663 | **145.412** | 49 | 1.45 |
| | 2 | 13 | 35.371 | **21.48** | 39 | 0.32 |
| ISO-5 Pass Authentication | 0 | 5 | 101.649 | **64.306** | 37 | 2.1 |
| Kao-Chow RA | 0 | 4 | 52.235 | **32.904** | 37 | 2 |
| Kao-Chow RAHK | 0 | 4 | **4.027** | 12.151 | -67 | 0.19 |
| Kao-Chow RAT | 0 | 4 | 114.414 | **77.463** | 32 | 1.94 |
| Otway-Rees | 0 | 4 | 72.516 | **44.91** | 38 | 2.16 |
| Secret 06 | 0 | 2 | **1.732** | 4.874 | -64 | 0.38 |
| Secret 07 | 0 | 4 | **2.589** | 6.662 | -61 | 0.28 |
| Wide Mouthed Frog | 0 | 3 | 16.11 | **15.563** | 3 | 1.92 |
| Woo and Lam Authentication | 0 | 4 | **1.371** | 5.83 | -76 | 0.28 |
| Yahalom | 0 | 4 | 45.216 | **28.838** | 36 | 1.91 |
| **2. Homomorphism Protocols** | | | | | | |
| Needham Schroeder Lowe ECB | 0 | 7 | 73.869 | **54.355** | 26 | 1.11 |
| **3. Exclusive OR Protocols** | | | | | | |
| Needham Schroeder Lowe XOR | 0 | 8 | **10.22** | 14.771 | -31 | 0.31 |
| SK3 | 0 | 3 | **4.162** | 10.171 | -59 | 0.17 |
| TMN ltv-F-tmn-asy | 0 | 5 | 157.442 | **110.264** | 30 | 0.88 |
| WIRED ltv-C-wep-asy | 0 | 5 | **14.392** | 21.335 | -33 | 0.15 |
| WIRED ltv-C-wep-variant | 0 | 5 | **15.571** | 22.781 | -32 | 0.15 |
| **4. API Protocols** | | | | | | |
| | 0 | 9 | **3.487** | 10.551 | -67 | 0.17 |
| YubiKey | 1 | 7 | 93824.875 | **65294.633** | 30 | 5.13 |
| | 21 | 8 | 341.529 | **228.15** | 33 | 0.8 |
| | 3 | 7 | 13092.864 | **10208.363** | 22 | 3 |
| YubiHSM attack(d) | 0 | 9 | 843.388 | **598.361** | 29 | 2.38 |

workers used in the experiments. When the average number of states at each layer for a worker is high, we have a chance to increase the number of workers to improve the running performance of our tool for the first two protocols, Amended Needham Schroeder and YukiKey. Up to a certain point, the more workers are used, the less busy workers are and the more burden the master needs to handle and communicate with workers that may not improve the running performance and even become worse as in the third case study, TLS attack. In addition, as mentioned above we parallelize only the first step in the second stage, but not the second step in the second stage. Hence, there is a limitation point for improvement of the first step by parallelization. Even if the number of workers is increased, it will not improve the running performance and may make the

**Table 2.** Results of Maude-NPA and Parallel Maude-NPA

| Protocol | Attack State | Depth | Maude-NPA (seconds) | Parallel Maude-NPA (seconds) | P(%) | States/ Layer/ Worker |
|---|---|---|---|---|---|---|
| **5. PKCS Protocols** | | | | | | |
| PKCS11 a1-noComp | 0 | 4 | 24.815 | **21.489** | 13 | 0.81 |
| PKCS11 a2-noComp | 0 | 6 | 69.532 | **46.842** | 33 | 0.75 |
| PKCS11 a3-noComp | 0 | 6 | 296.424 | **201.218** | 32 | 1.6 |
| PKCS11 a4-noComp | 0 | 7 | 62.886 | **45.154** | 28 | 0.88 |
| PKCS11 a5-noComp | 0 | 9 | 382.498 | **271.263** | 29 | 1.82 |
| **6. Choice Protocols** | | | | | | |
| encryption mode | 0 | 4 | **3.164** | 8.634 | -63 | 0.28 |
|  | 1 | 4 | **8.587** | 9.855 | -13 | 0.78 |
|  | 2 | 10 | 67.941 | **43.627** | 36 | 1.1 |
|  | 3 | 11 | 136.958 | **88.037** | 36 | 1.61 |
| rock paper scissors | 0 | 9 | 125.615 | **76.972** | 39 | 1.81 |
|  | 1 | 1 | **0.389** | 4.602 | -92 | 0.13 |
|  | 2 | 2 | **1.003** | 4.285 | -77 | 0.38 |
| TLS regular | 0 | 3 | **6.727** | 15.02 | -55 | 0.17 |
| TLS attack | 0 | 11 | 8695.211 | **6925.944** | 20 | 3.15 |
| **7. Distance-Bounding Protocols** | | | | | | |
| brands chaum | 1 | 4 | **6.236** | 8.817 | -29 | 0.25 |
|  | 2 | 6 | 16.186 | **16.088** | 1 | 0.29 |
| CRCS | 1 | 9 | 766.746 | **515.919** | 33 | 0.75 |
|  | 2 | 8 | 121.77 | **101.815** | 16 | 0.42 |
| H&K | 1 | 5 | 16.792 | **14.108** | 16 | 0.35 |
|  | 2 | 2 | **1.174** | 4.388 | -73 | 0.13 |
| MAD | 1 | 9 | 175.382 | **126.641** | 28 | 0.67 |
|  | 2 | 6 | 967.156 | **693.66** | 28 | 2.42 |
| Meadows v1-DH | 1 | 4 | **1.646** | 6.281 | -74 | 0.13 |
|  | 2 | 8 | **32.153** | 35.344 | -9 | 0.28 |
| Meadows v2-DH | 1 | 4 | **1.694** | 6.396 | -74 | 0.13 |
|  | 2 | 3 | **2.464** | 5.858 | -58 | 0.17 |
| Munilla | 1 | 7 | 186.24 | **96.735** | 48 | 1.45 |
|  | 2 | 4 | **6.279** | 9.017 | -30 | 0.19 |
| Swiss Knife | 1 | 4 | **6.69** | 9.384 | -29 | 0.25 |
|  | 2 | 4 | 26.862 | **24.539** | 9 | 0.38 |
| TREAD | 1 | 4 | **6.444** | 8.962 | -28 | 0.25 |
|  | 2 | 4 | **5.166** | 8.236 | -37 | 0.25 |

master busier to handle many workers at the same time. Note that there are no workers handling jobs when the second step is performed. Hence, it is significant to parallelize the second step as one piece of our future work.

In summary, the parallel version of Maude-NPA can improve the running performance of Maude-NPA effectively when dealing with most non-trivial case studies in which the number of states located at each layer is considerably large. The more states located at each layer and the deeper the search space is, the

**Table 3.** Parallel Maude-NPA with various numbers of workers

| Protocol | Attack State | Depth | Maude-NPA (seconds) | #Workers | Parallel Maude-NPA (seconds) | P(%) | States/ Layer/ Worker |
|---|---|---|---|---|---|---|---|
| | | | | 8 | 2821.933 | 39 | 11.11 |
| | | | | 16 | 2651.974 | 42 | 5.55 |
| Amended Needham Schroeder | 0 | 7 | 4588.61 | 24 | 2646.479 | 42 | 3.7 |
| | | | | 32 | 2622.72 | 43 | 2.78 |
| | | | | 40 | 2627.64 | 43 | 2.22 |
| | | | | 8 | 65294.633 | 30 | 5.13 |
| | | | | 16 | 61365.349 | 35 | 2.56 |
| YubiKey | 1 | 7 | 93824.875 | 24 | 60937.287 | 35 | 1.71 |
| | | | | 32 | 59652.934 | 36 | 1.28 |
| | | | | 40 | 60083.891 | 36 | 1.03 |
| | | | | 8 | 6997.392 | 20 | 3.15 |
| | | | | 16 | 6960.781 | 20 | 1.57 |
| TLS attack | 0 | 11 | 8695.211 | 24 | 7193.625 | 17 | 1.05 |
| | | | | 32 | 7220.197 | 17 | 0.79 |
| | | | | 40 | 7586.980 | 13 | 0.63 |

more improvement may be obtained by parallelization. For simple case studies, whose verification time is very small, for example, less than 40 s in our case studies, we do not need to use the parallel version of Maude-NPA, although we still can use it to obtain a reasonable result. We can see that the verification time for simple case studies is very small and so the use of the parallel version of Maude-NPA is not much different compared with Maude-NPA. Hence, it is sufficient to use solely the tool in the present paper for cryptographic protocol analysis with Maude-NPA.

## 6   Related Work

Our work is very close to parallel Breadth-first search algorithms. There are various parallel BFS algorithms that have been intensively studied [2,7,25,26, 35]. Some of these algorithms work efficiently compared to the classical serial BFS algorithm [9, Section 22.2]. PBFS [26] uses a multiset data structure called a bag instead of a queue (FIFO). The bag supports insertion essentially as fast as FIFO and can be split and combined efficiently. In addition, for efficient implementation, PBFS contains a benign race condition in their algorithm and uses a bag reducer that allows updating concurrently to a shared variable or data structure at the same time. A bag is a collection of pennants in which each pennant is a tree of $2^k$ nodes, where $k$ is a non-negative integer. Each node in this tree contains two pointers to denote its left and right children. The bag is a crucial data structure in PBFS that is implemented efficiently in C++, while we use a set data structure that can be defined in Maude. Both Maude-NPA and Parallel Maude-NPA are written in Maude, a specification language, which is not flexible to adapt various data structures able to be implemented efficiently in the low-level, however, the idea to parallelize BFS is shared. Furthermore, we have demonstrated that the breadth-first search in Maude-NPA can be reasonably parallelized.

In addition to Maude-NPA, there are several cryptographic security analysis tools, such as Athena [34], ProVerif [6], OFMC [5], Avispa [1], Scyther [10], Verifpal [22], and TAMARIN [28]. Among them, TAMARIN, which is a prover for the symbolic analysis of security protocols, is the closest to Maude-NPA that generalizes the backward search used by the Scyther tool to support the unbounded session model, reasoning modulo equational theories, and modeling complex control flow and mutable global state. In TAMARIN, protocol specification is specified in multiset rewriting rules, while property specification is written in a guarded fragment of first-order logic. Each protocol trace corresponds to a multiset rewriting derivation that is the sequences of the labels of the applied rules. TAMARIN performs an exhaustive backward search to look for a trace that does not satisfy the property and returns a counterexample as an attack. If no rule can be applied anymore and no counterexample is found, then the protocol satisfies the property. To the best of our knowledge, our tool is the first attempt to parallelize a dedicated cryptographic security tool, Maude-NPA. Although, there are many parallel model checking algorithms for LTL [3], such as DiVinE 3.0 [4], Garakabu2 [23,24], a multicore extension of SPIN [20], and Parallel $L + 1$-DCA2L2MC [11].

## 7    Conclusion

The paper has described a parallel version of Maude-NPA and a tool that supports it. The tool has been implemented in Maude by using a master-worker model with socket communication. The paper has also reported on some experiments of various kinds of protocols in which the tool can increase the running performance of Maude-NPA by 30% on average for most non-trivial case studies where one master and eight workers are used.

There are several lines of future work as mentioned in the paper. Furthermore, we would like to use the new meta-interpreter feature in Maude rather than sockets to reduce the time taken in verifying protocols by our tool to some extent. Meta-interpreters can be run in a separate process to handle jobs independently and processes can communicate to each other by using filesystem objects on the same host instead of sockets with the TCP/IP protocol. For our experiments, if we increase the number of jobs that will be sent simultaneously between workers to the master, the running performance becomes worse because each state in Maude-NPA carries more information after each state transition for optimizing and tracing back to the initial state from the state and each state is converted to a string before sending over sockets. Hence, using sockets to convey considerably large data between workers and the master in Maude is not a good way. Therefore, the use of processes in a shared memory machine may be better than sockets in terms of communication cost. Besides, it may be interesting to consider allocating workers dynamically instead of fixing its number beforehand as one line of our future work. Finally, we should conduct more case studies and use various numbers of workers with the tool to demonstrate its usefulness.

# A   The Number of States Located at Each Layer

The fourth column in Tables 4–5 shows the number of states located at each
layer starting from depth zero up to the depth bound for each protocol, which
is a list of natural numbers separated by commas. If the last value in the list is
$X$, it means that there are $X$ states located at the depth bound. Especially, if
$X$ is zero, it means that there is no state for the layer. If the last value in the
list is of the form $X + Y$, it means that there are $X + Y$ states located at the
depth bound while $Y$ is the number of initial states (counterexamples).

**Table 4.** The number of states located at each layer

| Protocol | Attack State | Depth | States located at layers $(0, ..., i)$ |
|---|---|---|---|
| **1. Symmetric Key Protocols** | | | |
| Amended Needham Schroeder | 0 | 7 | 1, 2, 4, 9, 26, 62, 152, 365 + 1 |
| Carlsen Secret Key Initiator | 0 | 5 | 1, 3, 8, 17, 40, 79 + 1 |
| Denning Sacco | 0 | 11 | 1, 1, 2, 3, 5, 7, 6, 5, 4, 3, 1, 0 |
| | 0 | 11 | 1, 4, 5, 9, 13, 18, 20, 22, 17, 12, 10, 5 + 1 |
| Diffie Hellman | 1 | 12 | 1, 6, 10, 11, 16, 20, 20, 21, 13, 9, 6, 3, 1 + 2 |
| | 2 | 13 | 1, 4, 6, 6, 7, 5, 3, 1, 0 |
| ISO-5 Pass Authentication | 0 | 5 | 1, 4, 4, 12, 23, 39 + 1 |
| Kao-Chow RA | 0 | 4 | 1, 3, 8, 17, 34 + 1 |
| Kao-Chow RAHK | 0 | 4 | 1, 1, 1, 2, 1, 0 + 1 |
| Kao-Chow RAT | 0 | 4 | 1, 2, 4, 14, 40 + 1 |
| Otway-Rees | 0 | 4 | 1, 2, 6, 15, 44 + 1 |
| Secret 06 | 0 | 2 | 1, 2, 2 + 1 |
| Secret 07 | 0 | 4 | 1, 4, 2, 1, 0 + 1 |
| Wide Mouthed Frog | 0 | 3 | 1, 5, 13, 26 + 1 |
| Woo and Lam Authentication | 0 | 4 | 1, 2, 2, 2, 0 + 2 |
| Yahalom | 0 | 4 | 1, 2, 8, 19, 30 + 1 |
| **2. Homomorphism Protocols** | | | |
| Needham Schroeder Lowe ECB | 0 | 7 | 1, 4, 9, 10, 5, 8, 14, 10 + 1 |
| **3. Exclusive OR Protocols** | | | |
| Needham Schroeder Lowe XOR | 0 | 8 | 1, 1, 2, 3, 3, 3, 2, 2, 2 + 1 |
| SK3 | 0 | 3 | 1, 2, 1, 0 |
| TMN ltv-F-tmn-asy | 0 | 5 | 1, 4, 7, 8, 8, 6 + 1 |
| WIRED ltv-C-wep-asy | 0 | 5 | 1, 2, 1, 1, 1, 0 |
| WIRED ltv-C-wep-variant | 0 | 5 | 1, 2, 1, 1, 1, 0 |
| **4. API Protocols** | | | |
| | 0 | 9 | 1, 1, 1, 2, 2, 1, 1, 1, 1, 0 + 1 |
| YubiKey | 1 | 7 | 1, 4, 4, 9, 21, 88, 160, 0 |
| | 21 | 8 | 1, 4, 7, 16, 14, 2, 2, 5, 0 |
| | 3 | 7 | 1, 4, 4, 6, 18, 55, 80, 0 |
| YubiHSM attack(d) | 0 | 9 | 1, 1, 2, 3, 4, 7, 13, 24, 40, 75 + 1 |

**Table 5.** The number of states located at each layer

| Protocol | Attack State | Depth | States located at layers (0, ..., i) |
|---|---|---|---|
| **5. PKCS Protocols** | | | |
| PKCS11 a1-noComp | 0 | 4 | 1, 3, 5, 7, 9 + 1 |
| PKCS11 a2-noComp | 0 | 6 | 1, 2, 2, 4, 11, 11, 4 + 1 |
| PKCS11 a3-noComp | 0 | 6 | 1, 3, 6, 13, 20, 21, 12 + 1 |
| PKCS11 a4-noComp | 0 | 7 | 1, 3, 7, 10, 10, 8, 6, 3 + 1 |
| PKCS11 a5-noComp | 0 | 9 | 1, 4, 11, 22, 31, 31, 15, 9, 5, 1 + 1 |
| **6. Choice Protocols** | | | |
| encryption mode | 0 | 4 | 1, 1, 1, 2, 3 + 1 |
| | 1 | 4 | 1, 2, 4, 8, 9 + 1 |
| | 2 | 10 | 1, 4, 9, 12, 15, 16, 13, 10, 6, 2, 0 |
| | 3 | 11 | 1, 4, 10, 18, 22, 24, 21, 18, 14, 8, 2, 0 |
| rock paper scissors | 0 | 9 | 1, 8, 16, 24, 27, 24, 18, 9, 3, 0 |
| | 1 | 1 | 1, 0 |
| | 2 | 2 | 1, 5, 0 |
| TLS regular | 0 | 3 | 1, 1, 1, 0 + 1 |
| TLS attack | 0 | 11 | 1, 4, 7, 10, 14, 18, 20, 24, 29, 35, 46, 69 |
| **7. Distance-Bounding Protocols** | | | |
| brands chaum | 1 | 4 | 1, 2, 3, 2, 0 |
| | 2 | 6 | 1, 3, 4, 3, 1, 1, 0 + 1 |
| CRCS | 1 | 9 | 1, 3, 8, 16, 26, 35, 28, 14, 4, 0 |
| | 2 | 8 | 1, 3, 3, 3, 6, 6, 3, 1, 0 + 1 |
| H&K | 1 | 5 | 1, 2, 4, 5, 2, 0 |
| | 2 | 2 | 1, 1, 0 |
| MAD | 1 | 9 | 1, 3, 7, 10, 10, 8, 5, 3, 1, 0 |
| | 2 | 6 | 1, 5, 10, 14, 18, 27, 40 + 1 |
| Meadows v1-DH | 1 | 4 | 1, 1, 1, 1, 0 |
| | 2 | 8 | 1, 2, 2, 3, 3, 3, 3, 1, 0 |
| Meadows v2-DH | 1 | 4 | 1, 1, 1, 1, 0 |
| | 2 | 3 | 1, 1, 1, 0 + 1 |
| Munilla | 1 | 7 | 1, 4, 7, 12, 22, 25, 10, 0 |
| | 2 | 4 | 1, 2, 2, 1, 0 |
| Swiss Knife | 1 | 4 | 1, 2, 3, 2, 0 |
| | 2 | 4 | 1, 4, 5, 2, 0 |
| TREAD | 1 | 4 | 1, 2, 3, 2, 0 |
| | 2 | 4 | 1, 3, 2, 1, 0 + 1 |

# References

1. Armando, A., et al.: The AVISPA tool for the automated validation of internet security protocols and applications. In: Etessami, K., Rajamani, S.K. (eds.) CAV 2005. LNCS, vol. 3576, pp. 281–285. Springer, Heidelberg (2005). https://doi.org/10.1007/11513988_27

2. Barnat, J., Brim, L., Chaloupka, J.: Parallel breadth-first search LTL model-checking. In: Proceedings of the 18th IEEE International Conference on Automated Software Engineering, pp. 106–115 (2003). https://doi.org/10.1109/ASE.2003.1240299
3. Barnat, J., et al.: Parallel model checking algorithms for linear-time temporal logic. In: Handbook of Parallel Constraint Reasoning, pp. 457–507. Springer, Cham (2018). https://doi.org/10.1007/978-3-319-63516-3_12
4. Barnat, J., et al.: DiVinE 3.0 – an explicit-state model checker for multithreaded C & C++ programs. In: Sharygina, N., Veith, H. (eds.) CAV 2013. LNCS, vol. 8044, pp. 863–868. Springer, Heidelberg (2013). https://doi.org/10.1007/978-3-642-39799-8_60
5. Basin, D., Mödersheim, S., Viganò, L.: OFMC: a symbolic model checker for security protocols. Int. J. Inf. Secur. **4**(3), 181–208 (2004). https://doi.org/10.1007/s10207-004-0055-7
6. Blanchet, B.: An efficient cryptographic protocol verifier based on Prolog rules. In: Proceedings of the 14th IEEE Computer Security Foundations Workshop, pp. 82–96 (2001). https://doi.org/10.1109/CSFW.2001.930138
7. Buluç, A., Madduri, K.: Parallel breadth-first search on distributed memory systems. In: Proceedings of 2011 International Conference for High Performance Computing, Networking, Storage and Analysis, SC 2011, New York, NY, USA. Association for Computing Machinery (2011). https://doi.org/10.1145/2063384.2063471
8. Clavel, M., et al.: All About Maude - A High-Performance Logical Framework. LNCS, vol. 4350. Springer, Heidelberg (2007). https://doi.org/10.1007/978-3-540-71999-1
9. Cormen, T.H., Leiserson, C.E., Rivest, R.L., Stein, C.: Introduction to Algorithms, 3rd edn. The MIT Press, Cambridge (2009)
10. Cremers, C.J.F.: The Scyther tool: verification, falsification, and analysis of security protocols. In: Gupta, A., Malik, S. (eds.) CAV 2008. LNCS, vol. 5123, pp. 414–418. Springer, Heidelberg (2008). https://doi.org/10.1007/978-3-540-70545-1_38
11. Do, C.M., Phyo, Y., Riesco, A., Ogata, K.: A parallel stratified model checking technique/tool for leads-to properties. In: 2021 7th International Symposium on System and Software Reliability (ISSSR), pp. 155–166 (2021). https://doi.org/10.1109/ISSSR53171.2021.00011
12. Dolev, D., Yao, A.: On the security of public key protocols. IEEE Trans. Inf. Theory **29**(2), 198–208 (1983). https://doi.org/10.1109/TIT.1983.1056650
13. Dong, L., Chen, K.: Introduction of cryptographic protocols. In: Cryptographic Protocol, pp. 1–12. Springer, Heidelberg (2012). https://doi.org/10.1007/978-3-642-24073-7_1
14. Escobar, S., Meadows, C., Meseguer, J.: A rewriting-based inference system for the NRL protocol analyzer and its meta-logical properties. Theor. Comput. Sci. **367**(1), 162–202 (2006). https://doi.org/10.1016/j.tcs.2006.08.035
15. Escobar, S., Meadows, C., Meseguer, J.: State space reduction in the Maude-NRL protocol analyzer. In: Jajodia, S., Lopez, J. (eds.) ESORICS 2008. LNCS, vol. 5283, pp. 548–562. Springer, Heidelberg (2008). https://doi.org/10.1007/978-3-540-88313-5_35
16. Escobar, S., Meadows, C., Meseguer, J.: Maude-NPA: cryptographic protocol analysis modulo equational properties. In: Aldini, A., Barthe, G., Gorrieri, R. (eds.) FOSAD 2007-2009. LNCS, vol. 5705, pp. 1–50. Springer, Heidelberg (2009). https://doi.org/10.1007/978-3-642-03829-7_1

17. Escobar, S., Meseguer, J.: Symbolic model checking of infinite-state systems using narrowing. In: Baader, F. (ed.) RTA 2007. LNCS, vol. 4533, pp. 153–168. Springer, Heidelberg (2007). https://doi.org/10.1007/978-3-540-73449-9_13

18. Fabrega, F., Herzog, J., Guttman, J.: Strand spaces: why is a security protocol correct? In: Proceedings of the 1998 IEEE Symposium on Security and Privacy, pp. 160–171 (1998). https://doi.org/10.1109/SECPRI.1998.674832

19. Goguen, J., Kirchner, C., Kirchner, H., Mégrelis, A., Meseguer, J., Winkler, T.: An introduction to OBJ 3. In: Kaplan, S., Jouannaud, J.-P. (eds.) CTRS 1987. LNCS, vol. 308, pp. 258–263. Springer, Heidelberg (1988). https://doi.org/10.1007/3-540-19242-5_22

20. Holzmann, G.J., Bosnacki, D.: The design of a multicore extension of the SPIN model checker. IEEE Trans. Software Eng. **33**(10), 659–674 (2007). https://doi.org/10.1109/TSE.2007.70724

21. Klop, J.W., Bezem, M., Vrijer, R.C.D.: Term Rewriting Systems. Cambridge University Press, Cambridge (2001)

22. Kobeissi, N., Nicolas, G., Tiwari, M.: Verifpal: cryptographic protocol analysis for the real world. In: Bhargavan, K., Oswald, E., Prabhakaran, M. (eds.) INDOCRYPT 2020. LNCS, vol. 12578, pp. 151–202. Springer, Cham (2020). https://doi.org/10.1007/978-3-030-65277-7_8

23. Kong, W., Hou, G., Hu, X., Ando, T., Hisazumi, K., Fukuda, A.: Garakabu2: an SMT-based bounded model checker for HSTM designs in ZIPC. J. Inf. Sec. Appl. **31**, 61–74 (2016). https://doi.org/10.1016/j.jisa.2016.08.001

24. Kong, W., Liu, L., Ando, T., Yatsu, H., Hisazumi, K., Fukuda, A.: Facilitating multicore bounded model checking with stateless explicit-state exploration. Comput. J. **58**(11), 2824–2840 (2015). https://doi.org/10.1093/comjnl/bxu127

25. Korf, R.E., Schultze, P.: Large-scale parallel breadth-first search. In: Proceedings of the 20th National Conference on Artificial Intelligence, AAAI 2005, vol. 3, pp. 1380–1385. AAAI Press (2005)

26. Leiserson, C.E., Schardl, T.B.: A work-efficient parallel breadth-first search algorithm (or how to cope with the nondeterminism of reducers). In: Proceedings of the Twenty-Second Annual ACM Symposium on Parallelism in Algorithms and Architectures, SPAA 2010, New York, NY, USA, pp. 303–314. Association for Computing Machinery (2010). https://doi.org/10.1145/1810479.1810534

27. Lowe, G.: An attack on the Needham-Schroeder public-key authentication protocol. Inf. Process. Lett. **56**(3), 131–133 (1995). https://doi.org/10.1016/0020-0190(95)00144-2

28. Meier, S., Schmidt, B., Cremers, C., Basin, D.: The TAMARIN prover for the symbolic analysis of security protocols. In: Sharygina, N., Veith, H. (eds.) CAV 2013. LNCS, vol. 8044, pp. 696–701. Springer, Heidelberg (2013). https://doi.org/10.1007/978-3-642-39799-8_48

29. Meseguer, J.: Conditional rewriting logic as a unified model of concurrency. Theor. Comput. Sci. **96**(1), 73–155 (1992). https://doi.org/10.1016/0304-3975(92)90182-F

30. Meseguer, J.: Membership algebra as a logical framework for equational specification. In: Presicce, F.P. (ed.) WADT 1997. LNCS, vol. 1376, pp. 18–61. Springer, Heidelberg (1998). https://doi.org/10.1007/3-540-64299-4_26

31. Meseguer, J.: Twenty years of rewriting logic. J. Log. Algebraic Methods Program. **81**(7–8), 721–781 (2012). https://doi.org/10.1016/j.jlap.2012.06.003

32. Meseguer, J., Thati, P.: Symbolic reachability analysis using narrowing and its application to verification of cryptographic protocols. Electron. Notes Theor. Comput. Sci. **117**, 153–182 (2005). https://doi.org/10.1016/j.entcs.2004.06.024. Proceedings of the Fifth International Workshop on Rewriting Logic and Its Applications (WRLA 2004)
33. Needham, R.M., Schroeder, M.D.: Using encryption for authentication in large networks of computers. Commun. ACM **21**(12), 993–999 (1978). https://doi.org/10.1145/359657.359659
34. Song, D.X.: Athena: a new efficient automatic checker for security protocol analysis. In: Proceedings of the 12th IEEE Computer Security Foundations Workshop, pp. 192–202 (1999). https://doi.org/10.1109/CSFW.1999.779773
35. Yoo, A., Chow, E., Henderson, K., McLendon, W., Hendrickson, B., Catalyurek, U.: A scalable distributed parallel breadth-first search algorithm on BlueGene/L. In: Proceedings of the 2005 ACM/IEEE Conference on Supercomputing, SC 2005, p. 25 (2005). https://doi.org/10.1109/SC.2005.4

# Maude as a Library: An Efficient All-Purpose Programming Interface

Rubén Rubio[(✉)] [iD]

Universidad Complutense de Madrid, Madrid, Spain
rubenrub@ucm.es

**Abstract.** We present a general and efficient programming interface to Maude from Python and other programming languages. All relevant Maude entities and operations are exposed in a documented object-oriented library to facilitate the integration of Maude into external programs and vice versa. This paper describes the design and implementation of the library, explains how to use it, and discusses some mature applications.

## 1  Introduction

Formal tools are more useful when they can cooperate and interact with the outside world through simple and well-defined interfaces. In addition to the traditional command-line interfaces, popular tools like the Z3 [10] and CVC4/5 [3,4] SMT solvers, the Storm [20] probabilistic model checker, or the Lean [11] theorem prover are offering programming interfaces to their functionality from languages like C++ and Python. Some are even conceived as libraries in the first place, like the Spot [16] platform for LTL and $\omega$-automata. This laudable trend also reaches mainstream programming languages like C/C++, whose compiler Clang can be used as a library to inspect the abstract syntax tree of programs and control the different compilation phases.

Maude [8] is a high-performance logical and semantic framework based on rewriting logic [27]. Maude programs are collections of modules corresponding to specifications in this logic, where states are terms in an equational logic that are transformed by the nondeterministic application of rewrite rules. Rewriting logic is reflective and Maude provides a universal theory where terms, modules, and other related concepts are represented as data that can be manipulated within the language. Several tools for analyzing Maude specifications and application-specific interactive interfaces have been written using these metaprogramming features. However, interacting with external tools and visualization is not so easy within Maude, and the interpreter has occasionally been extended with custom ad hoc extensions. Examples are the Maude Formal Environment [15], which interacts with external termination provers and libraries, and several analysis and visualization tools of the ELP group at Universitat Politècnica de València [1,2].

Maude is being used behind the scenes by some tools like the Tamarin prover [26] for security protocol verification, the $\mathbb{K}$ semantic framework [28] (until its fifth version), and the heterogeneous tool set Hets [9], among others. All this

© Springer Nature Switzerland AG 2022
K. Bae (Ed.): WRLA 2022, LNCS 13252, pp. 274–294, 2022.
https://doi.org/10.1007/978-3-031-12441-9_14

software includes ad hoc code to run an instance of the Maude interpreter as a separate process, issue commands to its standard input stream, and parse their answers. The IMaude agent of the InterOperability Platform (IOP) [25] follows the same approach to communicate with Maude, but then provides an abstraction for other user-defined agents of this framework to interact with the language. IMaude is used by the Pathway Logic Workbench [35], Mobile Maude [7], and the graphical interface to Maude-NPA [33], among others.

We present here an intuitive programming interface for Python and other programming languages that exposes almost all functionality of the Maude interpreter and some useful extensions. Moreover, the connection in the opposite direction, from Maude to the external language, is also supported. Unlike previous tools, these language bindings are directly linked with the Maude implementation, so several new possibilities and better performance are expected from this approach. The library comes with detailed documentation and API reference, and it has already been used in some relevant projects (see Sect. 6).

Its implementation relies on the Simple Wrapper and Interface Generator (SWIG) [13], so bindings can be produced for any language supported by this tool. However, only Python has been extensively tested and enhanced with language-specific adaptations to provide a more natural interface. The Python module is available at the Python Package Index (PyPI) and can be installed with the command `pip install maude`. Currently, the binding for Java has also been tested to a lesser degree and the those for other languages must be compiled from source, for which instructions are available. In the following, we will focus on the Python flavor of the bindings for simplicity, although most information can be generalized to other languages.

This paper starts with a quick overview of the library in Sect. 2, which is further illustrated by a simple example in Sect. 3. Some advanced features are introduced in Sect. 4, and the implementation is described in Sect. 5. Finally, Sects. 6 and 7 mention some applications and complete the discussion on related work in this introduction. More information can be found at github.com/fadoss/maude-bindings including documentation, examples, the API reference, and the source code of the library.

## 2    Overview of the Library

In this section, we describe the design and overall organization of the language bindings, which coincide for all supported languages. However, we will stick to the Python instance for simplicity, as explained before.

The `maude` library exhibits all relevant Maude entities and operations as objects and methods of the target language. There are classes `Term` for terms, `Module` for modules, `Sort` for sorts, `Symbol` for symbols (or operators), `Equation` for equations, `Substitution` for substitutions, and so on. Most commands in the Maude interpreter are gathered as methods of the `Term` class, like `reduce`, `rewrite`, `search`, `get_variants`, and `vu_narrow`. Some commands that are not applied to a singular term like `unify` are available through the `Module` class.

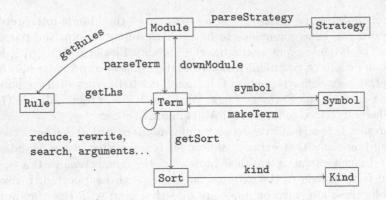

**Fig. 1.** Some relevant classes and methods in the library.

Operations that are reserved to the metalevel in the Maude interpreter are also implemented as regular methods, like iterating over the arguments of a term with `arguments`, obtaining its least sort with `getSort`, its root symbol with `symbol`, or applying a substitution with `instantiate`, among others. Figure 1 shows a selection of the basic classes along with some methods that relate them.

A simple program that reduces the term 2 * 3 with the `maude` Python package and prints its result 6 to the terminal would look as follows

```
import maude
maude.init()
m = maude.getModule('NAT')
t = m.parseTerm('2 * 3')
t.reduce()
print(t)
```

The first two instructions load the `maude` package and initialize it with the `init` function. This must be called before anything else in the library since it sets up some required resources and loads the Maude prelude. Everything in Maude takes place within modules, so a `Module` object is needed to begin with, and it can be obtained with the `getModule` or `getCurrentModule` functions. Typically, we will then parse a term with the `parseTerm` method and apply some operations to it. The `Module` class also includes several methods for inspecting its contents.

While the library offers enough resources to manipulate terms without resorting to the metalevel, moving through different levels of reflection is natively supported with the `upTerm` and `downTerm` methods of `Module`. For expressions in the Maude strategy language, these methods are called `upStrategy` and `downStrategy`. Moreover, a `Module` object can be obtained from its metarepresentation using the `downModule` function, while the converse operation can be achieved by simply reducing an `upModule` term in the `META-LEVEL` module.

In the next section, we illustrate the possibilities of the library through an example, giving further details on how to use it. Other advanced features are described in Sect. 4, and more information is available on the home page of the language bindings.

## 3    How to Use the Library, Illustrated by an Example

In this section, a toy interactive rewriter is implemented using the maude library, as an excuse to illustrate its usage and possibilities. Most of this example can be programmed directly in Maude using reflection, probably in a more verbose and complex manner, but the same procedures can be used when actual interaction with the outside world is pursued.

Our interactive prototype will repeatedly read commands from the terminal and reply to them. Implementing this kind of interface in Python is easy thanks to the standard cmd module. We only need to subclass the cmd.Cmd class and provide a method do_cmdname to handle the command cmdname. Its full source code is available in the inter.py file of the bindings repository. As already explained, we should start by importing the library with import maude and initializing it with maude.init(). The InteractiveRewriter class holding the implementation of all commands in the interpreter can then be defined[1].

```
import cmd
import maude

class InteractiveRewriter(cmd.Cmd):
  # A method will be added here for each command

if __name__ == '__main__':  # entry point
  maude.init()
  InteractiveRewriter().cmdloop()
```

For the moment, only two attributes are maintained, the current module and the term being rewritten, as specified in the class constructor.

```
def __init__(self):
  super().__init__()       # base class constructor
  self.module = None   # current module
  self.term = None     # current term
```

In order to bring modules to our scope, we need a load command to read them from Maude source files. Thus, we implement a method do_load that essentially delegates on the load function of the library.

```
def do_load(self, path):
  maude.load(path)
  self.module = maude.getCurrentModule()
  print('The current module is', self.module)
```

---

[1] The official documentation of the cmd module and other Python features that may appear is available at docs.python.org.

In addition, we set the current module using the `getCurrentModule` function, which gives the `Module` object for the last module that has been entered or explicitly selected in the file. Its name is printed in the screen by printing the object itself. However, we may want to select another module, for what we also provide commands to list the available modules and to select one of them.

```
def do_list(self, _):
  for module in maude.getModules():
    print(module)

def do_select(self, name):
  self.module = maude.getModule(name)
```

For example, assume we have a file `foo.maude` with the following module.

```
mod FOO-MODULE is                .
  sorts Foo Bar .
  subsort Bar < Foo .

  ops a b c :           -> Bar [ctor] .
  op  f     : Foo Foo -> Foo [ctor] .
  op  g     : Foo     -> Foo [ctor] .

  vars X Y : Foo .

  rl [swap] : f(X, Y) => f(Y, X) .
  rl [next] : a => b .
endm
```

After running the `inter.py` script with Python, the following command prompt will appear, where we can input `foo.maude` using the `load` command.

```
        *** Interactive rewriter for Maude ***

IRew> load foo
The current module is FOO-MODULE
```

*Term manipulation.* At this point, we need to choose a term to start rewriting.

```
def do_start(self, text):
  self.term = self.module.parseTerm(text)
```

Issuing the command `start` $t$ makes $t$ the current term in this session. We are not taking care about errors, but `self.term` would be `None` and error messages would have been printed if `text` could not be parsed as a term. For printing the syntax tree of this term, we can prepare a command `tree` by writing a method `do_tree` as before, which may simply call the following recursive function:

```
def print_tree(term, indent=''):
  print(f'{indent}{term.symbol()} : {term.getSort()}')

  for argument in term.arguments():
    print_tree(argument, indent + '  ')
```

The print_tree function starts by printing the top symbol of term and its sort with the appropriate indentation, and then proceeds recursively on the arguments via the arguments method. Notice that strings prefixed by f in Python are formatted by replacing the expressions between curly brackets with their values. For example, we can show the syntax tree of f(g(a), b) in FOO-MODULE by selecting this term with start and calling the tree command.

```
IRew> start f(g(a), b)
IRew> tree
f : Foo
  g : Foo
    a : Bar
  b : Bar
```

*Standard Commands.* One of the most useful commands in Maude is reduce.

```
def do_reduce(self, _):
  nrew = self.term.reduce()
  print(f'Reduced to {self.term} in {nrew} rewrites.')
```

Methods like reduce and rewrite modify the term to which they are applied and return the number of rewrites instead. Since the original term is overwritten, if desired, it can be copied before with its copy method. Another command with a straightforward implementation is the strategy-rewriting command srewrite:

```
def do_srewrite(self, text):
  strategy = self.module.parseStrategy(text)
  for result, nrew in self.term.srewrite(strategy):
    print(f'{result} in {nrew} rewrites')
```

Methods like srewrite, search, and vu_narrow that may produce multiple solutions return an iterator and do not alter the original term. As an example, we apply the strategy swap ; next to the current term with this command:

```
IRew> srewrite swap ; next
f(b, g(b)) in 2 rewrites
```

*Applying Rules.* For our interactive rewriter to honor its name, we should provide a command step to execute a single rewrite on the current term.

```
def do_step(self, label):
  results = []  # results of the rewriting step
```

```
for k, (result, subs, ctx, rl) in enumerate(
        self.term.apply(label if label else None)):
    where = self.print_context(ctx, rl.getLhs())
    results.append(result)

    print(f'({k}) {result} by applying {rl} '
          f'on {where} with {subs}')

self.select_one(results)
```

The apply method of Term calculates all possible rewrites with any rule labeled with the given string (or any rule at all if None is given instead). It provides an iterator over the rewritten terms (result), the matching substitutions (subs) and contexts (ctx), and the applied rules themselves (rl). Contexts designate a single position in a term, and we see them here as functions that fill that position with the given term. In other words, ctx(subs.instantiate(rl.getLhs())) is the original term, and ctx(subs.instantiate(rl.getRhs())) is result. In this case, we hide in the print_context method how the context is processed since we will come back to this soon. Every result is accumulated in a list that is later passed to another unspecified method select_one that lets the user choose the next term.

```
IRew> start f(f(b, c), a)
IRew> step swap
(0) f(a, f(b, c))
    by applying rl f(X, Y) => f(Y, X) [label swap] .
    on top with X=f(b, c), Y=a
(1) f(f(c, b), a)
    by applying rl f(X, Y) => f(Y, X) [label swap] .
    on f(@, a) with X=b, Y=c

Select one of the options (0-1):
```

*Matching and Substitutions.* In addition to the rules in the module, the interactive rewriter may be interested in experimenting with new rules, for what we add a command inline to apply inline rules.

```
IRew> start f(g(a), b)
IRew> inline g(X) => c
(0) f(c, b) in f(@, b) with X=a

There is a single option, done.
```

This command can be implemented by manually matching the left-hand side of => and replacing it with the right-hand side instantiated with the matching substitution. The match method of Term is the appropriate resource for this. It takes a pattern as an argument.

```
def do_inline(self, text):
  lhs, rhs = text.split('=>', maxsplit=1)
  lhs = self.module.parseTerm(lhs)
  rhs = self.module.parseTerm(rhs)

  results = []   # results of inline rewriting

  for k, (subs, ctx) in enumerate(self.term.match(lhs,
                          maxDepth=maude.UNBOUNDED)):
    result = ctx(subs.instantiate(rhs))
    where = self.print_context(ctx, lhs)

    print(f'({k}) {result} in {where} with {subs}')
    results.append(result)

  self.select_one(results)
```

The first block in the method separates the left- and right-hand sides of the inline rule and parses them in the current module. Then, `lhs` is matched against the current term `self.term`, obtaining the matching substitution `subs` and context `ctx`. By default, matching is limited to the top symbol without extension, but `minDepth` and `maxDepth` can be set to fix maximum and minimum depths. The auxiliary method `print_context` can be defined as follows.

```
def print_context(self, ctx, lhs):
  var_name = f'<<PH>>:{lhs.getSort()}'
  var_term = self.module.parseTerm(var_name)

  ctx = ctx(var_term)

  return 'top' if ctx.isVariable() \
    else str(ctx).replace(var_name, '@')
```

The context is instantiated with a placeholder variable `<<PH>>`. If the result is a variable, matching has happened on top. Otherwise, we replace the placeholder by the @ sign for aesthetic reasons.

*Building Terms and Modules.* Once convinced with the new rule, we may want to add it to the current module with a new command `add`. Since modules are immutable in Maude, the library does not provide any direct resource to modify them, but we can always draw on the metalevel. This requires a more complex processing that we will carefully explain. Given the command `add` $l$ => $r$, suppose both sides of the rule have been parsed into the variables `lhs` and `rhs`, like in the `inline` command. To modify the current module at the metalevel, we should obtain its metarepresentation by evaluating the `upModule` operator of the `META-LEVEL` module at the beginning of our `do_add` method.

```
ml = maude.getModule('META-LEVEL')
```

```
if self.metamodule is None:
  self.metamodule =
    ml.parseTerm(f"upModule('{self.module}, false)")
  self.metamodule.reduce()
```

The module term is stored in the `metamodule` attribute of the interpreter for the next time. Remember that the metarepresentation of a module in Maude is an operator with a set-like argument for each type of declaration or statement in it. Hence, we will construct the metarepresentation of the new rule and insert it in the slot of rule statements. The first ingredient is the operator

```
op rl_=>_[_]. : Term Term AttrSet -> Rule [ctor] .
```

for unconditional rules in the universal theory of `META-LEVEL`. The `findSymbol` method of `Module` allows finding operators in the module by their names and signatures, given as a sequence of domain kinds and a range kind. These kinds should be obtained first with the `findSort` and `kind` methods.

```
term_kind = ml.findSort('Term').kind()
rule_kind = ml.findSort('Rule').kind()
attr_kind = ml.findSort('Attr').kind()

rl_symb = ml.findSymbol('rl_=>_[_].', (term_kind,
                  term_kind, attr_kind), rule_kind)
```

Now, we only have to fill the gaps with the metarepresentations of `lhs` and `rhs`, and with the constant `none` for the attribute part of the statement. We parse this latter constant with the `parseTerm` as usual, but providing the additional argument `attr_kind` to restrict parsing to this kind and avoid ambiguities.

```
none_attr = ml.parseTerm('none', attr_kind)
```

Finally, `Symbol`'s `makeTerm` constructs a term with a given sequence of arguments.

```
rl_term = rl_symb.makeTerm((ml.upTerm(lhs),
                  ml.upTerm(rhs), none_attr))
```

Syntactic sugar is provided for invoking the `makeTerm` method when a `Symbol` object is applied as a function, so the previous is equivalent to

```
rl_term =
  rl_symb(ml.upTerm(lhs), ml.upTerm(rhs), none_attr)
```

Now, `rl_term` must be inserted into the seventh argument of the metamodule, which holds the set of rules in system and strategy modules. For simplicity, we assume that the module is not a functional one. In order to add the rule to this set, we must build a new term with the union operator `__` of `RuleSet`. The list of arguments of the metamodule is obtained into the `mm_args` variable.

```
rls_symb = ml.findSymbol('__', (rule_kind, rule_kind),
                          rule_kind)

mm_args = list(self.metamodule.arguments())
mm_args[7] = rls_symb(mm_args[7], rl_term)
```

Finally, the module is reassembled with the `makeTerm` method.

```
self.metamodule = self.metamodule.symbol()
                       .makeTerm(mm_args)
```

This new metamodule is converted to a `Module` object with the `downModule` function, then assigned to the `module` attribute of the rewriter.

```
self.module = maude.downModule(self.metamodule)
```

Term objects in the library belong to a fixed module and they cannot operate with entities from other modules, even if related by inclusion. Hence, if a term was already set, we must reparse it in the new module.

```
if self.term:
    self.term = self.module.parseTerm(str(self.term))
```

We can check that the new command works by executing the interpreter.

```
IRew> start a
IRew> add a => c
The rule has been inserted.
IRew> step
(0) b by applying rl a => b [label next] .
       on top with empty
(1) c by applying rl a => c . on top with empty

Select one of the options (0-1): 1
```

*Interoperability.* To conclude and connect with the interoperability goals of the library, we will implement a command `trs` that exports the rules in the module into the standard TRS format, used by multiple verification tools.

```
IRew> load foo
IRew> trs
(VAR X:Foo Y:Foo)
(RULES
  f(X:Foo, Y:Foo) -> f(Y:Foo, X:Foo)
  a -> b
)
```

Since the format includes a `VAR` entry specifying the set of variables in the rules, we must calculate this set with the following straightforward recursive function.

```
def find_vars(term, varset):
  if term.isVariable():
    varset.add(term)
  else:
    for argument in term.arguments():
      find_vars(argument, varset)
```

This find_vars function explores a term recursively accumulating its variables into the set varset of terms. Terms and most objects in the library can be safely used in dictionaries, sets, and other data structures since they support equality comparison and hashing. The implementation of the trs command simply iterates over the rules printing them. Instead of the default conversion of terms into strings, we use the prettyPrint method that permits finer control on the printing format. In particular, a zero argument causes terms to be printed in prefix form as required by the TRS format. Variables are also printed with an explicit type annotation.

```
def do_trs(self, _):
  varset = set()  # variables in the rules

  for rl in self.module.getRules():
    find_vars(rl.getLhs(), varset)
    find_vars(rl.getRhs(), varset)

  pv = lambda v: f'{v.getVarName()}:{v.getSort()}'
  print('(VARS', ' '.join(map(pv, varset)), ')')
  print('(RULES')

  for rl in self.module.getRules():
    lhs, rhs = rl.getLhs(), rl.getRhs()
    print(f'\t{lhs.prettyPrint(0)} -> '
          f'{rhs.prettyPrint(0)}')

  print(')')
```

In the general case, we should also ensure that identifiers respect the grammar of the TRS format and consider equations and structural axioms. The complete version of this example includes two more commands termination and confluence that automatically check these properties on the rules using the AProVE [19] and CSI [36] tools, with the generated TRS specification as input.

# 4   Advanced Features

This section introduces two features of the library with useful applications and no direct correspondence in the Maude interpreter.

## 4.1  Rewrite Graphs and Model Checking

Exploring the graph of all reachable states and transitions from a given initial
term is useful for debugging, visualizing, and model checking Maude specifications. We can recursively build this graph in the library using the `apply` method
or in Maude itself using the descent functions `metaSearch` or `metaXapply`, but
this does not work for strategy-controlled models and such a common operation deserves to be a builtin feature. The language bindings offer two classes
`RewriteGraph` and `StrategyRewriteGraph` to explore the rewrite graph of standard and strategy-controlled models, respectively. States are indexed by natural
numbers starting from zero, the state's term can be obtained with `getStateTerm`,
its successors can be enumerated with `getNextState`, and other methods can be
used to obtain the rule applied in each transition. This makes it easy to program
a search or any other algorithm in Python that directly operates with the graph
produced by Maude.

Moreover, a high-level interface to the Maude LTL model checker [18] and its
extension for strategy-controlled systems [29] is provided through these graphs.
This is more convenient than reducing, as usual, the `modelCheck` operator of
the `MODEL-CHECKER` module[2]. The `modelCheck` method of both graphs receives
a term of sort `Formula` and returns a record indicating whether the formula holds
and a counterexample that refutes if it does not. Counterexamples are described
by a cycle and a path to it from the initial state, both given as lists of indices
in the rewrite graph. One of the advantages of this approach is that the same
graph can be used to model check multiple properties, hence saving the work
required for the generation of the model in successive executions. Moreover, we
can further process the graph or the counterexample when model checking has
finished.

## 4.2  Custom Special Operators

Having overly shown that the `maude` module lets Python programmers evaluate
Maude code in their programs, the interaction in the opposite direction, calling
Python code from Maude, has not been explored yet.

User-defined and many predefined functions in Maude are specified with
equations, but the prelude also includes some *special* operators whose behavior is internally defined in the C++ code of the interpreter. Most operations on
the builtin types `Nat`, `Float`, `Qid`, and `String`, some polymorphic operators like
equality `==`, and most descent functions in the `META-LEVEL` module are examples of special operators. Moreover, the Maude implementation has occasionally
been extended ad hoc with new special operators, like in the Maude Formal
Environment [15].

---

[2] Even though the strategy language is part of the official releases of Maude [14], the
   strategy-aware model checker [29] is not yet, but we have included it in the Maude
   build used for this library.

The language bindings allow declaring custom special operators whose behavior against equational reduction and/or rule rewriting is defined in the target language. In the Maude side, the operator should be declared first with the `special` attribute and its `id-hook` `SpecialHubSymbol` option. For instance, the gamma function that extends the factorial to real (and complex) numbers can be declared as the following `gamma` operator within a module.

```
op gamma : Float -> Float [special (
  id-hook SpecialHubSymbol
)] .
```

On the Python side, we have to define and register the callback that is invoked when a term with `gamma` on top is reduced or rewritten. This is done by subclassing the `maude.Hook` class and implementing its `run` method, and then calling the functions `connectEqHook` and/or `connectRlHook` to register an object of the class as the handler for the special operator.

```
class GammaHook(maude.Hook):
  def run(self, term, data):
    module = term.symbol().getModule()
    argument, = term.arguments()

    value = math.gamma(float(argument))
    return module.parseTerm(str(value))
```

The `run` method receives the `term` that it should return reduced or rewritten. The implementation of `gamma` is directly provided by the `math` module of the Python standard library, so in this case we only need to convert the argument and result from a Maude term to a Python floating-point value and the other way around. Finally, we install the hook for equational reduction with the `connectEqHook` function.

```
hook = GammaHook()
maude.connectEqHook('gamma', hook)
```

After that, when we explicitly or implicitly reduce terms containing `gamma` in the library, `hook`'s `run` would be executed and we would obtain the desired number. For instance, if we program and run a REPL that parses and reduces every line from standard input, we can obtain the following:

```
Gamma> 1.2 + gamma(6.5)
2.8908527781504438e+2
```

In the signature of the `run` method, there is another argument `data` giving access to the `op-hook` and `term-hook` attributes of the special operator. Suppose we want to implement a custom predicate that tells whether a number is prime.

```
op isPrime : Nat -> Bool [special (
  term-hook trueTerm (true)
  term-hook falseTerm (false)
)] .
```

Using the above term hooks for the Boolean constants, we can define its `run` method by the expression

```
data.getTerm('trueTerm' if test_prime(argument)
             else 'falseTerm')
```

for some `test_prime` Python function. While the same can be achieved by parsing the constants with `parseTerm`, the advantage of hooks is that keep working even if truth values are renamed, for example to `tt` and `ff`, in a module importation within Maude. Further details are explained in the documentation.

## 5  Implementation

The language bindings for Maude are implemented on top of the official implementation of Maude using some additional C++ code and the Simple Wrapper and Interface Generator (SWIG) [13], as illustrated in Fig. 2. The desired programming interface is specified by selecting the classes, functions, and methods of the Maude implementation and the additional helper code that want to be exposed in the target language. Several languages like Python, Java, Lua, C#, Scheme, PHP, and JavaScript are supported, but only Python has been extensively tested and used in our case. From this specification, SWIG generates glue code in the selected language and in C, and this latter is then compiled into a binary module for the target language interpreter. This module is linked to the Maude implementation, which we have compiled as a shared library by adapting the build process. Indeed, we already did it to integrate Maude as a plugin for the language-independent model checker LTSmin [30]. Notice that Maude does not provide an official stable interface and the bindings are using its internal classes, so the implementation should be adapted on every new release of Maude. Moreover, instead of using the official Maude implementation as is, the language bindings are linked with our extension including a model checker for systems controlled by strategies [29], which does not alter any other aspect of the Maude implementation.

A large part of the classes and methods of the interface are direct wrappers to the homonym classes and methods of the Maude implementation, but some are implemented on purpose to facilitate the interaction. For example, terms are represented in Maude sometimes as trees and sometimes as nodes in a directed acyclic graph, but this particularity is hidden to the library user in the uniform `Term` class. This type is backed by an auxiliary C++ class `EasyTerm` that chooses the appropriate representation and manage the conversion between them. Custom special operators in Sect. 4.2 are supported by a `SpecialHubSymbol` subclass of the `Symbol` type of the Maude implementation written on purpose to allow registering C functions as callbacks for the equational reduction and rule rewrite handling methods of the symbol. The connection with the target language is based on the *directors* feature of SWIG and the `maude.Hook` class, whose `run` method implemented in the target language can be called from the registered callbacks of the special operator.

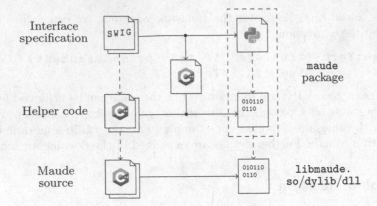

**Fig. 2.** Implementation structure.

When the Python interpreter executes the `import maude` statement, it loads the Python script generated by SWIG with the definition of all the classes and functions of the interface. This Python code loads the binary module that has been built from the SWIG-produced C code and the helper classes in the middle part of Fig. 2. This module is linked with the dynamic library `libmaude.so` (`.dylib` in macOS or `.dll` in Windows) that contains the Maude implementation. Every object of the library in the target language holds a pointer to an object living in the Maude implementation, whose methods are invoked when the equivalent methods of the library are called. However, arguments may need to be translated in the process, for example, from a Python list to a C++ vector. This is done by the glue code generated by the interface generator.

## 5.1 Performance Considerations

Since the language bindings replace text-based interprocess communication by direct procedure calls and despite the cost of the translations mentioned in the last paragraph, this approach is expectedly much more efficient than the classical interaction through the interpreter, especially when the results are frequently reused. We have executed some small experiments to compare the performance of reduction using (1) the `maude` Python library, (2) an I/O interaction that inputs `reduce` commands on a running Maude interpreter process and parse their results, and (3) a socket-based approach that communicates with a Maude-implemented TCP server that replies with the reduced forms of the terms it receives line by line. Reducing the constant 0 in the predefined module `CONVERSION` takes respectively (1) 3.21 μs, (2) 11.27 μs, and (3) 48.31 μs, so the best results are obtained with the `maude` Python library. Moreover, the last two options have been implemented in the simplest way possible and assuming unrealistic constraints, so production-ready implementations would likely be more costly.

Performance improvements are more noticeable when reusing the output of previous operations. For example, consider a toy Maude function `fibonacci`

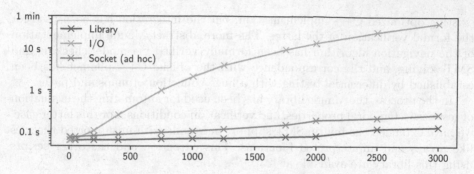

**Fig. 3.** Time spent in the iterative reduction of `fibonnaci` by number of iterations.

that expands a given list of integers by appending the sum of two leftmost numbers to the left. Repeatedly calling this function on the result of the previous call takes the amount of time depicted in Fig. 3 (in logarithmic scale) for an increasing number of iterations. In this experiment, the socket alternative has been improved to store and reuse the result of the previous call, which is already done by the bindings out of the box. While the language bindings and the socket approach show almost a constant execution time per iteration, the I/O alternative requires Maude to parse the list of integers again and again with a much higher cost. All these benchmarks are available at the bindings repository.

# 6 Some Applications

Since the first version of the library was released, almost two years ago, it has been applied from small quick scripts to more relevant projects. Examples of the latter are the integration of Maude into a robotic environment and a unified interface to several external verification tools.

## 6.1 Integration of Maude into the Robot Operating System

The Robot Operating System (ROS) [12] is a collaborative robotic framework organized as a collection of nodes that deal with the different robotic tasks and communicate with each other by message passing. One of its most prominent components is the navigation module. The officially supported languages for programming ROS nodes are C++ and Python, but in a recent work [24] Maude has been used for programming an alternative path-planning node and experimenting with the inclusion of declarative languages in this context. The `maude` Python library provides the required connection between the communication infrastructure of ROS and the actual path-planning algorithm. Even though random access to the map is enabled by a custom special operator (see Sect. 4.2), the efficiency of the Maude-based planner is not comparable to the existing optimized C++ implementation, but the integration has been used for the formal verification of the latter. The more abstract Maude implementation of the navigation algorithm has been formally verified via model checking and SMT solving, and the correspondence with the official C++ planner has been established by differential testing with a huge collection of maps and paths.

In the process, the Maude library has been used for automating the evaluation of test cases, temporal properties, and verification conditions. For this latter case, we have extended the builtin SMT support in Maude with unsupported theories like arrays and uninterpreted functions. This extension and the other scripts using this library are available at [23].

## 6.2   The Unified Maude Model Checker

The unified Maude model checking tool umaudemc [30] provides a uniform interface to the Maude LTL model checker [18] and several external model checkers for LTL, CTL, CTL\*, and $\mu$-calculus on standard and strategy-controlled Maude specifications. This interface reads the input data of the model-checking problem, builds the corresponding Kripke structure, calls the appropriate backend, and shows the results to the user. Among the supported backends, there are LTSmin, NuSMV [6], pyModelChecking [5], Spot [16], Spin [21], and a builtin $\mu$-calculus implementation written in Python. The maude library and the rewrite graphs discussed in Sect. 4.1 are used to generate the models, evaluate the atomic propositions, parse the temporal formula, and so on. More recently, we have extended umaudemc for specifying probabilities on top of Maude specifications, and checking properties and calculating quantitative values by probabilistic model checking using PRISM [22] and Storm [20] or by statistical model checking through simulation or the MultiVeSta tool [34]. By using external tools, we can efficiently support more logics and techniques while reducing the maintenance effort.

Moreover, umaudemc provides graphical and web-based interfaces for model checking, allows postprocessing the counterexamples, and generates visual representations of the rewrite graphs in different formats. This tool can also be used as a library for application-specific model-checking interfaces [31,32].

## 7   Related Work

As discussed in the introduction, several tools in the verification community maintain programming interfaces in addition to the traditional command-line ones, so that they can be used from other tools. Most applications interacting with Maude use ad hoc text-based communication with the interpreter, and the implementation of Maude has occasionally been extended to interact with external tools. The IMaude component of the IOP framework [25] is the closest precedent to this work in this context, since it provides a reusable and application-agnostic interface between Maude and external programs. However, our language bindings replace the textual communication with the interpreter with a more efficient binary connection with its implementation, extend the available functionality, simplify the installation process, and can be used from potentially more programming languages.

On the other hand, Maude itself is being extended for a richer connection to the outside world. The notion of external objects used for accessing Internet sockets since Maude 2.0 has been applied to read and write files and standard streams in 3.0, to external processes in 3.1, and to time and filesystem operations in 3.2. External tools have also been integrated into Maude 2.7.1 with limited support for SMT solving via the CVC4 [4] and Yices2 [17] tools.

## 8    Conclusions

We have introduced a general-purpose efficient programming interface to Maude from Python and other programming languages. Almost all functionality of the Maude interpreter is available through these language bindings along with some useful additions. Moreover, the connection in the opposite direction, calling external code from Maude, is also available via custom special operators. This work facilitates the interoperability between Maude and other tools, and tackles the claim for using Maude from external programs.

As future work, the library can be improved and extended in several directions, like adding native support for multiple interpreter sessions with separate databases through the infrastructure of metainterpreters, allowing the construction and manipulation of modules at the object level, or distributing compiled versions of the bindings for other languages. Moreover, there is currently no clear and explicit C/C++ interface, which can be very useful for applications where performance is a critical matter. Regarding applications, there are many possibilities for the library as we have suggested along the paper, from the elaboration of interfaces for specific frameworks to the development of more general tools.

**Acknowledgments.** I would like to thank Enrique Martin-Martin, Manuel Montenegro, Adrián Riesco, Juan Rodríguez-Hortalá, and Óscar Martín for the suggestions that brought this library into existence and their feedback to improve it. The first version of the bindings and extensions like custom operators were originally written for [24]. I also thank Narciso Martí-Oliet and Alberto Verdejo for their comments on this manuscript. This work was partially supported by the Spanish Ministry of Science and Innovation through projects TRACES (TIN2015-67522-C3-3-R) and ProCode (PID2019-108528RB-C22), and by the Spanish Ministry of Universities through the grant FPU17/02319.

## References

1. Alpuente, M., Ballis, D., Sapiña, J.: Efficient safety enforcement for Maude programs via program specialization in the ÁTAME system. Math. Comput. Sci. **14**(3), 591–606 (2020). https://doi.org/10.1007/s11786-020-00455-3
2. Alpuente, M., Escobar, S., Sapiña, J., Ballis, D.: Symbolic analysis of Maude theories with Narval. Theory Pract. Log. Program. **19**(5–6), 874–890 (2019). https://doi.org/10.1017/S1471068419000243

3. Barbosa, H. et al.: CVC5: A versatile and industrial-strength SMT solver. In: Fisman, D., Rosu, G. (eds.) TACAS 2022, Part I. LNCS, vol. 13243, pp. 415–442. Springer, Cham (2022). https://doi.org/10.1007/978-3-030-99524-9_24

4. Barrett, C., et al.: CVC4. In: Gopalakrishnan, G., Qadeer, S. (eds.) CAV 2011. LNCS, vol. 6806, pp. 171–177. Springer, Heidelberg (2011). https://doi.org/10.1007/978-3-642-22110-1_14

5. Casagrande, A.: pyModelChecking: a simple python model checking package (2020). https://pypi.org/project/pyModelChecking

6. Cimatti, A., et al.: NuSMV 2: an opensource tool for symbolic model checking. In: Brinksma, E., Larsen, K.G. (eds.) CAV 2002. LNCS, vol. 2404, pp. 359–364. Springer, Heidelberg (2002). https://doi.org/10.1007/3-540-45657-0_29

7. Clavel, M., Durán, F., Eker, S., Lincoln, P., Martí-Oliet, N., Meseguer, J., Talcott, C.: All About Maude - A High-Performance Logical Framework. LNCS, vol. 4350. Springer, Heidelberg (2007). https://doi.org/10.1007/978-3-540-71999-1

8. Clavel, M., et al.: Maude Manual v3.2.1 (2022)

9. Codescu, M., Mossakowski, T., Riesco, A., Maeder, C.: Integrating maude into hets. In: Johnson, M., Pavlovic, D. (eds.) AMAST 2010. LNCS, vol. 6486, pp. 60–75. Springer, Heidelberg (2011). https://doi.org/10.1007/978-3-642-17796-5_4

10. de Moura, L., Bjørner, N.: Z3: an efficient SMT solver. In: Ramakrishnan, C.R., Rehof, J. (eds.) TACAS 2008. LNCS, vol. 4963, pp. 337–340. Springer, Heidelberg (2008). https://doi.org/10.1007/978-3-540-78800-3_24

11. Moura, L., Ullrich, S.: The lean 4 theorem prover and programming language. In: Platzer, A., Sutcliffe, G. (eds.) CADE 2021. LNCS (LNAI), vol. 12699, pp. 625–635. Springer, Cham (2021). https://doi.org/10.1007/978-3-030-79876-5_37

12. The ROS developers. Robot Operating System (2020). https://www.ros.org/

13. The SWIG developers. Simplified Wrapper and Interface Generator (2020). http://www.swig.org/

14. Durán, F., et al.: Programming and symbolic computation in Maude. J. Log. Algebraic Methods Program. **110**, 100497 (2020). https://doi.org/10.1016/j.jlamp.2019.100497

15. Durán, F., Rocha, C., Álvarez, J.M.: Tool interoperability in the Maude Formal Environment. In: Corradini, A., Klin, B., Cîrstea, C. (eds.) CALCO 2011. LNCS, vol. 6859, pp. 400–406. Springer, Heidelberg (2011). https://doi.org/10.1007/978-3-642-22944-2_30

16. Duret-Lutz, A., Lewkowicz, A., Fauchille, A., Michaud, T., Renault, É., Xu, L.: Spot 2.0 — a framework for LTL and ω-automata manipulation. In: Artho, C., Legay, A., Peled, D. (eds.) ATVA 2016. LNCS, vol. 9938, pp. 122–129. Springer, Cham (2016). https://doi.org/10.1007/978-3-319-46520-3_8

17. Dutertre, B.: Yices 2.2. In: Biere, A., Bloem, R. (eds.) CAV 2014. LNCS, vol. 8559, pp. 737–744. Springer, Cham (2014). https://doi.org/10.1007/978-3-319-08867-9_49

18. Eker, S., Meseguer, J., Sridharanarayanan, A.: The Maude LTL model checker and its implementation. In: Ball, T., Rajamani, S.K. (eds.) SPIN 2003. LNCS, vol. 2648, pp. 230–234. Springer, Heidelberg (2003). https://doi.org/10.1007/3-540-44829-2_16

19. Giesl, J., et al.: Analyzing program termination and complexity automatically with AProVE. J. Autom. Reason. **58**(1), 3–31 (2017). https://doi.org/10.1007/s10817-016-9388-y

20. Hensel, C., Junges, S., Katoen, J.-P., Quatmann, T., Volk, M.: The probabilistic model checker Storm. Int. J. Softw. Tools Technol. Transf. **23**(4), 1–22 (2021). https://doi.org/10.1007/s10009-021-00633-z

21. Holzmann, G.J.: The SPIN Model Checker: Primer and Reference Manual. Addison-Wesley (2011)
22. Kwiatkowska, M., Norman, G., Parker, D.: PRISM 4.0: verification of probabilistic real-time systems. In: Gopalakrishnan, G., Qadeer, S. (eds.) CAV 2011. LNCS, vol. 6806, pp. 585–591. Springer, Heidelberg (2011). https://doi.org/10.1007/978-3-642-22110-1_47
23. Martin-Martin, E., Montenegro, M., Riesco, A., Rodríguez-Hortalá, J., Rubio, R.: Maude integration and verification for ROS Nav 2 (2021). https://github.com/demiourgoi/maudeROS
24. Martin-Martin, E., Montenegro, M., Riesco, A., Rodríguez-Hortalá, J., Rubio, R.: Verification of ROS Navigation using Maude. In: Martí-Oliet, N., (ed.) XX Jornadas de Programación y Lenguajes (PROLE). Sistedes (2021). http://hdl.handle.net/11705/PROLE/2021/008
25. Mason, I.A., Talcott, C.L.:. IOP: the interoperability platform & IMaude: an interactive extension of Maude. In: Martí-Oliet, N. (ed.) Proceedings of the Fifth International Workshop on Rewriting Logic and Its Applications, WRLA 2004, Barcelona, Spain, 27–28 March 2004. Electronic Notes in Theoretical Computer Science, vol. 117, pp. 315–333. Elsevier (2004). https://doi.org/10.1016/j.entcs.2004.06.016
26. Meier, S., Schmidt, B., Cremers, C., Basin, D.: The TAMARIN prover for the symbolic analysis of security protocols. In: Sharygina, N., Veith, H. (eds.) CAV 2013. LNCS, vol. 8044, pp. 696–701. Springer, Heidelberg (2013). https://doi.org/10.1007/978-3-642-39799-8_48
27. Meseguer, J.: Conditional rewriting logic as a unified model of concurrency. Theor. Comput. Sci. **96**(1), 73–155 (1992). https://doi.org/10.1016/0304-3975(92)90182-F
28. Rosu, G., Serbanuta, T.-F.: An overview of the K semantic framework. J. Log. Algebraic Methods Program. **79**(6), 397–434 (2010). https://doi.org/10.1016/j.jlap.2010.03.012
29. Rubio, R., Martí-Oliet, N., Pita, I., Verdejo, A.: Model checking strategy-controlled systems in rewriting logic. Autom. Softw. Eng. **29**(1), 1–62 (2021). https://doi.org/10.1007/s10515-021-00307-9
30. Rubio, R., Martí-Oliet, N., Pita, I., Verdejo, A.: Strategies, model checking and branching-time properties in Maude. J. Log. Algebr. Methods Program. **123**, 100700 (2021). https://doi.org/10.1016/j.jlamp.2021.100700
31. Rubio, R., Martí-Oliet, N., Pita, I., Verdejo, A.: Metalevel transformation of strategies. J. Log. Algebr. Methods Program. **124**, 100728 (2022). https://doi.org/10.1016/j.jlamp.2021.100728
32. Rubio, R., Martí-Oliet, N., Pita, I., Verdejo, A.: Simulating and model checking membrane systems using strategies in Maude. J. Log. Algebr. Methods Program. **124**, 100727 (2022). https://doi.org/10.1016/j.jlamp.2021.100727
33. Santiago, S., Talcott, C.L., Escobar, S., Meadows, C.A., Meseguer, J.: A graphical user interface for Maude-NPA. In: Lucio, P., Moreno, G., Peña, R., (eds.) Proceedings of the Ninth Spanish Conference on Programming and Languages (PROLE 2009), San Sebastián, Spain, 9–11 September, 2009, volume 258(1) of Electronic Notes Theory Computer Science, pp. 3–20. Elsevier (2009). https://doi.org/10.1016/j.entcs.2009.12.002
34. Sebastio, S., Vandin, A.: MultiVeStA: statistical model checking for discrete event simulators. In: Horváth, A., Buchholz, P., Cortellessa, V., Muscariello, L., Squillante, M.S., (eds.) 7th International Conference on Performance Evaluation

Methodologies and Tools, ValueTools '13, Torino, Italy, 10–12 December 2013, pp. 310–315. ICST/ACM (2013). https://doi.org/10.4108/icst.valuetools.2013.254377

35. Talcott, C.: Pathway logic. In: Bernardo, M., Degano, P., Zavattaro, G. (eds.) SFM 2008. LNCS, vol. 5016, pp. 21–53. Springer, Heidelberg (2008). https://doi.org/10.1007/978-3-540-68894-5_2

36. Zankl, H., Felgenhauer, B., Middeldorp, A.: CSI – a confluence tool. In: Bjørner, N., Sofronie-Stokkermans, V. (eds.) CADE 2011. LNCS (LNAI), vol. 6803, pp. 499–505. Springer, Heidelberg (2011). https://doi.org/10.1007/978-3-642-22438-6_38

# Author Index

Printed in the United States
by Baker & Taylor Publisher Services

Printed in the United States
by Baker & Taylor Publisher Services